EX LIBRIS
Kenneth Krieg

PROPERTIES OF ELECTRICAL ENGINEERING MATERIALS

HARPER'S SERIES IN ELECTRICAL ENGINEERING

Under the Editorship of Henry C. Bourne, Jr.

PROPERTIES OF ELECTRICAL ENGINEERING MATERIALS

G. C. Jain

DEPARTMENT OF ELECTRICAL ENGINEERING
RICE UNIVERSITY, HOUSTON, TEXAS

Harper & Row, Publishers

NEW YORK, EVANSTON, AND LONDON

Properties of Electrical Engineering Materials

Copyright © 1967 by G. C. Jain. Printed in the United States of America. All rights reserved. No part of this book may be used or reproduced in any manner whatsoever without written permission except in the case of brief quotations embodied in critical articles and reviews. For information address Harper & Row, Publishers, Incorporated, 49 East 33rd Street, New York, N.Y. 10016.

Library of Congress Catalog Card Number: 67-11656

Contents

Preface ix

Introduction xi

1 QUANTUM PHYSICS   1

    1.1 Bohr-Sommerfeld atom   1
    1.2 Wave equation of classical physics; Schrödinger equation   4
    1.3 Potential box   12
    1.4 Potential well   17
    1.5 Quantum oscillator   21
    1.6 Hydrogen atom   25
    1.7 Electron spin, exclusion principle, and the periodic table   36
    Problems   37

2 STATISTICAL PHYSICS AND ELECTRON EMISSION   39

    2.1 Gas kinetics   39
    2.2 Classical statistics   42
    2.3 Fermi-Dirac statistics   49
    2.4 Fermi-level   51
    2.5 Thermionic emission   57
    Problems   63

3 THERMAL PROPERTIES OF SOLIDS   66

    3.1 Classical theory of specific heats   67
    3.2 Einstein's theory of specific heat   69
    3.3 Vibrational modes of a continuous medium; Debye's approximations   72
    3.4 Equivalence of the elastic waves in an infinite one-dimensional array of identical atoms and a low-pass filter   79
    3.5 Diatomic linear lattice   83
    3.6 Heat capacity of conduction electrons   88

3.7 Thermal conductivity of solids 91
Problems 97

4 METALS, SEMICONDUCTORS, AND INSULATORS 99

4.1 Band theory of solids; Kronig-Penny model 99
4.2 Distinction between metals, semiconductors, and insulators 114
4.3 Free-electron model of electrons 115
4.4 Effect of electric field on metals; impurities in metals 117
4.5 Intrinsic semiconductors and insulators 127
4.6 Extrinsic semiconductors 138
4.7 Photoconductivity, excitons, and luminescence 153
Problems 161

5 CONTINUITY EQUATIONS AND P-N JUNCTIONS 163

5.1 Electrochemical potential 163
5.2 Diffusion; Modified Ohm's law 165
5.3 Continuity equations 170
5.4 P-N junction theory 177
5.5 Optical effects in P-N junctions 193
Problems 202

6 EFFECT OF ELECTRIC FIELDS ON INSULATORS 205

6.1 Results of field theory 206
6.2 Electronic, ionic, and orientational polarization in static fields 211
6.3 Ferroelectricity 231
6.4 Complex dielectric constant and lossy dielectrics 237
6.5 Ferroelectric energy conversion 249
Problems 252

7 MAGNETIC MATERIALS AND PROPERTIES 255

7.1 Magnetic dipole moment and magnetization 255
7.2 Diamagnetism 269
7.3 Paramagnetism 272
7.4 Ferromagnetism; Domain-wall motion 279
7.5 Antiferromagnetism 294
7.6 Ferrimagnetism and ferrites 298
7.7 Cooling by adiabatic demagnetization 299
Problems 301

8 TRANSPORT PROPERTIES 303

8.1 Irreversible processes; conjugate forces and fluxes 303
8.2 Thermoelectric energy conversion 315
8.3 Galvanomagnetic coefficients 320
8.4 Electrical and heat current densities; thermoelectric power 331
8.5 Boltzmann's transport equation 345
8.6 Thermomagnetic coefficients; Nernst-Ettingshausen generator 347
   Problems 352

9 QUANTUM ELECTRONICS 355

9.1 Why quantum electronics for engineers? 355
9.2 Energy levels in atomic systems 356
9.3 Operation of devices based on stimulated emission 363
9.4 The solid-state maser 366
9.5 The laser 368
9.6 Semiconductor lasers 371
   Problems 374

   Bibliography 377

   Index 379

Preface

More and more engineers are getting interested in solid-state theory. This book is designed to give them insight into the fundamentals of solid-state theory with practically no other background except an acquaintance with modern sophomore physics and brief knowledge of differential and integral calculus. No devices have been discussed in detail, but it is the contention of the author that a knowledge of the material covered in the book will help the student understand the devices more thoroughly.

The first eight chapters of the book originated from a course on the properties of electrical engineering materials which the author introduced for seniors and first-year graduate students at Purdue University and later taught at Rice University. The ninth chapter is based on the experiences, in the field of quantum electronics, of Professor T. A. Rabson of the Department of Electrical Engineering, Rice University, for which the author acknowledges his sincerest thanks.

The preparation of this book owes much to the works of other authors, including many listed in the Bibliography. The author would like to thank his valued colleague, Professor Henry C. Bourne, Jr., for his kind help and valuable comments and suggestions. My sincerest thanks to Mrs. Arlene McCourt, who typed the manuscript and without whose patience and consistent efforts it would have been impossible to bring out the book so soon.

G. C. JAIN

Houston, Texas
December, 1966

Introduction

The traditional role of the engineer has been to utilize and control the properties of nature to gain technological achievements for the benefit of mankind. This is a rather broad and pompous definition which can be interpreted to mean almost anything and to describe the endeavors of almost everyone. A more specific and pragmatic definition would be that the engineer is concerned with making use of the fundamental natural phenomena studied and revealed by natural scientists in order to design and create devices, systems, and techniques useful to mankind. It is implicit in applying the laws governing these natural phenomena that the engineer understand the laws. This book is concerned with giving the engineer an understanding of the physical laws which govern the behavior of matter with a particular emphasis on the behavior which is of importance in those devices of concern to an electrical engineer.

Without an understanding of the basic principles of quantum mechanics the device-oriented electrical engineer is lost. Even though many of the devices can be understood in terms of classical models, it is important to realize the limitations and realms of validity of these models in order to extend and improve device design. Such understanding can come only from a study of quantum-mechanical principles. Chapter 1, although by no means complete and rigorous, endeavors to provide the necessary background in quantum mechanics so that students can follow the deviations and calculations in succeeding chapters. Throughout the book an attempt is made to utilize concepts and mathematical techniques which have already become familiar to the engineering student.

Once one has learned the necessary laws governing the behavior of a single body, one soon realizes that in most problems of interest, many, many particles are involved, and one must resort to statistical techniques. Chapter 2 discusses the statistical distributions that arise in treating both classical and mechanical systems of particles. Specific examples of the applications of these statistical techniques to physical problems are given. In Chapter 3 use is again made of statistical techniques in understanding the thermal properties of solids.

Chapter 4 discusses the bulk electrical properties of matter. The main emphasis in this chapter is on semiconducting materials, because they are of special importance to electrical engineers and because both insulators and conductors are easily understood in terms of the model developed for semiconductors. The dynamic equations for electrons and holes in semiconductors are developed, and in the following chapter these equations are used to determine the electrical properties of a *P-N* junction.

A renewed interest has developed in the interaction of electromagnetic fields with insulators. Engineers are anxious to utilize the nonlinear and frequency-dependent relations between an externally applied electric field and the polarization present in the insulator for device development in the field of energy conversion and communication engineering. The theory explaining these phenomena is presented in Chapter 6.

Interactions of magnetic fields with matter have always been of paramount interest to the electrical engineer. Although the interactions are not completely understood, there are detailed microscopic theories which explain many of the observed magnetic properties of matter. These microscopic models account for such properties as diamagnetism, paramagnetism, ferromagnetism, antiferromagnetism, and ferrimagnetism. However, there are still some parts of the theory which remain phenomenological. Chapter 7 treats the above-mentioned phenomena in some detail.

Chapter 8 treats in detail those phenomena which involve the movement of large numbers of particles under the influence of external forces or as a result of being in a state of nonequilibrium. These movements must be treated on a statistical basis and are governed by the Boltzmann transport equation.

Chapter 9 treats one of the newer fields of science of interest to the electrical engineer, quantum electronics. This field includes the phenomenon of resonance interaction of an electromagnetic field with individual atomic systems. The process of stimulated emission of radiation from a nonequilibrium distribution of atoms is usually present in all devices classified under this heading of quantum electronics.

An attempt is made to present the main ideas in as simple and basic a form as possible. In many instances, special cases are treated first and the ideas thus developed are generalized into fundamental relations. In some places topics that might have been included for the sake of completeness have been omitted to leave room for more fundamentals without making the text unreasonably long. Although science and technology are rapidly changing, the fundamental knowledge necessary to understand the basic operation of most devices and systems of interest to the engineers is well developed, and although some refinements will occur any basic changes are unlikely. The major changes will probably be in the calculational techniques. It is the more stable fundamental principles that are presented in this book rather than more detailed descriptions of the devices whose operation is based on these principles.

G. C. J.

PROPERTIES OF ELECTRICAL ENGINEERING MATERIALS

1

Quantum Physics

1.1 Bohr-Sommerfeld atom

Many properties of electrical engineering materials may be understood only with the aid of quantum mechanics. It is however, pedagogically desirable and instructive to follow the developments of modern concepts of atomic structure from a historical point of view In 1913 Niels Bohr proposed his quantum model of the atom. In 1915–1916 Sommerfeld modified Bohr's model by adopting an approach similar to Bohr but with additional restrictions The theory is now called the *Bohr-Sommerfeld model* of the atom. The model was developed by applying the principles of classical mechanics but with certain quantum conditions placed on the system proposed. These conditions have no justification other than that the results obtained thereby agree with experimental results. It would be of interest to study the hydrogen atom as explained by the Bohr-Sommerfeld theory. The theory is made up of two *ad hoc* assumptions:

1. The electrons in Rutherford's atom exist only in circular stable orbits of fixed energy. Such orbits are fixed by the conditions that the angular momentum of the electron in its orbit is an integral multiple of $h/2\pi$, where h is Planck's constant:

$$h = 6.624 \times 10^{-34} \text{ J sec.}$$

It is worth mentioning that Ernest Rutherford established the general structure of atoms in 1911. He showed that the atom consisted of a positively charged nucleus, having nearly the mass of the entire atom surrounded by negatively charged electrons.

2. An electron may make transitions from one orbit to another and in doing so will emit or absorb radiation, whose frequency ν is given by the relationship

$$h\nu = E_f - E_i,$$

where E_f is the energy in the final stable orbit and E_i is the energy in the initial stable orbit.

In the model of the atom proposed, the mechanical and electrostatic forces of classical theory still hold. However, the model neglects the results of *Maxwell's theory of electromagnetic radiation*, which shows that an accelerating charge would continuously emit radiation with a corresponding loss in kinetic energy.

According to the model as applied to the hydrogen atom, the nucleus is considered fixed in space, and the electron, with charge e and mass m, is assumed to rotate about the nucleus at a distance r (Fig. 1.1). The

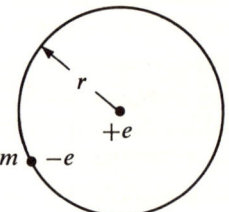

Fig. 1.1
Hydrogen atom

electron is maintained in its orbit of radius r by a balance between the centrifugal force and the electrostatic force of attraction between the nucleus, with charge $+e$, and the electron, with charge $-e$:

$$\frac{mv^2}{r} = \frac{e^2}{4\pi\epsilon_0 r^2}, \tag{1.1}$$

where $\epsilon_0 = 8.854 \times 10^{-12}$ F/m and v is the velocity of the electron.

The kinetic energy of the moving electron of mass m and velocity v is

$$\text{K.E.} = \tfrac{1}{2}mv^2,$$

and from Eq. (1.1),

$$\text{K.E.} = \frac{e^2}{8\pi\epsilon_0 r}.$$

Without the first *ad hoc* assumption, the balance in centrifugal and electrostatic force of attraction could be satisfied at any distance r. Considering the first assumption,

$$mvr = nh/2\pi, \quad \text{where } n \text{ is an integer } 1,2,3,4,\ldots. \quad (1.2)$$

Eliminating v from Eqs. (1.1) and (1.2),

$$r = \frac{n^2 h^2 \epsilon_0}{\pi e^2 m} \simeq 0.5\, n^2 \times 10^{-10}\, m. \quad (1.3)$$

Equation (1.3) indicates that only certain radii are possible for the orbiting of the electron. The second and third possible radii are approximately 2 and 4.5 Å, respectively.

The total energy of the electron can be determined by adding the kinetic energy (K.E.) and the potential energy (P.E.). K.E. can be expressed in terms of the radius of the orbit r by Eq. (1.1):

$$\text{K.E.} = \tfrac{1}{2}mv^2 = \frac{e^2}{8\pi\epsilon_0 r}.$$

To determine the potential energy, we have to assume an arbitrary zero. For this purpose it is customary to take the zero as being that of an electron at rest at an infinite distance from the nucleus. As the electron approaches the nucleus its potential energy will become increasingly negative, and a simple integration of the inverse-square law of force will show that the potential energy of the electron at a distance r is

$$\text{P.E.} = \int_\infty^r F\, dr = \int_\infty^r \frac{e^2}{4\pi\epsilon_0 r^2}\, dr = -\frac{e^2}{4\pi\epsilon_0 r}.$$

The total energy of an electron at a distance r is

$$E = \text{K.E.} + \text{P.E.} = \frac{e^2}{8\pi\epsilon_0 r} - \frac{e^2}{4\pi\epsilon_0 r} = -\frac{e^2}{8\pi\epsilon_0 r}. \quad (1.4)$$

It has already been shown by Eq. (1.3) that only certain values of r are possible, and substituting Eq. (1.3) in Eq. (1.4), the energy of the electron for a given value of n is

$$E_n = -\frac{me^4}{8\epsilon_0^2 h^2 n^2} \simeq -\frac{13.6}{n^2} \quad \text{electron volts (eV)}. \quad (1.5)$$

The electron volt is a unit of energy widely used in atomic or solid-state physics. It corresponds to the energy acquired by an electron accelerated through a potential difference of 1 V.

From Eq. (1.5),

$$\text{for } n = 1 \quad E_1 = -13.6 \text{ eV},$$
$$\text{for } n = 2 \quad E_2 = -3.4 \text{ eV},$$
$$\text{for } n = 3 \quad E_3 = -1.51 \text{ eV},$$

and so on.

The integer n determines the electronic energy. Apart from inadequacies of the theory, there are philosophical objections to the *ad hoc* manner in which the assumptions were introduced. Nonetheless, the great success of Bohr's theory lies in its going as far as it did to provide a great deal of supporting evidence for the fundamental significance in atomic structure of Planck's constant h.

1.2 Wave equation of classical physics; Schrödinger equation

A *harmonic force* is defined as a force whose magnitude is proportional to the distance from the point of application to a fixed point, called the *force center* and whose direction is such that the force always points toward the center. For example, Hooke's law tells us that the restoring force of a spring is proportional to the elongation, but opposite in direction, so that a spring exerts a harmonic force,

$$F = -fy,$$

where y is the distance from the force center to the point of application (Fig. 1.2) and f is Hooke's constant. The negative sign shows that the

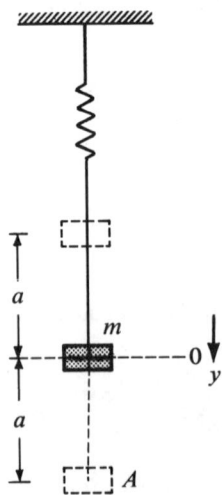

Fig. 1.2
Harmonic oscillator

elongation and restoring force are in opposite directions. Assume a frictionless spring with one end fastened and a mass m attached to the other. Let O be the starting point for the motion of mass m. If we pull the mass onto a distance a and release it with zero velocity, the mass will vibrate up and down, alternately compressing and extending the spring:

$$m \frac{d^2y}{dt^2} = -fy. \qquad (1.6)$$

Putting $f/m = \omega^2$, Eq. (1.6) can be rewritten

$$\frac{d^2y}{dt^2} + \frac{f}{m} y = 0$$

or

$$\frac{d^2y}{dt^2} + \omega^2 y = 0. \qquad (1.7)$$

A solution of the above differential equation is

$$y = A \cos(\omega t + B), \qquad (1.8)$$

where A and B are constants and can be determined with the help of boundary conditions.

At $t = 0$, $y = a$, and $v = 0$,

$$y = A \cos B = a \quad \text{at} \quad t = 0,$$

$$v = \frac{dy}{dt} = -A\omega \sin(\omega t + B),$$

and at $t = 0$

$$0 = -A\omega \sin B,$$

because $\quad A \neq 0, \quad \sin B = 0, \quad \text{or} \quad B = 0.$

$y = a \cos \omega t$ describes the motion of mass m attached to the spring. ω is the angular frequency in radians per second:

$$\omega = 2\pi \nu,$$

where ν is the frequency in cycles per second (Hz).

To get some insight into the properties of the solutions to the wave equation, we consider a group of identical springs and masses, with the masses arranged to lie on the curve:

$$y = a \cos kx \quad \text{at} \quad t = 0$$

and vibrating parallel to y axis (Fig 1.3). The displacement of the masses, each mass m, with reference to the axis OO', is $y = a$ at $x = 0$ and also $y = a$ at $x = \lambda$. Therefore,

$$k\lambda = 2\pi \quad \text{or} \quad k = 2\pi/\lambda.$$

The displacement y is
$$y = a \cos (2\pi/\lambda)x. \tag{1.9}$$

Let us assume that the masses attached to the springs can oscillate in such a way that the displacement cosine curve can move to the right with a velocity v, the displacement y along the x axis is shifted by vt in time t, and hence in Eq (1.9) x must be replaced by $x - vt$. Therefore,
$$y = a \cos (2\pi/\lambda)(x - vt) \tag{1.10}$$

is the equation of the traveling wave in the direction of x. Rewriting Eq. (1.10),
$$y = a \cos \left(kx - \frac{2\pi v}{\lambda} t\right).$$

Because the velocity v of the wave is the product of the number of cycles per second ν and the wavelength,
$$v = \lambda \nu \quad \text{and} \quad y = a \cos (kx - 2\pi \nu t),$$
or
$$y = a \cos (kx - \omega t), \tag{1.11}$$
where
$$\omega = 2\pi \nu.$$

Equation (1.11) is also a solution of the classical wave equation for wave motion in the direction of positive x in Fig. 1.3.

In general,
$$y = a \cos(kx \pm \omega t).$$

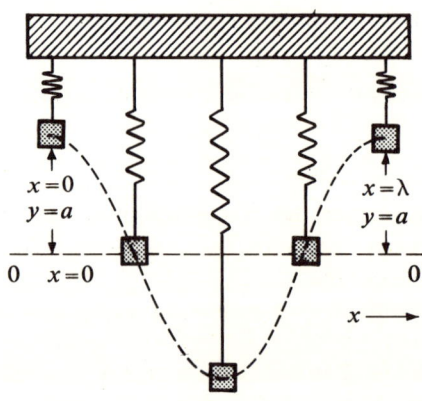

Fig. 1.3
Identical masses and springs, with the masses arranged to lie on a sinusoidal curve

SCHRÖDINGER EQUATION

The differential equation describing the one-dimensional wave motion is

$$\frac{\partial^2 y}{\partial t^2} = \frac{\omega^2}{k^2} \frac{\partial^2 y}{\partial x^2} = v^2 \frac{\partial^2 y}{\partial x^2}. \quad (1.12)$$

The cosine function we have investigated is obviously the solution of the wave equation because

$$\frac{d^2 y}{dt^2} = a\omega^2 \cos(kx - \omega t),$$

$$\frac{d^2 y}{dx^2} = ak^2 \cos(kx - \omega t).$$

Group velocity. Taking Eq. (1.11) as the classical wave equation, we shall determine the resultant displacement y of two waves of frequencies ω_1 and ω_2 and propagation constants k_1 and k_2 superimposed on each other.

$$y_1 = a \cos(k_1 x - \omega_1 t),$$
$$y_2 = a \cos(k_2 x - \omega_2 t),$$
$$y = y_1 + y_2 = a[\cos(k_1 x - \omega_1 t) + \cos(k_2 x - \omega_2 t)].$$

Let

$$\omega_1 = \omega_0 - \Delta\omega,$$
$$\omega_2 = \omega_0 + \Delta\omega,$$

where

$$\omega_0 = \tfrac{1}{2}(\omega_1 + \omega_2) \quad \text{and} \quad \Delta\omega = \tfrac{1}{2}(\omega_1 - \omega_2).$$

Similarly,

$$k_0 = \tfrac{1}{2}(k_1 + k_2) \quad \text{and} \quad \Delta k = \tfrac{1}{2}(k_1 - k_2).$$

Therefore,

$$y = a[\cos(k_0 x - \Delta k x - \omega_0 t + \Delta\omega t) + \cos(k_0 x + \Delta k x - \omega_0 t - \Delta\omega t)]$$

or, simplifying,

$$y = 2a[\cos(k_0 x - \omega_0 t) \cos(\Delta k x - \Delta\omega t]. \quad (1.13)$$

The velocity v, which represents the motion of the individual wave, is called the *phase velocity;* the velocity v_g of the motion of the group of waves is called the *group velocity*. Equation (1.13) suggests that the superimposed wave moves with an average phase velocity ω_0/k_0 and is modulated. The modulation which is representative of the group of waves of slightly different values of ω and k moves with a velocity $\Delta\omega/\Delta k$ and in the limit for differential values of ω and k,

$$v_g = \frac{d\omega}{dk} = \frac{d(2\pi\nu)}{d(2\pi/\lambda)} = \frac{d\nu}{d(1/\lambda)} = -\lambda^2 \frac{d\nu}{d\lambda}. \quad (1.14)$$

For the more general case, let us consider a wave made up of a continuous distribution of frequencies

$$\varphi(x,t) = \int_{-\infty}^{\infty} A(k)e^{i(\omega t - kx)}dk.$$

Assuming that $A(k)$ is appreciable only for k near k_0,

$$\omega(k) = \omega_0 + k'\frac{d\omega}{dk}\bigg|_{k=k_0} + \cdots,$$

$$\varphi(x,t) = \int_{-\infty}^{\infty} A(k_0 + k')\exp\left[i\left(\omega_0 t + \frac{d\omega}{dk}\bigg|_{k=k_0} k't - k_0 x - k'x\right)\right] dk'.$$

Because our limits of integration approach infinity we can let $A(k_0 + k') = B(k')$ and obtain

$$\varphi(x,t) = \exp[i(\omega_0 t - k_0 x)] \int_{-\infty}^{\infty} B(k')\exp\left[ik'\left(\frac{d\omega}{dk}\bigg|_{k=k_0} t - x\right)\right] dk'.$$

When $t = 0$ we have

$$\varphi(x,0) = e^{ik_0 x}f(x),$$

where

$$f(x) = \int_{-\infty}^{\infty} B(k')e^{-ik'x}dk'.$$

Because $B(k')$ is appreciable only where k' is small, $f(x)$ varies slowly with x:

$$\varphi(x,t) = \exp[i(\omega_0 t - k_0 x)]f\left(x - \frac{d\omega}{dk_0} t\right),$$

$$\text{group velocity} = v_g = \frac{d\omega}{dk}\bigg|_{k=k_0},$$

$$\text{phase velocity} = v_p = \frac{\omega_0}{k_0}.$$

Dual nature of electrons. In 1924 de Broglie suggested that small particles of matter such as electrons have a dual nature. The particles have the corresponding property of behaving like waves. Davisson and Germer in the United States and G. P. Thomson in England produced an experimental verification of de Broglie's hypothesis by showing that electron beams are diffracted by crystals in much the same manner as light is diffracted by ruled gratings.

The group velocity v_g, or simply v from now on, of the electron wave is the particle velocity of the electron also. The total energy E of the electron as a particle is the sum of kinetic and potential energy:

$$E = \text{K.E.} + \text{P.E.}$$

Let V be the potential energy. Therefore,

$$E = \tfrac{1}{2}mv^2 + V. \tag{1.15}$$

According to Planck's quantum theory, the energy E of the electron is given as

$$E = h\nu, \tag{1.16}$$

where ν is the frequency of the corresponding wave. According to de Broglie's hypothesis, the energy values given by Eqs. (1.15) and (1.16) must be equal. Therefore,

$$h\nu = \tfrac{1}{2}mv^2 + V. \tag{1.17}$$

Differentiating, with respect to λ,

$$h\frac{d\nu}{d\lambda} = mv\frac{dv}{d\lambda}.$$

Referring to Eq. (1.14),

$$v = v_g = -\lambda^2 \frac{d\nu}{d\lambda},$$

$$-\frac{h\nu}{\lambda^2} = mv\frac{dv}{d\lambda},$$

or

$$\frac{m}{h}\frac{dv}{d\lambda} = -\frac{1}{\lambda^2}.$$

Integrating with respect to λ, $mv/h = 1/\lambda$, assuming the constant of integration is zero:

$$\lambda = h/mv. \tag{1.18}$$

The result is known as *de Broglie's hypothesis* and is the statement of wave-particle duality. It connects the wavelength of the electron regarded as a wave with the mass of the electron as a particle.

The above-mentioned symmetry between radiation and matter led Schrödinger and Heisenberg to formulate a mechanics of material bodies in terms of equations normally used to describe wave phenomena.

The Schrödinger equation. The Schrödinger equation for an electron is a description of a particle in terms of wave equations. It has been shown that a single particle of mass m obeys the equation of motion:

$$\frac{d^2y}{dt^2} + \omega^2 y = 0$$

[Eq. (1.7)] under certain conditions. The wave equation described by a group of vibrating particles is

$$\frac{\partial^2 y}{\partial t^2} = v^2 \frac{\partial^2 y}{\partial x^2}$$

[Eq. (1.12)].

The Schrödinger equation is an analogous equation, which describes the behavior of a dual entity such as an electron. The displacement y for a classical wave is given by Eq. (1.11):

$$y = a \cos(kx - \omega t).$$

Putting it in the exponential form,

$$y = a e^{i(kx - \omega t)}. \tag{1.19}$$

Using de Broglie's hypothesis and Planck's relationships,

$$k = \frac{2\pi}{\lambda} = \frac{P}{\hbar},$$

where P is the momentum of the electron and

$$\hbar = h/2\pi, \qquad \omega = 2\pi \nu = E/\hbar$$

in Eq. (1.19), and replacing y by Ψ,

$$\Psi = a \exp\left[i\left(\frac{P}{\hbar} x - \frac{E}{\hbar} t\right)\right]. \tag{1.20}$$

Equation (1.20) is the equation of wave-particle motion because it takes into account the dual nature of the electron expressed by de Broglie's hypothesis and Planck's relationship. The interpretation of the function Ψ is discussed later.

From Eq. (1.20),

$$\frac{\partial^2 \Psi}{\partial x^2} = -a \frac{P^2}{\hbar^2} \exp\left[i\left(\frac{P}{\hbar} x - \frac{E}{\hbar} t\right)\right],$$

$$\frac{\partial \Psi}{\partial t} = -ai \frac{E}{\hbar} \exp\left[i\left(\frac{P}{\hbar} x - \frac{E}{\hbar} t\right)\right],$$

and, therefore,

$$-\frac{\partial^2 \Psi}{\partial x^2} \frac{\hbar^2}{P^2} = -\frac{\partial \Psi}{\partial t} \frac{\hbar}{Ei}$$

or

$$-\frac{\hbar^2}{P^2} E \frac{\partial^2 \Psi}{\partial x^2} - i\hbar \frac{\partial \Psi}{\partial t} = 0.$$

The total energy E of the electron is made up of

$$E = \text{K.E.} + \text{P.E.} = \frac{P^2}{2m} + V$$

and, therefore,

$$-\frac{\hbar^2}{2m}\frac{\partial^2 \Psi}{\partial x^2} - \frac{\hbar^2}{P^2}V\frac{\partial^2 \Psi}{\partial x^2} - i\hbar\frac{\partial \Psi}{\partial t} = 0. \tag{1.21}$$

Further, from Eq. (1.20),

$$\frac{\partial^2 \Psi}{\partial x^2} = -\frac{P^2}{\hbar^2}\Psi,$$

and, therefore, Eq. (1.21) can be rewritten

$$-\frac{\hbar^2}{2m}\frac{\partial^2 \Psi}{\partial x^2} + V\Psi - i\hbar\frac{\partial \Psi}{\partial t} = 0. \tag{1.22}$$

Equation (1.22) is the time-dependent Schrödinger equation. It governs the motion of the electron. It is a partial differential equation of Ψ involving x and t as independent variables. The equation can be simplified by using the standard procedure know as the *method of separation of variables*. Assuming Ψ can be expressed as the product of two functions $\varphi(x)$ and $T(t)$,

$$\Psi = \varphi(x)T(t), \tag{1.23}$$

$$\frac{\partial \Psi}{\partial t} = \varphi\frac{dT}{dt},$$

$$\frac{\partial^2 \Psi}{\partial x^2} = T\frac{d^2\varphi}{dx^2}.$$

Using these results in the time-dependent Schrödinger equation,

$$-\frac{\hbar^2}{2m}T\frac{d^2\varphi}{dx^2} + V\varphi T - i\hbar\varphi\frac{dT}{dt} = 0.$$

Dividing the above equation by φT,

$$-\frac{\hbar^2}{2m\varphi}\frac{d^2\varphi}{dx^2} + V = i\hbar\frac{1}{T}\frac{dT}{dt}. \tag{1.24}$$

In Eq. (1.24) the left side is a function of x only and the right side a function of t only. This could be possible only if both sides equal a constant:

$$i\hbar\frac{1}{T}\frac{dT}{dt} = \text{const} \quad \text{or} \quad T = e^{\text{const}(t/i\hbar)}.$$

Using the above result,

$$\Psi = \varphi(x)e^{\text{const}(t/i\hbar)}. \tag{1.25}$$

Also, from Eq. (1.20),

$$\Psi = ae^{iPx/\hbar}e^{-iEt/\hbar}. \tag{1.26}$$

A comparison of Eqs. (1.25) and (1.26) suggests that the constant must be E, the total energy of the electron. Hence, Eq. (1.24) can be rewritten

$$-\frac{\hbar^2}{2m}\frac{d^2\varphi}{dx^2} + V\varphi = E\varphi. \tag{1.27}$$

This is called the *Schrödinger amplitude equation* or simply the *time-independent Schrödinger equation*. In Eq. (1.27), V, the potential energy, has not been defined. Depending upon the value of V, the Schrödinger equation is really many sets of equations, one for each value of V.

The wave function φ governs the motion of electrons in the same way that light waves determine the motion of photons. Whenever φ vanishes, there are no electrons. If, then, a solution is sought that is to represent the motion of an electron in atom, φ must certainly vanish at infinity, for the bound electrons are all confined to a finite region. In addition, a physically useful solution should be a continuous and single-valued function of the position in the finite region. These conditions are simple but sufficient to determine characteristic values, which, in the old quantum mechanics, were singled out from among the many mechanically possible solutions only by imposing auxiliary conditions. The theory of differential equations shows that the conditions of single-valuedness and continuity can be satisfied in the solution of φ in Eq. (1.27) only for certain values of the parameter E. These values of E are called *proper values* or *characteristic values* or *eigenvalues*.

The electron is not a localized charge within the atom, but the charge and mass are "smeared out" over a certain region. The probability of finding an electron at a specific place is given by a mathematical function $\varphi\varphi^*$, where φ and φ^* are conjugate complexes. The probability of finding an electron in an element of volume ΔV is equal to $\varphi\varphi^* \, \Delta V$.

The solution of the wave equation (1.27) gives the probability of finding the one electron at different places. As already stated, the probability of finding the electron in an element of volume is proportional to the value of $\varphi\varphi^*$ at the place concerned. The value of the constant of proportionality is obtained by a process known as *normalization of the wave function*. If, for example, it is known that the electron is in an enclosure of volume V, the integral $\int \varphi\varphi^* \, dV$ over the volume V must equal 1, because the probability of finding the electron somewhere in the volume is unity.

1.3 Potential box

As the first example of the solution of the Schrödinger wave equation (1.27), we shall consider an electron confined to a potential box, a cube of side L (Fig. 1.4). It is assumed that the electron is a free electron $V = 0$

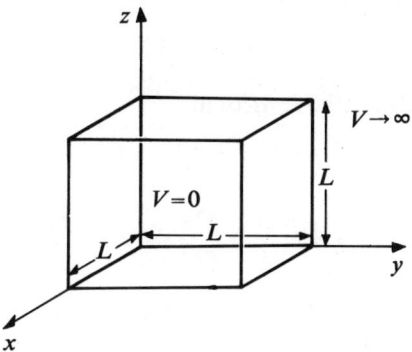

Fig. 1.4
Potential box

within the box. We are assuming the electron to be confined in the box, so the potential energy V is assumed to be infinite outside the box. Looking at (1.27),

$$-\frac{\hbar^2}{2m}\frac{d^2\varphi}{dx^2} + V\varphi = E\varphi,$$

we find that it governs the motion of the electron along the x axis. If the electron is allowed to move in any arbitrary direction, we would expect its motion to depend upon y and z in the same manner as it depends upon x, and the three-dimensional Schrödinger equation should be

$$-\frac{\hbar^2}{2m}\nabla^2\varphi + V\varphi = E\varphi, \tag{1.28}$$

where

$$\nabla^2 = \frac{\partial^2}{\partial x^2} + \frac{\partial^2}{\partial y^2} + \frac{\partial^2}{\partial z^2}.$$

Because it would take an infinite amount of work to get an electron outside the box, the probability density outside the box and at all six faces of the cube would be zero.

$V = 0$, so Eq. (1.28) reduces to

$$-\frac{\hbar^2}{2m}\nabla^2\varphi = E\varphi \tag{1.29}$$

or

$$\nabla^2\varphi + \frac{2mE}{\hbar^2}\varphi = 0.$$

The potential energy of the electron is zero; therefore, the total energy E is only the kinetic energy and hence

$$2mE = P^2.$$

Also
$$P^2/\hbar^2 = k^2.$$

Therefore, Eq. (1.29) can be rewritten

$$\nabla^2 \varphi + k^2 \varphi = 0 \tag{1.30}$$

or

$$\frac{\partial^2 \varphi}{\partial x^2} + \frac{\partial^2 \varphi}{\partial y^2} + \frac{\partial^2 \varphi}{\partial z^2} + k^2 \varphi = 0.$$

Using the method of separation of variables and assuming

$$\varphi(x,y,z) = X(x)Y(y)Z(z),$$

$$k^2 \varphi + \nabla^2 \varphi = YZ \frac{d^2 X}{dx^2} + ZX \frac{d^2 Y}{dy^2} + XY \frac{d^2 Z}{dz^2} + k^2 XYZ = 0.$$

Dividing by XYZ

$$\frac{1}{X}\frac{d^2 X}{dx^2} + \frac{1}{Y}\frac{d^2 Y}{dy^2} + \frac{1}{Z}\frac{d^2 Z}{dz^2} + k^2 = 0.$$

If the kinetic energy is factored into three directional components,

$$E = E_x + E_y + E_z = \frac{P_x^2}{2m} + \frac{P_y^2}{2m} + \frac{P_z^2}{2m}$$

or

$$k^2 = k_x^2 + k_y^2 + k_z^2,$$

and hence

$$\frac{1}{X}\frac{d^2 X}{dx^2} + \frac{1}{Y}\frac{d^2 Y}{dy^2} + \frac{1}{Z}\frac{d^2 Z}{dz^2} + k_x^2 + k_y^2 + k_z^2 = 0.$$

If the three parts, involving variation with x, y, and z, respectively, are to remain independent of each other,

$$\frac{1}{X}\frac{d^2 X}{dx^2} + k_x^2 = 0, \tag{1.31}$$

$$\frac{1}{Y}\frac{d^2 Y}{dy^2} + k_y^2 = 0, \tag{1.32}$$

$$\frac{1}{Z}\frac{d^2 Z}{dz^2} + k_z^2 = 0. \tag{1.33}$$

Considering Eq. (1.31),

$$\frac{d^2 X}{dx^2} + k_x^2 X = 0.$$

Solving the above differential equation,

$$X = Ae^{ik_zx} + Be^{-ik_zx}, \qquad (1.34)$$

where A and B are constants of integration:

$$X(x) = (A + B)\cos k_x x + i(A - B)\sin k_x x.$$

The constants A and B could be determined from the boundary conditions. Along the x axis $X = 0$ at $x = 0$ and $x = L$ because there is no probability of finding the electron at the faces. Therefore,

$$0 = (A + B)$$

and

$$0 = (A + B)\cos k_x L + i(A - B)\sin k_x L$$

or

$$0 = i(A - B)\sin k_x L = 2iA \sin k_x L.$$

This is only possible if

$$k_x L = n_x \pi,$$

where n_x is a nonzero integer, because if $n_x = 0$, $X(x)$ would be zero throughout the box.

Now $k_x = (2mE_x)^{1/2}\hbar^{-1}$.
Therefore,

$$\frac{2mE_x}{\hbar^2} L^2 = n_x^2 \pi^2$$

or

$$E_x = \frac{n_x^2 \pi^2 \hbar^2}{2mL^2}. \qquad (1.35)$$

Equation (1.35) suggests that not all values of the kinetic energy in the x axis are possible. The energy is quantized according to the equation. The wave function in the x axis is

$$X(x) = i(A - B) \sin(n_x \pi / L)x.$$

To determine the coefficient $A - B$, we use the fact that since the electron is localized in the box, the probability of finding it anywhere from $x = 0$ to $x = L$ is unity:

$$\int_0^L X(x)X^*(x)\, dx = (A - B)^2 \int_0^L \sin^2 \frac{n_x \pi x}{L} dx = 1$$

or

$$(A - B)^2 = 2/L.$$

Hence

$$X(x) = i(2/L)^{1/2} \sin(n_x \pi x / L).$$

In a similar way,

$$Y(y) = i(2/L)^{1/2} \sin(n_y \pi y/L) \quad \text{and} \quad Z(z) = i(2/L)^{1/2} \sin(n_z \pi z/L),$$

where n_y and n_z are nonzero integers.

The total energy of the electron E is

$$E = E_x + E_y + E_z = \frac{\pi^2 \hbar^2}{2mL^2}(n_x^2 + n_y^2 + n_z^2). \tag{1.36}$$

Each set of integers n_x, n_y, and n_z in this case determines both φ and E and is said to specify a possible state of the electron. Because n_x, n_y, and n_z can take infinite values, there is no limit to the number of possible states of the electron. It would be of interest to calculate the number of states between energy interval E and $E + \Delta E$.

It can be seen from Eq. (1.36) that the same energy value can be obtained for various combinations of the same three integers n_x, n_y, and n_z, for example, 3, 2, 1; 2, 3, 1; and 1, 2, 3. Each combination of integers represents the same energy but a different wave function. Such energy states are called *threefold-degenerate*.

According to Eq. (1.36) the energy states are discrete, but because the adjacent energy states differ only by a negligibly small amount, the energy states may be considered as continuously distributed.

Density of states. *Density of states* is defined as the number of states per unit volume in an energy interval. We would therefore first calculate the number of states between energy interval E and $E + \Delta E$ in the box and then divide the number of states by the volume of the box to obtain the density of states. Drawing a coordinate system with n_x, n_y, and n_z, the energy E of the electron for a given value of n_x, n_y, and n_z depends upon the distance R of n_x, n_y, and n_z values from the origin ($n_x = n_y = n_z = 0$):

$$R^2 = n_x^2 + n_y^2 + n_z^2.$$

Therefore, Eq. (1.36) reduces to

$$E = \frac{\pi^2 \hbar^2}{2mL^2} R^2. \tag{1.37}$$

Any change in the values of the three quantum numbers changes R and hence E. The problem of finding the number of states between energy interval E and $E + \Delta E$ is therefore equivalent to the problem of finding the number of states in a shell of thickness ΔR at a distance R. Because n_x, n_y, and n_z are only positive values, the number of points we are trying to calculate is the volume of an octant of thickness ΔR.

Solving the above differential equation,
$$X = Ae^{ik_xx} + Be^{-ik_xx}, \tag{1.34}$$
where A and B are constants of integration:
$$X(x) = (A + B)\cos k_x x + i(A - B)\sin k_x x.$$

The constants A and B could be determined from the boundary conditions. Along the x axis $X = 0$ at $x = 0$ and $x = L$ because there is no probability of finding the electron at the faces. Therefore,
$$0 = (A + B)$$
and
$$0 = (A + B)\cos k_x L + i(A - B)\sin k_x L$$
or
$$0 = i(A - B)\sin k_x L = 2iA \sin k_x L.$$
This is only possible if
$$k_x L = n_x \pi,$$
where n_x is a nonzero integer, because if $n_x = 0$, $X(x)$ would be zero throughout the box.

Now $k_x = (2mE_x)^{1/2}\hbar^{-1}$.
Therefore,
$$\frac{2mE_x}{\hbar^2}L^2 = n_x^2 \pi^2$$
or
$$E_x = \frac{n_x^2 \pi^2 \hbar^2}{2mL^2}. \tag{1.35}$$

Equation (1.35) suggests that not all values of the kinetic energy in the x axis are possible. The energy is quantized according to the equation. The wave function in the x axis is
$$X(x) = i(A - B)\sin(n_x\pi/L)x.$$

To determine the coefficient $A - B$, we use the fact that since the electron is localized in the box, the probability of finding it anywhere from $x = 0$ to $x = L$ is unity:
$$\int_0^L X(x)X^*(x)\,dx = (A - B)^2 \int_0^L \sin^2 \frac{n_x\pi x}{L}\,dx = 1$$
or
$$(A - B)^2 = 2/L.$$
Hence
$$X(x) = i(2/L)^{1/2} \sin(n_x\pi x/L).$$

In a similar way,

$$Y(y) = i(2/L)^{1/2} \sin(n_y \pi y/L) \quad \text{and} \quad Z(z) = i(2/L)^{1/2} \sin(n_z \pi z/L),$$

where n_y and n_z are nonzero integers.

The total energy of the electron E is

$$E = E_x + E_y + E_z = \frac{\pi^2 \hbar^2}{2mL^2}(n_x^2 + n_y^2 + n_z^2). \tag{1.36}$$

Each set of integers n_x, n_y, and n_z in this case determines both φ and E and is said to specify a possible state of the electron. Because n_x, n_y, and n_z can take infinite values, there is no limit to the number of possible states of the electron. It would be of interest to calculate the number of states between energy interval E and $E + \Delta E$.

It can be seen from Eq. (1.36) that the same energy value can be obtained for various combinations of the same three integers n_x, n_y, and n_z, for example, 3, 2, 1; 2, 3, 1; and 1, 2, 3. Each combination of integers represents the same energy but a different wave function. Such energy states are called *threefold-degenerate*.

According to Eq. (1.36) the energy states are discrete, but because the adjacent energy states differ only by a negligibly small amount, the energy states may be considered as continuously distributed.

Density of states. Density of states is defined as the number of states per unit volume in an energy interval. We would therefore first calculate the number of states between energy interval E and $E + \Delta E$ in the box and then divide the number of states by the volume of the box to obtain the density of states. Drawing a coordinate system with n_x, n_y, and n_z, the energy E of the electron for a given value of n_x, n_y, and n_z depends upon the distance R of n_x, n_y, and n_z values from the origin ($n_x = n_y = n_z = 0$):

$$R^2 = n_x^2 + n_y^2 + n_z^2.$$

Therefore, Eq. (1.36) reduces to

$$E = \frac{\pi^2 \hbar^2}{2mL^2} R^2. \tag{1.37}$$

Any change in the values of the three quantum numbers changes R and hence E. The problem of finding the number of states between energy interval E and $E + \Delta E$ is therefore equivalent to the problem of finding the number of states in a shell of thickness ΔR at a distance R. Because n_x, n_y, and n_z are only positive values, the number of points we are trying to calculate is the volume of an octant of thickness ΔR.

If ΔG is the number of states in the interval between R and $R + \Delta R$,

$$\Delta G = \tfrac{1}{8} 4\pi R^2 \Delta R = \frac{\pi R^2}{2} \Delta R.$$

From Eq. (1.37),

$$\frac{1}{2} \frac{2mL^2}{\pi^2 \hbar^2} dE = R\, dR,$$

$$\frac{(2mL^2)^{1/2}}{\pi \hbar} E^{1/2} = R.$$

Therefore,

$$dG = \frac{\pi}{2} R R\, dR = \left(\frac{2mL^2}{\pi^2 \hbar^2}\right)^{3/2} \frac{\pi}{4} E^{1/2}\, dE$$

or

$$dG = \frac{4\pi}{h^3} m^{3/2} L^3 (2E)^{1/2}\, dE.$$

$$\frac{dG}{\text{unit vol.}} = \frac{4\pi}{h^3} m^{3/2} (2E)^{1/2}\, dE = G(E) dE. \quad (1.38)$$

The energy and momentum for a free electron are related by

$$E = P^2/2m,$$

so it can be easily shown that the density of states in the momentum interval P and $P + dP$ is given by

$$\frac{dG}{\text{unit vol.}} = \frac{4\pi P^2}{h^3} dP. \quad (1.39)$$

1.4 Potential well

In the case of a potential box we were concerned with the solution of the Schrödinger equation. We shall now take up the case of a potential well. Considering the one-dimensional motion of the electron in a potential

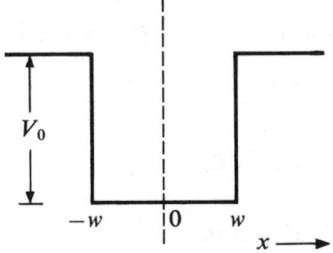

Fig. 1.5
Potential well

configuration given by Fig. 1.5, we have the following values of the potential energy:

$$V = V_0 \quad \text{for } x \leq -w,$$
$$V = 0 \quad \begin{cases} \text{for } x \leq w, \\ \text{for } -w \leq x, \end{cases}$$
$$V = V_0 \quad \text{for } x \geq w.$$

Let us assume that the total energy of the particle is less than the height V_0 of the well, so the particle would not have the energy to escape the wall and hence the well. As a result of our analysis we shall see that in spite of this restriction, there is a finite probability of finding the particle outside the well. It is a direct consequence of the quantum-mechanical treatment of the problem.

According to Eq. (1.27),

$$\frac{d^2\varphi}{dx^2} - \frac{2m}{\hbar^2}(V - E)\varphi = 0.$$

Dividing the values of x from $-\infty$ to ∞ in three regions and calling wave functions

φ_l in the region $-\infty$ to $-w$,

φ_c in the region $-w$ to w,

φ_r in the region w to ∞,

the Schrödinger equation could be split into three differential equations for the three regions:

$$\frac{d^2\varphi_l}{dx^2} - \frac{2m}{\hbar^2}(V_0 - E)\varphi_l = 0,$$

$$\frac{d^2\varphi_c}{dx^2} + \frac{2mE}{\hbar^2}\varphi_c = 0,$$

$$\frac{d^2\varphi_r}{dx^2} - \frac{2m}{\hbar^2}(V_0 - E)\varphi_r = 0.$$

Substituting,

$$\frac{2m}{\hbar^2}(V_0 - E) = \beta^2 \quad (V_0 > E \text{ assumed}),$$

$$\frac{2mE}{\hbar^2} = \alpha^2,$$

$$\frac{d^2\varphi_l}{dx^2} - \beta^2\varphi_l = 0,$$

$$\frac{d^2\varphi_c}{dx^2} + \alpha^2\varphi_c = 0,$$

$$\frac{d^2\varphi_r}{dx^2} - \beta^2\varphi_r = 0,$$

The solutions of the above three equations are

$$\varphi_l = Ae^{\beta x} + Be^{-\beta x},$$
$$\varphi_c = Ce^{i\alpha x} + De^{-i\alpha x},$$
$$\varphi_r = Ee^{\beta x} + Fe^{-\beta x},$$

where A, B, C, D, E, and F are arbitrary constants.

Because the probability of finding a particle at infinity is not infinite and $e^{\beta x}$ would get very large if $x \to \infty$, E must be zero. Similarly, B must be zero. As such,

$$\varphi_l = Ae^{\beta x},$$
$$\varphi_c = Ce^{i\alpha x} + De^{-i\alpha x},$$
$$\varphi_r = Fe^{-\beta x}.$$

The general character of the above equations is shown in Fig. 1.6 on the assumption that φ is real and A and F are positive. Now, to arrive at a

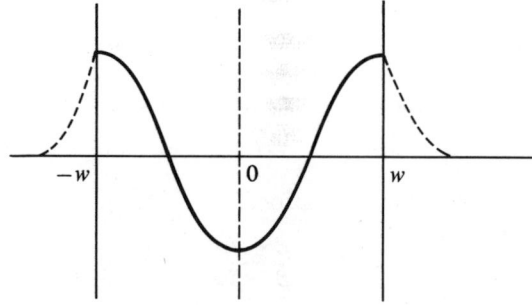

Fig. 1.6
Wave function for a potential well

wave function describing the motion of the electron, the wave function must be single-valued and continuous. As such,

$$\varphi_l(-w) = \varphi_c(-w), \qquad \varphi_r(w) = \varphi_c(w).$$

Because there are no infinite discontinuities in V,

$$\frac{d\varphi_l}{dx}(-w) = \frac{d\varphi_c}{dx}(-w), \qquad \varphi'_r(w) = \varphi'_c(w).$$

To satisfy the above boundary conditions, we take φ_l in the equation as it stands and adjust φ_c to satisfy the single-valuedness and continuity equations. This can be easily done without difficulty, because φ_c contains two adjustable constants, and there are just two conditions to satisfy:

$$\varphi_l(-w) = Ae^{-\beta w} = \varphi_c(-w) = Ce^{-i\alpha w} + De^{i\alpha w},$$

$$\varphi'_l(-w) = \beta Ae^{-\beta w} = \varphi'_c(-w) = Ci\alpha e^{-i\alpha w} - Di\alpha e^{i\alpha w}.$$

Therefore,

$$Ae^{-\beta w} = Ce^{-i\alpha w} + De^{i\alpha w},$$

$$-A\beta e^{-\beta w} = -Ci\alpha e^{-i\alpha w} + Di\alpha e^{i\alpha w}.$$

Eliminating A from the above equations,

$$D = C\left(-\frac{\beta - i\alpha}{\beta + i\alpha} e^{-2i\alpha w}\right). \tag{1.40}$$

We next try to satisfy the continuity and single-valuedness condition at w by adjusting φ_r:

$$\varphi_r = Fe^{-\beta x},$$

$$\varphi_c = Ce^{i\alpha w} + De^{-i\alpha w}.$$

At $x = w$,

$$\varphi_r(w) = Fe^{-\beta w}, \qquad \varphi'_r(w) = -F\beta e^{-\beta w},$$

$$\varphi_c(w) = Ce^{i\alpha w} + De^{-i\alpha w}, \qquad \varphi'_c(w) = Ci\alpha e^{i\alpha w} - Di\alpha e^{-i\alpha w},$$

$$Fe^{-\beta w} = Ce^{i\alpha w} + De^{-i\alpha w}, \qquad -F\beta e^{-\beta w} = Ci\alpha e^{i\alpha w} - Di\alpha e^{-i\alpha w}.$$

Eliminating F from the above equations,

$$D = C\left(-\frac{\beta + i\alpha}{\beta - i\alpha} e^{2i\alpha w}\right). \tag{1.41}$$

Equations (1.40) and (1.41) are in general inconsistent, and therefore it is impossible to find a value of φ which satisfies the conditions of single-valuedness and continuity in the three regions. In other words, the Schrödinger equation does not possess a well-behaved function for any arbitrary value of the energy of the particle E lying between 0 and V_0. Conclusions can be drawn that not every energy lying between 0 and V_0 is a possible value of the energy of the particle moving in the field. It now remains to investigate the circumstance in which we obtain consistency. From Eqs. (1.40) and (1.41), equating D/C,

$$\frac{\beta + i\alpha}{\beta - i\alpha} e^{2i\alpha w} = \frac{\beta - i\alpha}{\beta + i\alpha} e^{-2i\alpha w}, \tag{1.42}$$

and we obtain the relationship between β, α, and w. Only if the above relationship is satisfied would we obtain a solution of Schrödinger's equation over the three regions which would be single-valued and continuous. From Eq. (1.42),

$$(\beta + i\alpha)^2 e^{2i\alpha w} = (\beta - i\alpha)^2 e^{-2i\alpha w},$$
$$(\beta^2 + 2i\alpha\beta - \alpha^2)e^{2i\alpha w} = (\beta^2 - \alpha^2 - 2i\alpha\beta)e^{-2i\alpha w},$$
$$(\beta^2 - \alpha^2)(e^{2i\alpha w} - e^{-2i\alpha w}) = -(2i\alpha\beta e^{-2i\alpha w} + 2i\alpha\beta e^{2i\alpha w}).$$

Simplifying,

$$i(\beta^2 - \alpha^2)\sin 2\alpha w = -2i\alpha\beta \cos 2\alpha w$$

or

$$\tan 2\alpha w = -\frac{2\alpha\beta}{\beta^2 - \alpha^2}.$$

Substituting the values of α and β,

$$\tan\left[2w\left(\frac{2mE}{\hbar^2}\right)^{1/2}\right] = -\frac{2(2m/\hbar^2)[E(V_0 - E)]^{1/2}}{(2m/\hbar^2)V_0 - (2mE/\hbar^2) - (2mE/\hbar^2)}$$

or

$$\tan\left[2w\left(\frac{2mE}{\hbar^2}\right)^{1/2}\right] = \frac{2[E(V_0 - E)]^{1/2}}{2E - V_0}. \quad (1.43)$$

The values of E lying between 0 and V_0 for which our Schrödinger equation possesses well-behaved functions as solutions are the roots of this transcendental equation lying between 0 and V_0. The roots that turn out to be real, discrete, and finite in number are the energy levels of our dynamical system lying between 0 and V_0. They are $E_1, E_2, E_3, \ldots$.

A discussion of the potential well would not be complete without finding the explicit form of the Schrödinger function φ_n belonging to an energy value E_n lying between 0 and V_0. From Eq. (1.42),

$$\left(\frac{\beta + i\alpha}{\beta - i\alpha}\right)^2 e^{4i\alpha w} = 1 \quad \text{or} \quad \frac{\beta + i\alpha}{\beta - i\alpha}e^{2i\alpha w} = \pm 1.$$

As such,

$$\varphi_n = \begin{cases} C(e^{-i\alpha_n w} \pm e^{i\alpha_n w})e^{\beta_n(w+x)} \cdots\cdots\cdot \text{for } x < -w, \\ C(e^{i\alpha_n x} \pm e^{-i\alpha_n x}) \cdots\cdots\cdots \text{for } -w < x < w, \\ C(e^{i\alpha_n w} \pm e^{-i\alpha_n w})e^{\beta_n(w-x)} \cdots\cdots\cdot \text{for } w \le x. \end{cases}$$

It can be shown that if $E > V_0$, all energies would be possible.

1.5 Quantum oscillator

It has already been discussed in Sec. 1.2 that a vibrating mass m on the end of a spring of Hooke's-law constant f is called *an oscillator;* its motion is governed by the Schrödinger equation. A harmonic force $F = -fx$ and the potential energy $V = -\int F\,dx = \int fx\,dx = fx^2/2$.

We are discussing a one-dimensional oscillator, so the motion of the one-dimensional quantum oscillator would be given by the one-dimensional Schrödinger equation:

$$\frac{d^2\varphi}{dx^2} + \frac{2m}{\hbar^2}\left(E - \frac{fx^2}{2}\right)\varphi = 0. \tag{1.44}$$

Let $x = sa$, where s is the new variable and a is a constant to be chosen later. Equation (1.44) can be rewritten

$$-\frac{\hbar^2}{2m}\frac{d^2\varphi}{dx^2} + \frac{fx^2}{2}\varphi = E\varphi,$$

$$-\frac{\hbar^2}{2ma^2}\frac{d^2\varphi}{ds^2} + \frac{fa^2s^2}{2}\varphi = E\varphi,$$

or

$$-\frac{d^2\varphi}{ds^2} + \frac{mfa^4s^2}{\hbar^2}\varphi = E\varphi\frac{2ma^2}{\hbar^2}.$$

Let $a = (\hbar^2/mf)^{1/4}$

and define

$$\lambda = \frac{2mE}{\hbar^2}a^2 = \frac{2mE}{\hbar^2}\left(\frac{\hbar^2}{mf}\right)^{1/2} = \frac{2}{\hbar}\left(\frac{m}{f}\right)^{1/2}E$$

or

$$\lambda = \frac{4\pi}{h}\left(\frac{m}{f}\right)^{1/2}E.$$

The frequency of the classical oscillators is

$$\nu = \frac{1}{2\pi}\left(\frac{f}{m}\right)^{1/2}$$

and, therefore,

$$\lambda = 2E/h\nu.$$

The Schrödinger equation for this case reduces to

$$-\frac{d^2\varphi}{ds^2} + s^2\varphi = \lambda\varphi. \tag{1.45}$$

Let $\varphi(s) = v(s)\,e^{-s^2/2}$ be a solution of Eq. (1.45). Substituting the solution in Eq. (1.45),

$$\frac{d\varphi}{ds} = \frac{dv(s)}{ds}e^{-s^2/2} - v(s)s\,e^{-s^2/2} = \frac{dv(s)}{ds}e^{-s^2/2} - \varphi(s)s,$$

$$\frac{d^2\varphi}{ds^2} = \frac{d^2v(s)}{ds^2}e^{-s^2/2} - \frac{dv(s)}{ds}se^{-s^2/2} - \frac{d\varphi(s)}{ds}s - \varphi(s),$$

or

$$\frac{d^2\varphi}{ds^2} = \frac{d^2v(s)}{ds^2}e^{-s^2/2} - \frac{dv(s)}{ds}se^{-s^2/2} + s^2v(s)e^{-s^2/2} - v(s)e^{-s^2/2}.$$

Therefore, if $\varphi(s) = v(s) e^{-s^2/2}$ is a solution of the Schrödinger equation (1.45), the following condition must be satisfied:

$$\frac{d^2v(s)}{ds^2} e^{-s^2/2} - 2s \frac{dv(s)}{ds} e^{-s^2/2} + (\lambda - 1)v(s)e^{-s^2/2} = 0$$

or

$$e^{-s^2/2}\left[\frac{d^2v(s)}{ds^2} - 2s\frac{dv(s)}{ds} + (\lambda - 1)v(s)\right] = 0,$$

$e^{-s^2/2} \neq 0$ or otherwise $\varphi(s) = 0$.

Therefore,

$$\frac{d^2v(s)}{ds^2} - 2s\frac{dv(s)}{ds} + (\lambda - 1)v(s) = 0. \tag{1.46}$$

Equation (1.46) is a differential equation of the second order. Let $v(s)$ be represented by a power series

$$v(s) = \sum_n a_n s^n.$$

As such,

$$\frac{dv(s)}{ds} = \sum_n a_n n s^{n-1},$$

$$\frac{d^2v(s)}{ds^2} = \sum_n a_n n(n-1) s^{n-2}.$$

Therefore,

$$\sum_n a_n n(n-1) s^{n-2} + \sum_n a_n(\lambda - 1 - 2n) s^n = 0. \tag{1.47}$$

If Eq. (1.47) is to hold for all values of s, the coefficient of each power of s must vanish. The coefficient of s^n would be given by

$$a_{n+2}(n+2)(n+1) + a_n(\lambda - 1 - 2n),$$

and equating the coefficient to zero,

$$a_{n+2} = \frac{2n + 1 - \lambda}{(n+2)(n+1)} a_n. \tag{1.48}$$

Equation (1.48) is a recursion relation for the coefficients. The initial term of the series can be obtained by assuming the series to start with $n = \tau$ or $a_\tau \neq 0$. The first part of the equation contributes terms lower than τ, and therefore the coefficient of the term $s^{\tau-2}$ must be zero. Hence

$$a_\tau \tau(\tau - 1) = 0$$

if $\tau = 0$ or $\tau = 1$.

The series starts either with zero or unity values of τ. From Eq. (1.48) we would conclude that $\tau = 0$ would lead to an even power series and $\tau = 1$ to an odd power series.

$$v_0(s) = \sum_{n=0}^{\text{even powers of } s} a_n s^n,$$

$$v_1(s) = \sum_{n=1}^{\text{odd powers of } s} a_n s^n.$$

Therefore, the general solution for φ is

$$\varphi = A\varphi_0(s) + B\varphi_1(s),$$

where A and B are constant.

For certain values of λ, the numerator of the recursion relation (1.48) is zero and the power series terminate. If the series terminate for $n = i$,

$$2i + 1 - \lambda = 0,$$

$$\lambda = \frac{2E}{h\nu} = (2i + 1),$$

or

$$E = \frac{h\nu}{2}(2i + 1) = h\nu(i + \tfrac{1}{2}), \tag{1.49}$$

where $i = 0, 1, 2, 3, \ldots$. The corresponding acceptable wave functions are

$$\varphi_0 = e^{-s^2/2} \sum_{n=1}^{i} a_n s^n \quad \text{when } i \text{ is even,}$$

$$\varphi_1 = e^{-s^2/2} \sum_{n=1}^{i} a_n s^n \quad \text{when } i \text{ is odd.}$$

Equation (1.49) suggests that the possible values of energy of a one-dimensional quantum oscillator are quantized. The energy levels are $h\nu/2$, $3h\nu/2$, $5h\nu/2, \ldots$. This is in contrast to the classical theory, in which an oscillator could have all possible values of energies. $\varphi_n^2(x)$, where $n = 0, 1, 2, \ldots$, would be the probability per unit length of finding an electron of energy $E_n = (n + \tfrac{1}{2})h\nu$ in a small length dx located at x. We defined this as the probability density.

1.6 Hydrogen atom

In Sec. 1.1 the classical theory of the structure of a hydrogen atom, the Bohr-Sommerfeld theory, was discussed. It would be very interesting to study the hydrogen atom on a quantum-mechanical basis. This study is essential, as it helps to explain various electrical properties of materials. The hydrogen atom consists of a proton of charge $+e$ and an electron of

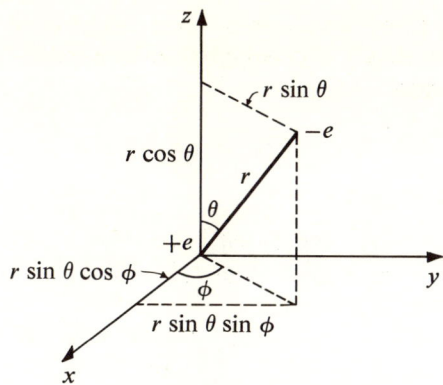

Fig. 1.7
Spherical coordinates

charge $-e$ and mass m. We shall regard the position of the proton as fixed from the origin and the electron as at a point whose spherical coordinates are r, θ, and ϕ (Fig. 1.7). The necessity of expressing the position of the electron with spherical coordinates originated from the fact that the potential energy of the electron in the field of force of the nucleus is a function of r or $V = V(r)$.

The three-dimensional Schrödinger equation is given by Eq. (1.28):

$$-\frac{\hbar^2}{2m}\nabla^2\varphi + V\varphi = E\varphi. \qquad (1.28)$$

This equation has to be solved for φ for a hydrogen atom. From Fig. 1.7,

$x = r\sin\theta\cos\phi,$
$y = r\sin\theta\sin\phi,$
$z = r\cos\theta,$
$r^2 = x^2 + y^2 + z^2,$

$$\nabla^2 = \frac{\partial^2}{\partial x^2} + \frac{\partial^2}{\partial y^2} + \frac{\partial^2}{\partial z^2} = \frac{1}{r^2}\frac{\partial}{\partial r}\left(r^2\frac{\partial}{\partial r}\right) + \frac{1}{r^2\sin\theta}\frac{\partial}{\partial\theta}\left(\sin\theta\frac{\partial}{\partial\theta}\right) + \frac{1}{r^2\sin^2\theta}\frac{\partial^2}{\partial\phi^2}.$$

Therefore the Schrödinger equation (1.28) can be rewritten in terms of r, θ, and ϕ as

$$\frac{1}{r^2}\frac{\partial}{\partial r}\left(r^2\frac{\partial\varphi}{\partial r}\right) + \frac{1}{r^2\sin\theta}\frac{\partial}{\partial\theta}\left(\sin\theta\frac{\partial\varphi}{\partial\theta}\right) + \frac{1}{r^2\sin^2\theta}\frac{\partial^2\varphi}{\partial\phi^2}$$
$$+ \frac{2m}{\hbar^2}[E - V(r)]\varphi = 0, \qquad (1.50)$$

where

$$V(r) = -\int F\,dr = \int_\infty^r -\frac{e^2}{4\pi\epsilon_0 r^2}\,dr = -\frac{e^2}{4\pi\epsilon_0 r}.$$

Equation (1.50) is a partial differential equation of φ involving r, θ, and ϕ as independent variables. The equation can be used by the method of separation of variables.

Let
$$\varphi = f_1(r)f_2(\theta)f_3(\phi). \tag{1.51}$$

Substituting Eq. (1.51) in Eq. (1.50),

$$f_2 f_3 \frac{1}{r^2}\frac{d}{dr}\left(r^2\frac{df_1}{dr}\right) + \frac{f_1 f_3}{r^2 \sin\theta}\frac{d}{d\theta}\left(\sin\theta\frac{df_2}{d\theta}\right) + \frac{f_1 f_2}{r^2 \sin^2\theta}\frac{d^2 f_3}{d\phi^2}$$
$$+ \frac{2m}{\hbar^2}[E - V(r)]f_1 f_2 f_3 = 0.$$

Dividing by $f_1 f_2 f_3$ and multiplying by $r^2 \sin^2\theta$,

$$\frac{\sin^2\theta}{f_1}\frac{d}{dr}\left(r^2\frac{df_1}{dr}\right) + \frac{\sin\theta}{f_2}\frac{d}{d\theta}\left(\sin\theta\frac{df_2}{d\theta}\right) + \frac{2m}{\hbar^2}[E - V(r)]$$
$$\times r^2 \sin^2\theta = -\frac{1}{f_3}\frac{d^2 f_3}{d\phi^2}. \tag{1.52}$$

Because the right side of Eq. (1.52) is only a function of ϕ and the left side only a function of r and θ, both equal a constant. Let this constant be m_l^2. Therefore,

$$-\frac{1}{f_3}\frac{d^2 f_3}{d\phi^2} = m_l^2$$

or
$$\frac{d^2 f_3}{d\phi^2} + m_l^2 f_3 = 0. \tag{1.53}$$

Solving Eq. (1.53),
$$f_3 = \text{const } e^{+im_l\phi}. \tag{1.54}$$

f_3 is a part of φ given by Eq. (1.51) and as such must be single-valued if φ is to be single-valued. By observation of Eq. (1.54) it can be seen that f_3 is single-valued at $\phi = 0$ and $\phi = 2\pi$ if m_l is an integer. m_l is a quantum number; it is an integer and can have values ± 1, ± 2, ± 3, ... The quantum number is called the *magnetic number* m_l. We shall discuss it further later.

Equation (1.52) can be rewritten

$$\frac{\sin^2\theta}{f_1}\frac{d}{dr}\left(r^2\frac{df_1}{dr}\right) + \frac{\sin\theta}{f_2}\frac{d}{d\theta}\left(\sin\theta\frac{df_2}{d\theta}\right) + \frac{2m}{\hbar^2}[E - V(r)]r^2 \sin^2\theta = m_l^2. \tag{1.55}$$

Dividing both sides of Eq. (1.55) by $\sin^2\theta$,

$$\frac{1}{f_1}\frac{d}{dr}\left(r^2\frac{df_1}{dr}\right) + \frac{2mr^2}{\hbar^2}[E - V(r)] = \frac{m_l^2}{\sin^2\theta} - \frac{1}{f_2 \sin\theta}\times\frac{d}{d\theta}\left(\sin\theta\frac{df_2}{d\theta}\right). \tag{1.56}$$

The left side of Eq. (1.56) is a function of r and the right side only a function of θ. Both sides could be equated to a constant C:

$$-\frac{1}{f_2 \sin\theta}\frac{d}{d\theta}\left(\sin\theta \frac{df_2}{d\theta}\right) + \frac{m_l^2}{\sin^2\theta} = C$$

or

$$\frac{1}{\sin\theta}\frac{d}{d\theta}\left(\sin\theta \frac{df_2}{d\theta}\right) + \left(C - \frac{m_l^2}{\sin^2\theta}\right)f_2 = 0. \quad (1.57)$$

Expressing θ in terms of η, a new variable,

$$\cos\theta = \eta, \qquad -\sin\theta\, d\theta = d\eta, \qquad \sin\theta = (1-\eta^2)^{1/2},$$

Eq. (1.57) can be rewritten

$$\frac{d}{d\eta}\left[(1-\eta^2)\frac{df_2}{d\eta}\right] + \left(C - \frac{m_l^2}{1-\eta^2}\right)f_2 = 0.$$

Expanding,

$$(1-\eta^2)\frac{d^2 f_2}{d\eta^2} - 2\eta\frac{df_2}{d\eta} + \left(C - \frac{m_l^2}{1-\eta^2}\right)f_2 = 0. \quad (1.58)$$

Equation (1.58) is the *Legendre equation*. The equation has solutions which satisfy the condition of single-valuedness and continuity if the constant C has the form

$$C = (K + |m_l|)(K + |m_l| + 1),$$

where K and $|m_l|$ are whole numbers.

Setting

$$K + |m_l| = l,$$
$$C = l(l+1) \qquad \text{where } l \geq |m_l|.$$

l is referred to as the *orbital angular momentum quantum number* and $|m_l|$ is the magnitude of m_l.

The constant $C = l(l+1)$ results in the solution of f_2 as polynomials:

$$f_2(\eta) = P_l^{|m_l|}(\cos\theta).$$

For example,

$l = 0, \qquad m_l = 0, \qquad P_0^0 = 1,$

$l = 1, \qquad m_l = 0, \qquad P_1^0 = \cos\theta,$

$l = 2, \qquad m_l = 0, \qquad P_2^0 = \frac{3}{2}\cos^2\theta - \frac{1}{2},$

$l = 3, \qquad m_l = 0, \qquad P_3^0 = \frac{5}{3}\cos^3\theta - \frac{3}{2}\cos\theta,$

and so on.

If $m_l = 0$, the polynomials are called *Legendre polynomials*, and if $m_l \neq 0$, the polynomials are called *associated Legendre polynomials*.

Referring back to Eq. (1.56), the right side equals $l(l+1)$, and, as such,

$$\frac{1}{f_1}\frac{d}{dr}\left(r^2 \frac{df_1}{dr}\right) + \frac{2mr^2}{\hbar^2}[E - V(r)] = l(l+1). \qquad (1.59)$$

Before solving Eq. (1.59), it would be worthwhile mentioning that the quantum numbers m_l and l are valid for all potential energies as long as they are functions of r. To determine f_1 as a function of r, let

$$f_1(r) = \frac{\chi(r)}{r},$$

$$\dot{f}_1(r) = \frac{r\dot{\chi}(r) - \chi(r)}{r^2},$$

or

$$r^2 \dot{f}_1(r) = r\dot{\chi}(r) - \chi(r)$$

and

$$\frac{d}{dr}(r^2 \dot{f}_1(r)) = r\ddot{\chi}(r) + \dot{\chi}(r) - \dot{\chi}(r) = r\ddot{\chi}(r).$$

Substituting these values in Eq. (1.59),

$$r^2 \frac{\ddot{\chi}(r)}{\chi(r)} + \frac{2mr^2}{\hbar^2}[E - V(r)] = l(l+1)$$

or

$$\ddot{\chi}(r) + \left\{\frac{2m}{\hbar^2}[E - V(r)] - \frac{l(l+1)}{r^2}\right\}\chi(r) = 0. \qquad (1.60)$$

$\chi(r)$ can be obtained from Eq. (1.60) and depends upon the specific form of $V(r)$:

$$\varphi(r, \theta, \phi) = \text{const } e^{\pm i m_l \phi} P_l{}^{m_l}(\cos \theta)[\chi(r)/r]$$

is a general solution of the Schrödinger equation.

Putting $V(r) = -e^2/4\pi\epsilon_0 r$ in Eq. (1.60),

$$\frac{d^2\chi}{dr^2} + \left[\frac{2mE}{\hbar^2} + \frac{2m}{\hbar^2}\frac{e^2}{4\pi\epsilon_0 r} - \frac{l(l+1)}{r^2}\right]\chi = 0.$$

Substituting,

$$\lambda = \frac{e^2}{2}\left(-\frac{2m}{\hbar^2 E}\right)^{1/2},$$

$$\rho = 2\left(-\frac{2mE}{\hbar^2}\right)^{1/2} r,$$

$$d\rho = 2\left(-\frac{2mE}{\hbar^2}\right)^{1/2} dr,$$

$$\frac{d\chi}{dr} = \frac{d\chi}{d\rho}\frac{d\rho}{dr} = 2\left(-\frac{2mE}{\hbar^2}\right)^{1/2}\frac{d\chi}{d\rho},$$

$$\frac{d^2\chi}{dr^2} = \frac{d}{dr}\left[\frac{d\chi}{d\rho} 2\left(-\frac{2mE}{\hbar^2}\right)^{1/2}\right],$$

or

$$\frac{d^2\chi}{dr^2} = 4\left(-\frac{2mE}{\hbar^2}\right)\frac{d^2\chi}{d\rho^2}.$$

Therefore,

$$4\left(-\frac{2mE}{\hbar^2}\right)\frac{d^2\chi}{d\rho^2} + \left[\frac{2mE}{\hbar^2} + \frac{2m}{\hbar^2}\frac{1}{4\pi\epsilon_0}\frac{e^2}{\rho} 2\left(-\frac{2mE}{\hbar^2}\right)^{1/2} - 4\left(-\frac{2mE}{\hbar^2}\right)\frac{1}{\rho^2}\right.$$

$$\left. \times l(l+1)\right]\chi = 0.$$

Dividing by $4(-2mE/\hbar^2)$,

$$\frac{d^2\chi}{d\rho^2} + \left[-\frac{1}{4} + \frac{\lambda}{4\pi\epsilon_0 \rho} - \frac{l(l+1)}{\rho^2}\right]\chi = 0. \qquad (1.61)$$

Let Eq. (1.61) have a solution

$\chi(\rho) = e^{-\rho/2}\sum a_s \rho^{s+l+1},$
$\dot{\chi}(\rho) = e^{-\rho/2}\sum (s+l+1)a_s\rho^{l+s} - \frac{1}{2}e^{-\rho/2}\sum a_s\rho^{s+l-1},$
$\ddot{\chi}(\rho) = e^{-\rho/2}\sum (s+l)(s+l+1)a_s\rho^{l+s-1} - e^{-\rho/2}\sum(s+l+1)a_s\rho^{s+l}$
$\quad + \frac{1}{4}e^{-\rho/2}\sum a_s\rho^{s+l+1}.$

Substituting the solution in Eq. (1.61),

$$e^{-\rho/2}\left[\sum (s+l)(s+l+1)a_s\rho^{s+l-1} - \sum(s+l+1)a_s\rho^{s+l}\right.$$
$$\left. + \frac{\lambda}{4\pi\epsilon_0}\sum a_s\rho^{s+l} - l(l+1)\sum a_s\rho^{s+l-1}\right] = 0.$$

Since $e^{-\rho/2} \neq 0$,

$$\sum[(s+l)(s+l+1) - l(l+1)]a_s\rho^{s+l-1} - \sum\left[(s+l+1) - \frac{\lambda}{4\pi\epsilon_0}\right]$$
$$\times a_s\rho^{s+l} = 0. \quad (1.62)$$

Suppose the first term in the series $\chi(\rho)$ is $s = \tau$; i.e., $a_\tau \neq 0$ and $a_\tau \cdot \rho^{\tau+l+1}$. $e^{-\rho/2}$ is the first term in the series. The term with $\rho^{\tau+l-1}$ power must therefore have the coefficient zero. Thus

$$(s+l)(s+l+1) - l(l+1) = s(2l+s+1) = 0,$$

and with $s = \tau$,

$$\tau(2l + \tau + 1) = 0.$$

This means that either $\tau = 0$ or $\tau = -2l - 1$. The condition $\tau = -2l - 1$ is not acceptable, because this would mean that the initial term in the solution of $\chi(\rho)$ is $e^{-\rho/2} a_\tau \rho^{-l}$, and for $\rho \to 0$ the value of $\chi(\rho) \to \infty$, and hence the wave function, would misbehave. Therefore,

$$\chi(\rho) = e^{-\rho/2} \sum_{s=0} a_s \rho^{s+l+1} \qquad \text{starting with } \tau = 0.$$

If Eq. (1.62) is to hold for all values of ρ, the coefficient of each power of ρ must vanish. The coefficient of ρ^{s+l} is given by

$$a_{s+1}(s+1)(s+2l+2) - a_s\left(s+l+1 - \frac{\lambda}{4\pi\epsilon_0}\right) = 0$$

or

$$a_{s+1} = a_s \frac{s+l+1 - (\lambda/4\pi\epsilon_0)}{(s+1)(s+2l+2)}. \qquad (1.63)$$

Similar to Eq. (1.48) for the quantum oscillator, Eq. (1.63) is a recursion formula for the coefficients.

The equation

$$\chi(\rho) = e^{-\rho/2} \sum_{s=0} a_s \rho^{s+l+1}$$

would behave as e^ρ as $s \to \infty$. However, it can be seen that the series terminates before $s \to \infty$. The condition for termination of the series can be obtained from Eq. (1.63). The numerator of the right side of Eq. (1.63) is zero if

$$\lambda/4\pi\epsilon_0 = \nu + l + 1,$$

assuming the series to terminate if $s = \nu$. This condition suggests that only for certain values of λ do well-behaved solutions exist. It may be mentioned that ν is the νth term, and as such an integer.

Therefore,

$$(\nu + l + 1)^2 = \frac{\lambda^2}{(4\pi\epsilon_0)^2} = -\frac{me^4}{2\hbar^2 E} \frac{1}{16\pi^2\epsilon_0^2}$$

or

$$E = -\frac{me^4}{8\epsilon_0^2 h^2(\nu + l + 1)^2}.$$

Putting

$$n = \nu + l + 1,$$

$$E = -\frac{me^4}{8\epsilon_0^2 h^2 n^2}. \qquad (1.64)$$

HYDROGEN ATOM

n is called the *principal quantum number*. The possible values of n are integers 1, 2, 3, ... The minimum value of $n = 1$ is obtained if $l = 0$, $\nu = 0$. Because $l \geq m_l$, $m_l = 0$, $n = 1$ is called the *ground state*.

Equation (1.64) suggests that the possible values of energy of an electron in a hydrogen atom are quantized. Equation (1.64) is the same as Eq. (1.5) calculated for an electron in a hydrogen atom according to the Bohr-Sommerfeld theory.

Wave function for a hydrogen atom in the ground state. It has been said in the quantum-mechanical treatment of the electron in a hydrogen atom that the wave function φ of the electron is the product of three functions f_1, f_2, and f_3:

$$\varphi = f_1(r)f_2(\theta)f_3(\phi),$$

$$f_1(r) = \frac{\chi(r)}{r},$$

$$f_2(\theta) = P_l^{m_l}(\cos\theta),$$

$$f_3(\phi) = \text{const } e^{\pm i m_l \phi}.$$

For the ground state $n = 1$ or $\nu + l + 1 = 1$. Therefore, $\nu = 0$, $l = 0$, and $m_l = 0$. Hence

$$f_3(\phi) = \text{const},$$

$$f_2(\theta) = P_0^0(\cos\theta) = 1,$$

$$f_1(r) = \frac{\chi(r)}{r} = \frac{\chi(\rho)}{\rho},$$

$$\chi(\rho) = e^{-\rho/2} \sum_{s=0} a_s \rho^{s+l+1}.$$

For $s = \nu = 0$,

$$\chi(\rho) = e^{-\rho/2} a_0 \rho,$$

$$\frac{\chi(\rho)}{\rho} = a_0 e^{-\rho/2}.$$

The relationship of ρ and r is

$$\rho = 2\left(-\frac{2mE}{\hbar^2}\right)^{1/2} r,$$

$$\frac{\rho}{2} = \left(-\frac{2mE}{\hbar^2}\right)^{1/2} r.$$

The energy E of the electron in the ground state ($n = 1$) is given by Eq. (1.64) and is

$$E_1 = -\frac{me^4}{8\epsilon_0^2 h^2}.$$

Therefore,
$$\frac{\rho}{2} = \left(\frac{2m}{\hbar^2}\frac{me^4}{8\epsilon_0^2 h^2}\right)^{1/2} r = \frac{\pi me^2}{\epsilon_0 h^2} r.$$

Referring to Eq. (1.3), the first Bohr radius r is given by
$$r_1 = \frac{h^2 \epsilon_0}{\pi me^2}.$$

Hence
$$\rho/2 = r/r_1.$$

Therefore $f_1(r) = a_0 e^{-r/r_1}$ and hence the wave function for the hydrogen atom in the ground state is

$$\varphi = A e^{-r/r_1}, \quad \text{where } A \text{ is a constant involving } a_0,$$
$$\varphi\varphi^* = A^2 e^{-2r/r_1}.$$

$\varphi\varphi^*$ is the probability density. Because the probability of finding the electron over all space is unity,

$$\int \varphi\varphi^* \, dv = 1.$$

This normalization condition helps to determine the constant A^2. Assuming the electron to be in an element of volume between r and $r + dr$ distance from the nucleus (Fig. 1.7),

$$dv = 4\pi r^2 \, dr.$$

The normalization condition would be

$$\int_0^\infty \varphi\varphi^* 4\pi r^2 \, dr = 1,$$

$$\int_0^\infty A^2 e^{-2r/r_1} 4\pi r^2 \, dr = 1,$$

or

$$4\pi A^2 \int_0^\infty r^2 e^{-2r/r_1} \, dr = 1.$$

The integral $\int_0^\infty r^2 e^{-2r/r_1} \, dr$ can be solved by integrating by parts. It can be easily shown that

$$\int_0^\infty r^2 e^{-2r/r_1} \, dr = \frac{r_1^3}{4},$$

and, as such, $A^2 = 1/\pi r_1^3$. Hence the probability density of the electron in a hydrogen atom is

$$\varphi\varphi^* = (\pi r_1^3)^{-1} e^{-2r/r_1}.$$

HYDROGEN ATOM

The charge on the electron is $-e$, and therefore the product of charge and probability density would result in the charge density:

$$\text{charge density} = -\frac{e}{\pi r_1{}^3} e^{-2r/r_1}. \tag{1.65}$$

In electrical engineering the volume charge density is represented by ρ and according to Eq. (1.65) turns out to be an exponential function of r. Therefore,

$$\rho(r) = -\frac{e}{\pi r_1{}^3} e^{-2r/r_1}. \tag{1.66}$$

Equation (1.66) and Fig. 1.8 show that the charge density is a maximum at $r = 0$ and decays exponentially with increasing r.

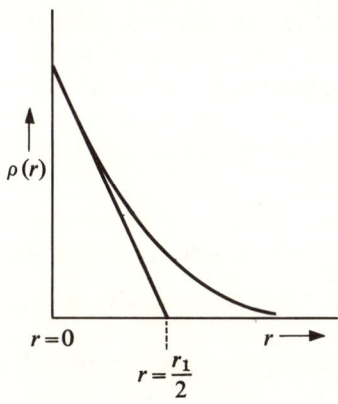

Fig. 1.8
Charge density of the electron for a hydrogen atom in the ground state

Most probable position of the electron. The most probable position of the electron is given by maximizing the probability function and determining the position at which the probability function would be a maximum. The probability dP of finding the electron between r and $r + dr$ is

$$dP = \varphi \varphi^* 4\pi r^2 dr.$$

Substituting the value of $\varphi \varphi^*$,

$$dP = (\pi r_1{}^3)^{-1} e^{-2r/r_1} 4\pi r^2 \, dr$$

or

$$dP = \frac{4}{r_1{}^3} e^{-2r/r_1} r^2 \, dr = F(r) \, dr,$$

where $F(r) = (4/r_1^3)e^{-2r/r_1}r^2$ is called the *probability function*. The probability is a maximum if

$$\frac{d[F(r)]}{dr} = 0.$$

Differentiating $F(r)$ and equating to zero,

$$\frac{4}{r_1^3}\left(2re^{-2r/r_1} - \frac{2}{r_1}r^2e^{-2r/r_1}\right) = 0$$

or $r = r_1$. The probability of finding the electron is a maximum at $r = r_1$ for a hydrogen atom in the ground state. r_1 is the first Bohr radius. Needless to say, although the Bohr-Sommerfeld assumptions were *ad hoc*, the quantum theory with all its rigorous treatment of electron gives results that conform with the classical theory.

Physical interpretation of quantum numbers n, l, and m_l. In this section we have seen that the state of motion of the electron in the hydrogen atom can be described by the quantum numbers n, l, and m_l, and a set of these numbers is said to define the state of the electron:

$$n = 1, 2, 3, \ldots,$$
$$l = 0, 1, 2, 3, \ldots, n-1,$$
$$m_l = l, l-1, l-2, \ldots, 0, \ldots, -(l-2), -(l-1), -l.$$

The total number of states for a given value of n is

$$\sum_{l=0}^{n-1}(2l+1) = 1 + 3 + 5 + \cdots [2(n-1)+1] = n^2.$$

Principal quantum number n. The *principal quantum number n* is numerically equal to the n of the Bohr theory, and it is again a measure of the total energy of the electron in the state concerned. The larger the value of n, the greater the energy, and in the hydrogen atom the total energy is proportional to $-1/n^2$. With infinitely large n, the total energy is zero, and this corresponds to an electron at an infinite distance from the nucleus.

Orbital angular momentum quantum number l. *Oribital angular momentum quantum number l*, as the name suggests, is a measure of the angular momentum of the state concerned. It will be shown that the angular momentum is of magnitude $(h/2\pi)[l(l+1)]^{1/2}$. It may be mentioned that the postulate of the Bohr theory was that the angular momentum was equal to $nh/2\pi$, so Bohr's guess was not exactly right. One of the difficult things to visualize is that with $n = 1$, $l = 0$, and therefore, according to the quantum theory, $(h/2\pi)[l(l+1)]^{1/2} = 0$. The name *orbital angular momentum* corresponds with the angular momentum of the orbital motion in the old theory. It is an unfortunate expression, because we must no longer think of the electron as being a little particle running around an orbit. Considering the motion of a particle with a potential energy dependent upon r, the total energy E is

$$E = \text{K.E.} + \text{P.E.} = \tfrac{1}{2}mv^2 + V(r).$$

In terms of polar coordinates,

$$v^2 = \left(\frac{dr}{dt}\right)^2 + r^2\left(\frac{d\theta}{dt}\right)^2.$$

Therefore,

$$E = \frac{m}{2}\left(\frac{dr}{dt}\right)^2 + \frac{m}{2}r^2\left(\frac{d\theta}{dt}\right)^2 + V(r).$$

By definition the angular momentum

$$\mathbf{M}_a = \mathbf{r} \times m\mathbf{v}$$

or

$$M_a = mr^2\frac{d\theta}{dt},$$

and, as such, the total energy of a particle traversing a linear path:

$$E = \frac{m}{2}\left(\frac{dr}{dt}\right)^2 + \frac{M_a^2}{2mr^2} + V(r).$$

For a linear path,

$$\text{K.E.} = \frac{m}{2}\left(\frac{dr}{dt}\right)^2,$$

and

$$\text{effective potential energy} = \frac{M_a^2}{2mr^2} + V(r). \tag{1.67}$$

This has an analogy in the quantum-mechanical treatment of the motion of electrons in a hydrogen atom [Eq. (1.59)]:

$$\frac{1}{f_1}\frac{d}{dr}\left(r^2\frac{df_1}{dr}\right) + \frac{2mr^2}{\hbar^2}[E - V(r)] = l(l+1). \tag{1.59}$$

If $f_1 = \chi(r)/r$,

$$\ddot{\chi}(r) + \frac{2m}{\hbar^2}\left[E - V(r) - \frac{\hbar^2}{2mr^2}l(l+1)\right]\chi(r) = 0. \tag{1.68}$$

Equation (1.68) is a form of the Schrödinger equation in one dimension — with the difference that it is in a potential field:

$$V_{\text{effective}} = V(r) + \frac{\hbar^2 l(l+1)}{2mr^2}.$$

By analogy,
$$M_a^2 = \hbar^2 l(l+1)$$
or
$$M_a = \hbar[l(l+1)]^{1/2}.$$

Physical interpretation of m_l. There is a magnetic moment associated with the rotation of a charge. The magnetic field is set up perpendicular to the plane of the orbit. The angular momentum vector M_a is parallel to the magnetic field set up by the electron. The angular momentum vector M_a will have a component parallel to the external magnetic field. Because of the reaction between the two magnetic fields, the angular momentum vector will precess about the external magnetic field. This motion is quantized and
$$m_l = 0, \pm 1, \pm 2, \ldots, \pm l.$$

m_l is called the *magnetic orbital quantum number* and is an index of the splitting of the energy levels with a given n and l in a magnetic field.

1.7 Electron spin, exclusion principle, and the periodic table

Electron spin. In 1925–1926 Uhlenbeck and Goudschmidt postulated that the electron is spinning about its own axis and as a result has an angular momentum and magnetic field. The motion or spin of the electron is quantized just as energy in the preceding various solutions of the Schrödinger equation. In all the examples we have discussed, the energy could have an infinite number of energy levels, whereas it was suggested that the angular momentum vector of the spin could have only one of the two possible values of its projection. The values are equal and opposite in sign. One speaks of spin up and spin down, calling the spin quantum number m_s:
$$m_s = \pm \tfrac{1}{2}.$$

Exclusion principle. Pauli enunciated that no more than one electron can occupy a state described by the four quantum numbers n, l, m_l, and m_s. This is known as *Pauli's exclusion principle*. The periodic table follows from the four quantum numbers and the exclusion principle.

Periodic table. It is convenient in discussing the periodic table to use a notation developed by spectroscopists. The quantum numbers l's are denoted by letters as follows:
$$l = 0, 1, 2, 3, 4, \ldots,$$
$$s, p, d, f, g, \ldots.$$

These symbols are based on the historial names sharp, principal, diffused, fundamental, etc., for spectral series.

ELECTRON SPIN, EXCLUSION PRINCIPLE

The periodic table can be easily constructed with the help of quantum numbers n, l, m_l, and m_s and Pauli's exclusion principle.

For $n = 1$, l must equal 0, m_l must equal 0, and $m_s = \pm\frac{1}{2}$. Thus there are two possible states for $n = 1$.

For $n = 2$, l may equal 1 and 0.
For $l = 0$, $m_l = 0$ and $m_s = \pm\frac{1}{2}$, giving two possible states.
For $l = 1$, $m_l = -1$ and $m_s = \pm\frac{1}{2}$, giving two possible states.
For $l = 1$, $m_l = 0$ and $m_s = \pm\frac{1}{2}$, giving two possible states.
For $l = 1$, $m_l = 1$ and $m_s = \pm\frac{1}{2}$, giving two possible states.

Thus for $n = 2$, a total of eight states is possible. Similarly, for $n = 3$, there are eighteen possible states.

The principal quantum number n is indicated by an integer in front of the letter l. The number of electrons having the same value of l is expressed as an index of l.

Hydrogen has 1 electron $1s$.
Helium has 2 electrons $1s^2$.
Lithium has 3 electrons $1s^2 2s^1$.
Neon has 10 electrons $1s^2 2s^2 2p^6$.

The electron configurations of light elements are simple, and follow the rule of filling one shell after another. The electrons belonging to a given value of n are said to form shells. The hydrogen-like model does not hold for more complicated atoms. Structures of C, Si, and Ge:

C has atomic number 6 and hence 6 electrons $1s^2 2s^2 2p^2$.
Si has atomic number 14 and hence 14 electrons $1s^2 2s^2 2p^6 3s^2 3p^2$.
Ge has atomic number 32 and hence 32 electrons $1s^2 2s^2 2p^6 3s^2 3p^6 3d^{10} 4s^2 4p^2$.

The three elements C, Si, and Ge have similar structures. They all have four electrons in their last unfilled shell. These are valence electrons and are responsible for binding.

Problem 1

a. Show that the group velocity of a plane wave of amplitude $\exp\left[i\left(kx - \frac{Et}{\hbar}\right)\right]$ is $v_g = 1/\hbar\, (dE/dK)$. Assume $E = \hbar w$.
b. From this fact prove the de Broglie's relationship $\lambda = h/mv$.
c. What is the wavelength of an electron beam accelerated through a 100-V potential. Express the answer in angstroms.   Ans.   1.23 Å

Problem 2

Show that the following two expressions are solutions of the time-dependent Schrödinger equation, provided the total energy $E = \hbar\omega$:

$$\varphi_1 = Ae^{i(Kx - \omega t)},$$
$$\varphi_2 = Ae^{i(-Kx - \omega t)}.$$

Problem 3

a. Suppose a potential well has a width $2W$ and the walls of the one-dimensional well are infinitely high ($V_0 \to \infty$). Would you expect the probability of finding an electron outside the well to be finite? Why?

b. Show that the possible values of energy of an electron in the infinite well are

$$E_n = \frac{n^2 h^2}{32 m W^2},$$

where n is an integer.

Problem 4

a. Show that the spherically symmetric wave function for the state of energy $n = 2, l = 0$ for a hydrogen atom is given by

$$\varphi_2 = \frac{1}{(8\pi)^{1/2}} \frac{1}{r_1^{3/2}} \left(1 - \frac{r}{2r_1}\right) e^{-r/2r_1},$$

where r_1 is the radius of the first Bohr orbit.

b. Plot the probability per unit radius of finding an electron as a function of r for this state and find the most probable value of r in this state. Express r in terms of r_1.

Problem 5 (Tunnel Effect)

Solve in detail the problem of partial reflection and transmission of a rectangular potential such as is shown in Fig. A. Find the probability of transmission for a 1-MeV proton through a 4-MeV-high 10^{-12}-cm-thick rectangular potential-energy barrier. Ans. 0.0015.

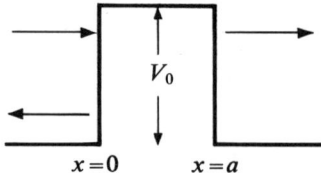

2
Statistical Physics and Electron Emission

2.1 Gas kinetics

Gas kinetics deals with the dynamics of gas, molecules. The gas molecules in the kinetic theory are noninteracting free molecules, and the results obtained from the theory are applicable to electrons in solids when the electrons behave as noninteracting potential free electrons. Electrons in such a case are often referred to as an *electron gas*. It is therefore worthwhile reviewing a few results of gas kinetics.

In the kinetic theory of gases, gas molecules are considered as spheres free to move in straight lines until they collide with another molecule or the boundary walls of the container of the gas. Considering the gas to be monoatomic and assuming potential free molecules, the total energy of the molecules is the kinetic energy of translational motion. The gas is supposed to be at a temperature T degrees Kelvin. Consider that the gas is enclosed in a cube of side L, and that each gas molecule has a mass m. Under equilibrium conditions, equal pressure is exerted by the gas on all six surfaces of the container. The force exerted on surface A must be equal and opposite in direction to the force exerted on B (Fig. 2.1). Let us assume that there are N molecules in the container of volume L^3. Different molecules are randomly moving with different velocity. Let the component of velocity of an ith molecule in the x direction be v_{ix}. Assuming only translational motion of the molecule, this molecule has a momentum mv_{ix}, and upon striking the surface will rebound with a velocity $-v_{ix}$ and a momentum $-mv_{xi}$. The momentum transferred to the wall is $2mv_{ix}$ in one collision with the wall B.

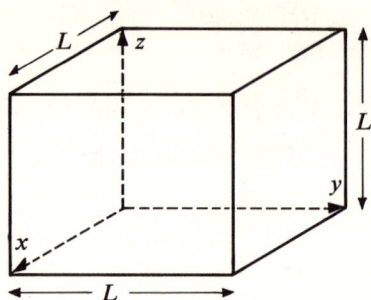

Fig. 2.1
Cube of side L containing gas

The average force due to a molecule of velocity component v_{ix} in a cube of side L at the wall B is

$$F_{ix} = \frac{\text{change in momentum}}{\text{collision}} \times \text{collisions/sec.}$$

The distance covered by the molecule between two consecutive collisions is $2L$, as the molecule after the first collision with wall B must travel up a wall A a distance L, strike the wall, and rebound to travel another length L before striking surface B again.

Therefore,

$$\text{collisions/sec} = v_{ix}/2L.$$

Hence

$$F_{ix} = 2mv_{ix}\frac{v_{ix}}{2L} = \frac{mv_{ix}^2}{L}.$$

Total force exerted on surface B is the sum of the forces exerted by all the N particles:

$$F_x = F_{1x} + F_{2x} + F_{3x} + F_{4x} + \cdots$$

or

$$F_x = \sum_i F_{ix} = \sum_{i=1}^{i=N} \frac{m(v_{ix})^2}{L}.$$

The pressure exerted on the plane B is

$$P_x = \frac{F_x}{L^2} = \sum_{i=1}^{i=N} \frac{m(v_{ix})^2}{L^3}.$$

Similarly, it can be shown that the pressure exerted on a plane at a distance $y = L$ from the origin in Fig. 2.1 is

$$P_y = \sum_{i=1}^{i=N} \frac{m(v_{iy})^2}{L^3}$$

and, by analogy,

$$P_z = \sum_{i=1}^{i=N} \frac{m(v_{iz})^2}{L^3}.$$

As already stated, the pressure exerted on all the six faces of the cube is equal. Therefore,

$$P_x = P_y = P_z = P,$$

$$3P = P_x + P_y + P_z = \frac{m}{L^3}[(v_{ix})^2 + (v_{iy})^2 + (v_{iz})^2],$$

or

$$P = \frac{m}{3L^3} \sum_{i=1}^{i=N} (v_{ix})^2 + (v_{iy})^2 + (v_{iz})^2.$$

The mean-square velocity $\overline{v^2}$ is the sum of the mean-square-velocity components,

$$\overline{v^2} = \overline{v_x^2} + \overline{v_y^2} + \overline{v_z^2}.$$

Also,

$$\overline{v_x^2} = \sum_{i=1}^{N} \frac{(v_{ix})^2}{N},$$

$$\overline{v_y^2} = \sum_{i=1}^{N} \frac{(v_{iy})^2}{N},$$

$$\overline{v_z^2} = \sum_{i=1}^{N} \frac{(v_{iz})^2}{N},$$

and, as such,

$$P = \frac{m}{3L^3} N\overline{v^2}.$$

Define n as the number of molecules per unit volume:

$$n = N/L^3.$$

Therefore,

$$P = \tfrac{1}{3}mn\overline{v^2}. \tag{2.1}$$

According to the ideal gas equation, the pressure P is given by

$$P = nkT, \tag{2.2}$$

where k is Boltzmann's constant. Equating the right sides of Eqs. (2.1) and (2.2),

$$\overline{v^2} = 3kT/m. \tag{2.3}$$

The average energy of the molecule is the average kinetic energy $\frac{1}{2}mv^2$ and equals $\frac{3}{2}kT$.

The thermal-energy density of the gas, the kinetic-energy density, depends upon temperature alone and not on pressure and volume.

2.2 Classical statistics

In Sec. 2.1 the velocity of the molecule was considered, and from consideration of the velocity of the individual particle, the expression for the mean-square velocity of the molecule was defined. In the discussion of electrical engineering materials and their properties we are mainly concerned with electrons. For microscopic particles such as electrons, it would be more appropriate to speak of the probability of particles having velocities within a certain range of velocities rather than exact velocities, as there is no way of measuring the velocity of the microscopic particle.

Suppose there are five microscopic particles in a box having velocities 10, 50, 70, 81, and 92 m/sec. By definition, the probability of finding a particle that possesses a velocity between 0 and 20 m/sec is

$$\Delta P = \Delta N/N = \tfrac{1}{5} = 0.2.$$

Increasing the velocity range to 80 m/sec,

$$\Delta P = \Delta N/N = \tfrac{3}{5} = 0.6.$$

If the velocity range is extended from 0 to 100 m/sec,

$$\Delta P = \Delta N/N = \tfrac{5}{5} = 1.0.$$

Generalizing, if ΔP is the probability of finding a particle within the velocity range v and $v + \Delta v$,

$$\Delta P = \Delta n/n,$$

where Δn is the number of particle within the required velocity range v and $v + \Delta v$ and n is the total number of particles. n is used in place of N, as the carrier concentration is normally expressed as per unit volume. From the above numerical example, ΔP is a function of the velocity $f(v)$ and depends upon the width of the interval:

$$\Delta P = \Delta n/n = f(v)\,\Delta v.$$

$f(v)$ is called the *velocity distribution function* or in general the *probability density function*.

By definition,

$$\sum \Delta P = 1 = \sum \Delta N/N = \sum \Delta n/n = \sum f(v)\Delta v.$$

In a system of microscopic particles $f(v)$ may be assumed continuous and the summation sign may be replaced by an integration:

$$\int_0^\infty f(v)\, dv = 1. \tag{2.4}$$

Equation (2.4) is known as the *normalization condition* and follows from the basic physical fact that the particle must have some velocity between 0 and ∞, and, as such, the probability of finding a particle having any velocity between 0 and ∞ is unity.

Equation (2.4) can be expanded into the component form if the velocity distribution function is $f(v_x, v_y, v_z)$. The normalization condition would now be

$$\int_{-\infty}^{\infty} \int_{-\infty}^{\infty} \int_{-\infty}^{\infty} f(v_x, v_y, v_z)\, dv_x\, dv_y\, dv_z = 1. \tag{2.5}$$

The particle could have velocity components anywhere between $-\infty$ and $+\infty$.

Average velocities. The probability of finding a particle of velocity magnitude v in the interval v and $v + \Delta v$ is $f(v)\, \Delta v$. Therefore, the average velocity v_a could be defined as

$$v_a = \int_0^\infty v f(v)\, dv. \tag{2.6}$$

It may once more be stressed that Eq. (2.6) is true if and only if Eq. (2.4) satisfied. If, however, Eq. (2.4) is not satisfied,

$$v_a = \frac{\int_0^\infty v f(v)\, dv}{\int_0^\infty f(v)\, dv}.$$

Henceforth we shall assume that the normalization condition is satisfied by the form of $f(v)$.

Similarly, the average velocity component in the x direction is

$$v_{x_a} = \int_{-\infty}^{\infty} \int_{-\infty}^{\infty} \int_{-\infty}^{\infty} v_x f(v_x, v_y, v_z)\, dv_x\, dv_y\, dv_z, \tag{2.7}$$

$$v_{y_a} = \int_{-\infty}^{\infty} \int_{-\infty}^{\infty} \int_{-\infty}^{\infty} v_y f(v_x, v_y, v_z)\, dv_x\, dv_y\, dv_z, \tag{2.8}$$

$$v_{z_a} = \int_{-\infty}^{\infty} \int_{-\infty}^{\infty} \int_{-\infty}^{\infty} v_z f(v_x, v_y, v_z)\, dv_x\, dv_y\, dv_z. \tag{2.9}$$

The mean-square velocity is

$$\overline{v^2} = \int_0^\infty v^2 f(v)\,dv, \tag{2.10}$$

and the x component of the mean-square velocity is

$$\overline{v_x^2} = \iint_{-\infty}^{\infty}\!\!\int v_x^2 f(v_x,v_y,v_z)\,dv_x\,dv_y\,dv_z. \tag{2.11}$$

Total energy. In the previous subsection we considered the velocity distribution function and the average velocities. In the case of potential free molecules of a monoatomic gas or electrons, the energy of the particles is the kinetic energy of translational motion. A correlation between energy and velocity of translational motion is

$$E = \tfrac{1}{2}mv^2 = \tfrac{1}{2}m(v_x^2 + v_y^2 + v_z^2).$$

The probability of finding an electron of energy E in the interval E and $E + \Delta E$ is defined in a fashion similar to velocity distribution:

$$\Delta P = \Delta n/n = P(E)\,\Delta E.$$

$P(E)$ is the probability of an electron having an energy E and is the probability energy function. The total energy E_t is given by

$$E_t = n\int_0^\infty E P(E)\,dE,$$

and the average energy is

$$\overline{E} = \frac{E_t}{n} = \int_0^\infty E P(E)\,dE.$$

It now remains to determine the form of $f(v)$ or $P(E)$, which we shall discuss in the following subsection.

Probability of an electron having an energy E, P(E). We assume that two or more electrons could have the same energy. We are thus ignoring the results of quantum-mechanical theory and Pauli's exclusion principle. The case is referred to as a *classical* one. To determine $P(E)$ we shall consider collisions between electrons of energies E_1 and E_2. After collision let the energies of the electrons be E_1' and E_2'. From the laws of conservation of energy,

$$E_1 + E_2 = E_1' + E_2'.$$

Let $E_1' = E_1 - \Delta E$. Therefore, $E_2' = E_2 + \Delta E$. ΔE is the change in energy.

Let (on the average) α collisions per second occur in a system of electrons in equilibrium in which electrons of energy E_1 and E_2 collide and change their energies to E_1' and E_2'. α is proportional to the product of $p(E_1)$ and $p(E_2)$, where $p(E_1)$ is the probability of occupation of a state of energy E_1 and $p(E_2)$ is the probability of occupation of a state of energy E_2:

$$p(E) = p(E) \cdot G(E)n, \text{ where } G(E) \text{ is defined on page 49}$$

$$\alpha = \text{const } p(E_1)p(E_2).$$

Let (on the average) α' collisions per second occur in the same system of electrons in which electrons of energies E_1' and E_2' collide and change their energies to E_1 and E_2:

$$\alpha' = \text{const } p(E_1')p(E_2').$$

In equilibrium $\alpha = \alpha'$, and, therefore,

$$p(E_1)p(E_2) = p(E_1')p(E_2')$$

or

$$p(E_1)p(E_2) = p(E_1 - \Delta E)p(E_2 + \Delta E). \tag{2.12}$$

An examination of Eq. (2.12) and a Taylor-series expansion shows that there is only one simple mathematical expression for $p(E)$ that satisfies Eq. (2.12):

$$p(E) = Ae^{-\beta E}, \tag{2.13}$$

where A is a constant and β is a universal constant.

Equation (2.13) leads to the velocity distribution function $f(v)$, from the basic fact that the total energy E of a free electron is the kinetic energy:

$$E = \tfrac{1}{2}m(v_x^2 + v_y^2 + v_z^2) = \tfrac{1}{2}mv^2.$$

Therefore,

$$f(v) = Ae^{-1/2m\beta v^2} \cdot 4\pi v^2 \tag{2.14}$$

$$f(v_x, v_y, v_z) = A \exp[-\tfrac{1}{2}m\beta(v_x^2 + v_y^2 + v_z^2)]. \tag{2.15}$$

In Eqs. (2.13), (2.14), and (2.15) the constant A has different values and its value is determined by normalization conditions:

$$\int_0^\infty P(E)\,dE = 1, \tag{2.16}$$

$$\int_0^\infty f(v)\,dv = 1, \tag{2.17}$$

$$\int\int\int_{-\infty}^\infty f(v_x, v_y, v_z)\,dv_x\,dv_y\,dv_z = 1. \tag{2.18}$$

In the following we shall determine the constant A for the function $f(v_x, v_y, v_z)$ in Eq. (2.15) and hence use the normalization condition (2.18).

Gamma functions. Gamma functions are defined by functions of the type

$$\int_0^\infty x^{n-1}e^{-x}\,dx = \Gamma(n) \quad \text{where } n > 0. \tag{2.19}$$

A knowledge of gamma functions is very useful in solving integrals of the type given by Eq. (2.18).

It is shown in books of calculus that

$$\Gamma(n+1) = n\Gamma n,$$

$$\Gamma(1) = 1,$$

$$\Gamma(\tfrac{1}{2}) = \pi^{1/2}.$$

Determination of constant A in Eq. (2.15). According to the normalization condition (2.18),

$$A \iiint_{-\infty}^{\infty} \exp[-\tfrac{1}{2}m\beta(v_x^2 + v_y^2 + v_z^2)]dv_x\,dv_y\,dv_z = 1. \tag{2.20}$$

To carry out the above integration [Eq. (2.20)] we set

$$X = \tfrac{1}{2}m\beta v_x^2,$$
$$Y = \tfrac{1}{2}m\beta v_y^2,$$
$$Z = \tfrac{1}{2}m\beta v_z^2,$$
$$dX = m\beta v_x\,dv_x,$$

and, therefore,

$$dv_x = \frac{1}{2}\left(\frac{m\beta}{2}\right)^{-1/2} X^{-1/2}\,dX,$$

$$dv_y = \frac{1}{2}\left(\frac{m\beta}{2}\right)^{-1/2} Y^{-1/2}\,dY,$$

$$dv_z = \frac{1}{2}\left(\frac{m\beta}{2}\right)^{-1/2} Z^{-1/2}\,dZ.$$

Hence

$$A(2m\beta)^{-3/2} \iiint_{-\infty}^{\infty} (e^{-X}X^{-1/2}\,dX)(e^{-Y}Y^{-1/2}\,dY) \times (e^{-Z}Z^{-1/2}\,dZ) = 1.$$

Assuming the velocity distribution to be symmetrical about zero-velocity components, the limits could be changed to 0 to ∞. As such,

$$8A(2m\beta)^{-3/2} \iiint_0^{\infty} (e^{-X}X^{-1/2}\,dX)(e^{-Y}Y^{-1/2}\,dY) \times (e^{-Z}Z^{-1/2}\,dZ) = 1.$$

From the results of gamma functions, it can be concluded that

$$\int_0^\infty e^{-X} X^{-1/2}\, dX = \int_0^\infty e^{-Y} Y^{-1/2}\, dY = \int_0^\infty e^{-Z} Z^{-1/2}\, dZ = \pi^{1/2}.$$

Hence

$$8A(2m\beta)^{-3/2} \pi^{3/2} = 1 \quad \text{or} \quad A = \left(\frac{m\beta}{2\pi}\right)^{3/2}.$$

The distribution function would be

$$f(v_x, v_y, v_z) = \left(\frac{m\beta}{2\pi}\right)^{3/2} \exp\left[-\frac{m\beta}{2}(v_x^2 + v_y^2 + v_z^2)\right]. \quad (2.21)$$

Having determined the constant A, it now remains to evaluate the universal constant β in Eq. (2.15). This can be done by calculating the average energy per unit volume in the system of electrons from the distribution function using the definitions of average velocities and equating to the average energy from the simple gas laws discussed in Sec. 2.1.

If W is the energy per unit volume and n the number of electrons per unit volume,

$$W = \tfrac{1}{2} n m \overline{v^2} = \tfrac{1}{2} mn(\overline{v_x^2} + \overline{v_y^2} + \overline{v_z^2})$$

from the distribution function. Also

$$W = \tfrac{3}{2} nkT \quad \text{from simple gas laws,}$$

$$\overline{v_x^2} = \int\int\int_{-\infty}^{\infty} v_x^2 f(v_x, v_y, v_z)\, dv_x\, dv_y\, dv_z.$$

Substituting the value of $f(v_x, v_y, v_z)$ from Eq. (2.21) in the above equation,

$$\overline{v_x^2} = \left(\frac{m\beta}{2\pi}\right)^{3/2} \int\int\int_{-\infty}^{\infty} \exp\left[-\frac{m\beta}{2}(v_x^2 + v_y^2 + v_z^2)\right] v_x^2\, dv_x\, dv_y\, dv_z.$$

Solving the right side with the help of gamma functions,

$$\overline{v_x^2} = 1/m\beta.$$

Similarly,

$$\overline{v_y^2} = 1/m\beta = \overline{v_z^2}$$

Hence

$$W = \frac{1}{2} mn \frac{3}{m\beta} = \frac{3}{2} \frac{n}{\beta}.$$

Also from gas laws,

$$W = \tfrac{3}{2} nkT.$$

The above relations result in a value of $\beta = 1/kT$, where k is the Boltzmann factor and T is the temperature of the electron gas in degrees Kelvin.

The complete expression for $f(v_x,v_y,v_z)$ is

$$f(v_x,v_y,v_z) = \left(\frac{m}{2\pi kT}\right)^{3/2} \exp\left[-\frac{1}{2}\frac{m(v_x^2 + v_y^2 + v_z^2)}{kT}\right]. \quad (2.22)$$

The term $e^{-E/kT}$ in the distribution or probability equation, the Boltzmann factor, is extremely important. Having determined the velocity distribution function, it will be helpful later to determine the probability function.

The probability of finding an electron having velocity components between v_x and $v_x + \Delta v_x$, $v_y + \Delta v_y$, v_z and $v_z + \Delta v_z$ is

$$\Delta P = f(v_x,v_y,v_z)\, \Delta v_x\, \Delta v_y\, \Delta v_z$$

or

$$\Delta P = \left(\frac{m}{2\pi kT}\right)^{3/2} \exp\left[-\frac{1}{2}\frac{m}{kT}(v_x^2 + v_y^2 + v_z^2)\right] \Delta v_x\, \Delta v_y\, \Delta v_z. \quad (2.23)$$

In terms of absolute velocity, the probability of finding an electron having a velocity between v and $v + \Delta v$ can be determined by Eq. (2.23). The elemental volume $\Delta v_x\, \Delta v_y\, \Delta v_z = 4\pi v^2\, \Delta v$ and

$$v_x^2 + v_y^2 + v_z^2 = v^2.$$

Therefore,

$$\Delta P = \left(\frac{m}{2\pi kT}\right)^{3/2} \exp\left(-\frac{1}{2}\frac{mv^2}{kT}\right) 4\pi v^2\, \Delta v. \quad (2.24)$$

In terms of energy, the probability of finding an electron having an energy value between E and $E + \Delta E$ can be obtained from Eq. (2.24):

$$E = \tfrac{1}{2}mv^2$$

and hence

$$\Delta P = P(E)\, \Delta E = \frac{2}{(\pi kT)^{1/2}} E^{1/2} e^{-E/kT} \frac{1}{kT}\, \Delta E$$

or

$$P(E)\, \Delta E = \frac{2}{\pi^{1/2}} \frac{1}{(kT)^{3/2}} e^{-E/kT} E^{1/2}\, \Delta E \quad (2.25)$$

The consistency can be easily proved by trying the normalization condition on Eq. (2.25) and showing that

$$\int_0^\infty P(E)\, dE = 1.$$

Without normalizing one could easily say that the probability of finding an electron of energy E is

$$P(E) = Ae^{-E/kT}. \quad (2.26)$$

The above expression will be used often later.

From the results of gamma functions, it can be concluded that

$$\int_0^\infty e^{-X} X^{-1/2}\, dX = \int_0^\infty e^{-Y} Y^{-1/2}\, dY = \int_0^\infty e^{-Z} Z^{-1/2}\, dZ = \pi^{1/2}.$$

Hence

$$8A(2m\beta)^{-3/2}\pi^{3/2} = 1 \quad \text{or} \quad A = \left(\frac{m\beta}{2\pi}\right)^{3/2}.$$

The distribution function would be

$$f(v_x, v_y, v_z) = \left(\frac{m\beta}{2\pi}\right)^{3/2} \exp\left[-\frac{m\beta}{2}(v_x^2 + v_y^2 + v_z^2)\right]. \qquad (2.21)$$

Having determined the constant A, it now remains to evaluate the universal constant β in Eq. (2.15). This can be done by calculating the average energy per unit volume in the system of electrons from the distribution function using the definitions of average velocities and equating to the average energy from the simple gas laws discussed in Sec. 2.1.

If W is the energy per unit volume and n the number of electrons per unit volume,

$$W = \tfrac{1}{2}nm\overline{v^2} = \tfrac{1}{2}mn(\overline{v_x^2} + \overline{v_y^2} + \overline{v_z^2})$$

from the distribution function. Also

$$W = \tfrac{3}{2}nkT \quad \text{from simple gas laws,}$$

$$\overline{v_x^2} = \int\!\!\int\!\!\int_{-\infty}^{\infty} v_x^2 f(v_x, v_y, v_z)\, dv_x\, dv_y\, dv_z.$$

Substituting the value of $f(v_x, v_y, v_z)$ from Eq. (2.21) in the above equation,

$$\overline{v_x^2} = \left(\frac{m\beta}{2\pi}\right)^{3/2} \int\!\!\int\!\!\int_{-\infty}^{\infty} \exp\left[-\frac{m\beta}{2}(v_x^2 + v_y^2 + v_z^2)\right] v_x^2\, dv_x\, dv_y\, dv_z.$$

Solving the right side with the help of gamma functions,

$$\overline{v_x^2} = 1/m\beta.$$

Similarly,

$$\overline{v_y^2} = 1/m\beta = \overline{v_z^2}$$

Hence

$$W = \frac{1}{2} mn \frac{3}{m\beta} = \frac{3}{2}\frac{n}{\beta}.$$

Also from gas laws,

$$W = \tfrac{3}{2}nkT.$$

The above relations result in a value of $\beta = 1/kT$, where k is the Boltzmann factor and T is the temperature of the electron gas in degrees Kelvin.

The complete expression for $f(v_x,v_y,v_z)$ is

$$f(v_x,v_y,v_z) = \left(\frac{m}{2\pi kT}\right)^{3/2} \exp\left[-\frac{1}{2}\frac{m(v_x^2 + v_y^2 + v_z^2)}{kT}\right]. \tag{2.22}$$

The term $e^{-E/kT}$ in the distribution or probability equation, the Boltzmann factor, is extremely important. Having determined the velocity distribution function, it will be helpful later to determine the probability function.

The probability of finding an electron having velocity components between v_x and $v_x + \Delta v_x$, $v_y + \Delta v_y$, v_z and $v_z + \Delta v_z$ is

$$\Delta P = f(v_x,v_y,v_z) \, \Delta v_x \, \Delta v_y \, \Delta v_z$$

or

$$\Delta P = \left(\frac{m}{2\pi kT}\right)^{3/2} \exp\left[-\frac{1}{2}\frac{m}{kT}(v_x^2 + v_y^2 + v_z^2)\right] \Delta v_x \, \Delta v_y \, \Delta v_z. \tag{2.23}$$

In terms of absolute velocity, the probability of finding an electron having a velocity between v and $v + \Delta v$ can be determined by Eq. (2.23). The elemental volume $\Delta v_x \, \Delta v_y \, \Delta v_z = 4\pi v^2 \, \Delta v$ and

$$v_x^2 + v_y^2 + v_z^2 = v^2.$$

Therefore,

$$\Delta P = \left(\frac{m}{2\pi kT}\right)^{3/2} \exp\left(-\frac{1}{2}\frac{mv^2}{kT}\right) 4\pi v^2 \, \Delta v. \tag{2.24}$$

In terms of energy, the probability of finding an electron having an energy value between E and $E + \Delta E$ can be obtained from Eq. (2.24):

$$E = \tfrac{1}{2}mv^2$$

and hence

$$\Delta P = P(E) \, \Delta E = \frac{2}{(\pi kT)^{1/2}} E^{1/2} e^{-E/kT} \frac{1}{kT} \Delta E$$

or

$$P(E) \, \Delta E = \frac{2}{\pi^{1/2}} \frac{1}{(kT)^{3/2}} e^{-E/kT} E^{1/2} \, \Delta E \tag{2.25}$$

The consistency can be easily proved by trying the normalization condition on Eq. (2.25) and showing that

$$\int_0^\infty P(E) \, dE = 1.$$

Without normalizing one could easily say that the probability of finding an electron of energy E is

$$P(E) = Ae^{-E/kT}. \tag{2.26}$$

The above expression will be used often later.

2.3 Fermi-Dirac statistics

In Sec. 2.2 we studied classical statistics. Electrons were assumed to be particles just as molecules in a gas. The results of quantum mechanics and Pauli's exclusion principle were not applied to the motion of the electron. Fermi-Dirac statistics takes into consideration the following two aspects when considering the motion of electrons.

1. The wave characteristic of the electron restricts the number of allowed energy levels between E and $E + \Delta E$.
2. According to Pauli's exclusion principle the number of electrons with an allowed energy between E and $E + \Delta E$ is drastically restricted.

It was shown in Sec. 1.3 that the motion of an electron in a potential box as a free electron is characterized by three quantum numbers n_x, n_y, and n_z of the electron, and as a result of the numbers the density of electronic states between energy interval E and $E + dE$ is given by Eq. (1.38):

$$\frac{dG}{\text{unit vol.}} = \frac{4\pi}{h^3} m^{3/2}(2E)^{1/2}\, dE = G(E)\, dE. \tag{1.38}$$

It was discussed in Sec. 1.7 that electron motion is not completely defined by these three quantum numbers. The electron also possesses a quantum-mechanical property called *spin*, which has associated with it a certain magnetic moment. The term "spin" is used to describe this property because the classical property that corresponds most closely to spin is a rotation on the electron about its axis. If a magnetic field is applied, this magnet can have two possible orientations given by spin quantum numbers n_s. Pauli's exclusion principle states that no two electrons have identical quantum numbers. This restricts the number of electrons per wave number to 2 and the number of electrons possible per unit volume between energy E and $E + dE$ to

$$g(E)\, dE = 2 \times \frac{4\pi}{h^3} m^{3/2}(2E)^{1/2}\, dE. \tag{2.27}$$

If all possible states are occupied,

$$g(E)\, dE = dn, \tag{2.28}$$

where dn is the number of electrons available between energy E and $E + dE$. $g(E)$ is called the *density of states function*.

If however, all possible energy states between energy E and $E + dE$ are not occupied and there exists a probability of occupation of a state of energy E, $p(E)$, which may or may not be unity, the total number of electrons in the energy interval between E and $E + dE$ is

$$dn = g(E)p(E)\, dE, \tag{2.29}$$

because dn = density of states × probability of occupation of a state.

We shall now proceed to calculate $p(E)$ on the assumption of a quantum-mechanical behavior of the electrons. To obtain $p(E)$ we shall again consider collisions between two electrons with energies E_1 and E_2 before the collision and E_1' and E_2' after the collision. Following the law of conservation and energy,

$$E_1' + E_2' = E_1 + E_2.$$

In equilibrium the average number of collisions with initial energies E_1 and E_2 and final energies E_1' and E_2' must equal the average number of collisions with initial energies E_1' and E_2' and final energies E_1 and E_2. Looking at transitions $E_1, E_2 \rightarrow E_1'$, and E_2', we would expect that the number of such collisions is proportional to the probabilities of occupation of initial states E_1 and E_2 and to the probabilities of the final energy states E_1' and E_2' being empty.

Probability of occupation of an energy state $E_1 = p(E_1)$.
Probability of occupation of an energy state $E_2 = p(E_2)$.
Probability of an energy state E_2' being empty $= [1 - p(E_2')]$.
Probability of an energy state E_1' being empty $= [1 - p(E_1')]$.

Therefore the average number of transitions $E_1, E_2 \rightarrow E_1'$, and E_2' is $Cp(E_1)p(E_2)[1 - p(E_1')][1 - p(E_2')]$, where C is the constant of probability. Following a similar argument, the average number of transitions $E_1', E_2' \rightarrow E_1$, and E_2 equals $Cp(E_1')p(E_2')[1 - p(E_1)][1 - p(E_2)]$. In equilibrium

$$Cp(E_1)p(E_2)[1 - p(E_1')][1 - p(E_2')] = Cp(E_1')p(E_2')[1 - p(E_1)][1 - p(E_2)].$$

Dividing both sides of the above equation by $Cp(E_1)p(E_2)p(E_1')p(E_2')$,

$$\left[\frac{1}{p(E_1)} - 1\right]\left[\frac{1}{p(E_2)} - 1\right] = \left[\frac{1}{p(E_1')} - 1\right]\left[\frac{1}{p(E_2')} - 1\right]. \quad (2.30)$$

Putting $E_1' = E_1 - \Delta E$ and $E_2' = E_2 + \Delta E$,

$$\left[\frac{1}{p(E_1)} - 1\right]\left[\frac{1}{p(E_2)} - 1\right] = \left[\frac{1}{p(E_1 + \Delta E)} - 1\right]\left[\frac{1}{p(E_2 - \Delta E)} - 1\right]. \quad (2.31)$$

An examination of Eq. (2.31) shows that there is only one simple mathematical expression for $p(E)$ that satisfies Eq. (2.31),

$$\left[\frac{1}{p(E)} - 1\right] = Be^{\beta E}$$

or

$$p(E) = (Be^{\beta E} + 1)^{-1}. \quad (2.32)$$

β is the universal constant $1/kT$ derived in Sec. 2.2. It has a positive sign.

B is also positive and is a constant. The signs follow from the fact that the states with lower energies are more likely to be occupied.

If $Be^{\beta E} \gg 1$,

$$p(E) = \frac{1}{B}e^{-\beta E} = \frac{1}{B}e^{-E/kT},$$

which is the probability of a state of energy being occupied according to the Boltzmann statistic.

Introducing $B = e^{-E_f/kt}$ in Eq. (2.32),

$$p(E) = \left[\exp\left(\frac{E-E_f}{kT}\right) + 1\right]^{-1}, \qquad (2.33)$$

where E_f is called the *Fermi energy level*. In Sec. 2.4 we shall discuss Fermi levels. It may be mentioned here that having determined $p(E)$ [Eq. (2.33)] and knowing $g(E)\,dE$ [Eq. (2.28)], dn (the number of electrons between energy level E and $E + dE$) can be determined by Eq. (2.29).

2.4 Fermi levels

The factor $[\exp(E - E_f/kT) + 1]^{-1}$ in the equation of $p(E)$ depends upon the energy in a peculiar way. At $T = 0$ the factor is unity for $E < E_f$ and zero for $E > E_f$. Hence at $T = 0$,

$$p(E) = \begin{cases} \left[\exp\left(\frac{E-E_f}{kT}\right) + 1\right]^{-1} = 1 & \text{for } E < E_f, \\ \left[\exp\left(\frac{E-E_f}{kT}\right) + 1\right]^{-1} = 0 & \text{for } E > E_f. \end{cases}$$

This would mean that at $T = 0$ the probability of occupation of all states with energy values lower than the Fermi energy level is unity, and the probability of occupation of states higher than the Fermi energy level is zero; in other words, no electron has an energy higher than E_f. Hence at $T = 0$,

$$\int dn = \int_0^{E_f} g(E)\,dE \qquad p(E) = 1,$$

$$n = \begin{cases} \int_0^{E_f} \frac{8\pi}{h^3} m^{3/2} 2^{1/2} E^{1/2}\,dE, \\ \frac{8\pi}{h^3} m^{3/2} 2^{1/2} \tfrac{2}{3} E_f^{3/2}. \end{cases}$$

Calling the Fermi level at absolute zero E_{f0},

$$E_{f0} = \left(\frac{h^3}{8\pi 2^{1/2}} m^{-3/2} \tfrac{3}{2}\right)^{2/3} = \frac{h^2}{2m}\left(\frac{3n}{8\pi}\right)^{2/3},$$

where n is the number of electrons per unit volume. In the expression

$$E_{f0} = \frac{h^2}{2m}\left(\frac{3n}{8\pi}\right)^{2/3}, \tag{2.34}$$

E_{f0} can be calculated if n is known, because h and m are constants.

At $T > 0$, the factor $[\exp(E - E_f/kT) + 1]^{-1}$ is close to unity for $E \ll E_f$ and is zero for $E \gg E_f$. At $E = E_f$ the factor is $\frac{1}{2}$. At any temperature other than zero, the probability of occupation of the energy state $E = E_f$ is $\frac{1}{2}$. Therefore the Fermi energy level or Fermi level is that level of energy whose probability of occupation by an electron at $T > 0$ is $\frac{1}{2}$. The function $p(E)$ is plotted against E for $T = 0$ and $T > 0$ in Fig. 2.2.

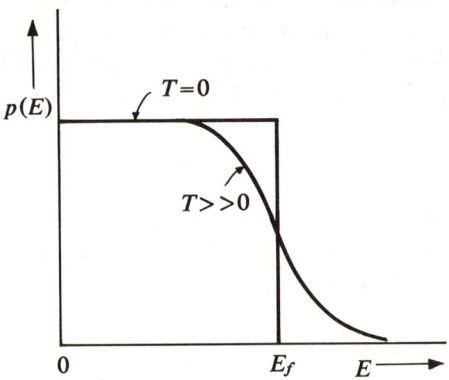

Fig. 2.2
Fermi distribution function $p(E)$ at absolute zero and at a temperature $T \gg 0$

From Eq. (2.27), which gives the density of states in the energy interval E and $E + dE$,

$$g(E)\,dE = 2 \times \frac{4\pi}{h^3} m^{3/2}(2E)^{1/2}\,dE. \tag{2.27}$$

$g(E)$, the density of states function, can be plotted as a function of E:

$$g(E) = 2 \times \frac{4\pi}{h^3} m^{3/2} 2^{1/2} E^{1/2}. \tag{2.35}$$

Figure 2.3 shows the variation of $g(E)$ against E, the half-power variation. The definition of the density of states suggests that on the $g(E)$ versus E curve, the area of a small strip of thickness dE at any arbitrary value of E would be the number of states available in the energy interval E and $E + dE$. The area under the curve from $E = 0$ to any value of E would be the total density of states available from $E = 0$ to the value of E.

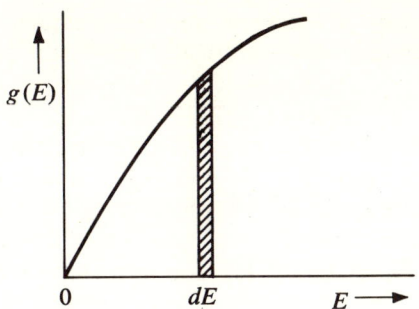

Fig. 2.3
Variation of density of states $g(E)$ with E

The product $g(E)p(E)$ can be obtained by multiplying Eqs. (2.35) and (2.33). Looking at Figs. 2.2 and 2.3 one can conclude that the product is not only a function of E but also depends upon T, because $p(E)$ depends strongly upon T. The product $g(E)p(E)$ is plotted against E in Fig. 2.4 for $T = 0$ and for $T \gg 0$. In Fig. 2.4 the area of a small strip of thickness at any arbitrary value of E would be the number of electrons occupying the states, in other words, the number of electrons present in the energy interval E and $E + dE$. The area under the curve from $E = 0$ to any value of E would give the total density of electrons available from $E = 0$ to the value of E.

From Fig. 5.6 it can be seen that the rise in temperature above absolute zero affects only high-energy electrons. Electrons with energy near E_f can undergo slight excitation in the neighboring unoccupied states as the temperature is raised above absolute zero.

In Fermi-Dirac statistics an electron gas has quite different properties from those of a classical gas at or near absolute zero temperature. At these

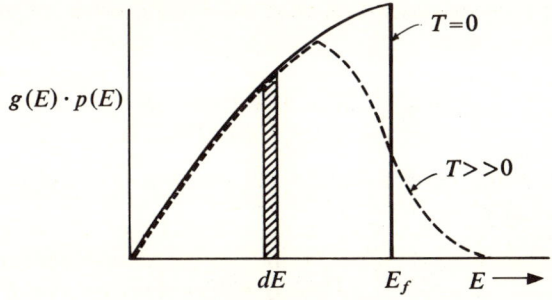

Fig. 2.4
Energy distribution

low temperatures the electron gas is said to be degenerate. As the temperature is raised, degeneracy is gradually removed. Looking at the expression of dn,

$$dn = g(E)p(E)\,dE.$$

Also

$$dn = 2 \times \frac{4\pi}{h^3} m^{3/2} 2^{1/2} E^{1/2} \left[\exp\left(\frac{E - E_f}{kT}\right) + 1 \right]^{-1}. \tag{2.36}$$

If $E \gg E_f$ and $e^{E/kT} \gg 1$, Eq. (2.36) reduces to Eq. (2.37):

$$dn = 2 \times \frac{4\pi}{h^3} m^{3/2} 2^{1/2} E^{1/2} e^{-E/kT}. \tag{2.37}$$

Comparing the right sides of Eqs. (2.37) and (2.25) we find that the two expressions are similar in E, the difference being in constants. This indicates that if $E \gg E_f$ and $e^{E/kT} \gg 1$, a Fermi-Dirac electron gas which is degenerate becomes a classical electron gas or nondegenerate. It can be shown easily that at a given temperature if $E - E_f \geq 4kT$,

$$\exp\left(\frac{E - E_f}{kT}\right) \gg 1 \quad \text{and} \quad p(E) = \exp\left(\frac{E_f - E}{kT}\right) = e^{E_f/kT} e^{-E/kT}.$$

We shall make use of this fact in our discussion of semiconductors.

Calculation of Fermi level E_f at $T \neq 0$. The constant E_f, the Fermi energy level (or Fermi level) which occurs in the Fermi-Dirac distribution function, is

$$p(E) = \left[\exp\left(\frac{E - E_f}{kT}\right) + 1 \right]^{-1},$$

is equal only to the maximum energy of the electron at $T = 0$. At all other temperatures the value of E_f may be different from E_{f0} and is calculated from the basic physical definition that the total number of electrons per unit volume n is given by

$$n = \int dn = \int g(E)p(E)\,dE, \tag{2.38}$$

the integration being carried out for all energies. To solve the integral (2.38), let us consider the integral

$$I = \int_0^\infty p(E) \frac{d}{dE}[F(E)]\,dE, \tag{2.39}$$

where $F(E)$ is any function that vanishes for $E = 0$.

By partial integration of Eq. (2.39),

$$I = \left[p(E)F(E)\right]_0^\infty - \int_0^\infty F(E) \frac{d}{dE}[p(E)]\, dE.$$

In the term $p(E)F(E)\Big|_0^\infty$, the first term would be zero for $E \to \infty$, because the probability of finding an electron with infinite energy is zero. The second term is zero for $E = 0$ by the basic assumption of the function $F(E)$. Hence at both the limits the product $p(E)F(E) = 0$. Therefore,

$$I = -\int_0^\infty F(E) \frac{d}{dE}[p(E)]\, dE. \tag{2.40}$$

It can be seen from Fig. 2.2 that $p(E)$ changes only in the neighborhood of E_f and hence $(d/dE)[p(E)]$ is a function that vanishes except in the neighborhood of $E = E_f$. In the integral of Eq. (2.40) the term $(d/dE)[p(E)]$ appears as the product, so the whole integral would be appreciable only in the neighborhood of E_f. We therefore expand $F(E)$ by a Taylor series at $E = E_f$:

$$F(E) = F(E_f) + (E - E_f)F'(E_f) + \frac{(E - E_f)^2}{2} F''(E_f) + \cdots.$$

Substituting the expansion in Eq. (2.40),

$$I = -F(E_f) \int_0^\infty \frac{dp(E)}{dE}\, dE - F'(E_f) \int_0^\infty (E - E_f) \frac{dp(E)}{dE}\, dE$$

$$- F''(E_f) \int_0^\infty \frac{1}{2}(E - E_f)^2 \frac{d}{dE}[p(E)]\, dE - \cdots.$$

It can be easily shown that

$$-\int_0^\infty \frac{dp(E)}{dE}\, dE = 1,$$

$$-\int_0^\infty (E - E_f) \frac{dp(E)}{dE}\, dE = 0,$$

$$-\frac{1}{2}\int_0^\infty (E - E_f)^2 \frac{d}{dE}[p(E)]\, dE \cong \frac{\pi^2}{6}(kT)^2.$$

Therefore,

$$I = F(E_f) + \frac{\pi^2}{6}(kT)^2 F''(E_f).$$

Let us now put

$$F(E) = \int_0^E g(E)\, dE.$$

Therefore,
$$\frac{d}{dE}[F(E)] = g(E),$$
and hence
$$I = \int_0^\infty p(E)g(E)\,dE = F(E_f) + \frac{\pi^2}{6}(kT)^2 F''(E_f)$$
or
$$n = F(E_f) + \frac{\pi^2}{6}(kT)^2 F''(E_f).$$

We have already discussed that E_{f0} is the Fermi level at absolute zero,
$$n = \int_0^{E_{f0}} g(E)\,dE,$$
and hence
$$\int_0^{E_{f0}} g(E)\,dE = F(E_f) + \frac{\pi^2}{6}(kT)^2 F''(E_f). \tag{2.41}$$

The first term on the right side can be written
$$F(E_f) = \int_0^{E_f} g(E)\,dE.$$

The second term on the right side can be rewritten
$$\frac{\pi^2}{6}(kT)^2 F''(E_f) = \frac{\pi^2}{6}(kT)^2 g'(E_f).$$

Hence
$$\int_0^{E_{f0}} g(E)\,dE = \int_0^{E_f} g(E)\,dE + \frac{\pi^2}{6}(kT)^2 g'(E_f)$$
or
$$g(E_f)[E_f - E_{f0}] + \frac{\pi^2}{6}(kT)^2 g'(E_f) = 0.$$

Now from Eq. (2.35),
$$g(E) = 2 \times \frac{4\pi}{h^3} m^{3/2} 2^{1/2} E^{1/2}. \tag{2.35}$$

Differentiating Eq. (2.35),
$$g'(E) = \frac{4\pi}{h^3} m^{3/2} 2^{1/2} E^{-1/2}. \tag{2.42}$$

From Eqs. (2.35) and (2.42),

$$g(E_f) = 2 \times \frac{4\pi}{h^3} m^{3/2} 2^{1/2} E_f^{1/2},$$

$$g'(E_f) = \frac{4\pi}{h^3} m^{3/2} 2^{1/2} E_f^{-1/2},$$

and hence

$$(E_f - E_{f0}) \times 2 \times \frac{4\pi}{h^3} m^{3/2} 2^{1/2} E_f^{1/2} = -\frac{\pi^2}{6} \frac{4\pi}{h^3} m^{3/2} 2^{1/2} E_f^{-1/2},$$

$$(E_f - E_{f0})E_f = -\frac{\pi^2}{12} (kT)^2,$$

or

$$E_f - E_{f0} = -\frac{\pi^2}{12} (kT)^2 \frac{1}{E_f}.$$

Because

$$E_f \simeq E_{f0},$$

$$E_f = E_{f0} \left[1 - \frac{\pi^2}{12} \frac{(kT)^2}{E_{f0}^2} \right]. \tag{2.43}$$

The Fermi level at $T = 0$ is E_{f0}, which decreases with increasing temperature.

2.5 Thermionic emission

The process of thermionic emission consists of providing sufficient thermal energy to the electrons in the material that the conduction electron near the surface of the material can be emitted from the material. The term *conduction electron* will be discussed in detail later. At this stage we may understand by conduction electrons those electrons that are not tightly bound to the nucleus and as such may be treated as free electrons for motion. Conduction electrons are mostly available in metals in large numbers, and hence we would be concerned with thermionic emission in metals. The thermal energy in the metal is transferred to the electrons and some of the electrons can overcome a natural potential-energy barrier that exists at the surface and can be emitted. The potential barrier is known as the *surface work function*. The energy transferred at high temperature is in the form of violent thermal lattice vibration in the metal. Conduction electrons get this energy transferred to them. The metals in consideration are elevated to temperatures as high as 1500 to 2700°K.

Let the metal under consideration be at a temperature T and let us cal-

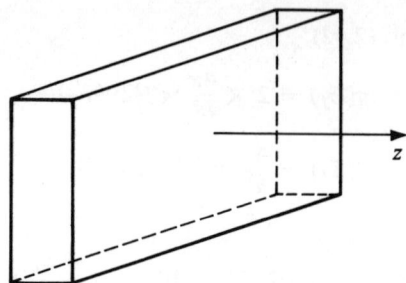

Fig. 2.5
Metal block for illustrating the thermionic emission process

culate the number of electrons emitted from the unit surface normal to the z axis in Fig. 2.5, J_z.

Let dJ_z be the differential current density due to electrons dn having velocity components v_x, v_y, and v_z. Therefore,

$$dJ_z = ev_z\,dn, \tag{2.44}$$

where e is the electronic charge 1.6×10^{-19} C.

The differential charge-carrier density dn for electrons having momentum between P_x and $P_x + dP_x$, P_y and $P_y + dP_y$, and P_z and $P_z + dP_z$ is given by

$$dn = 2 \times \frac{dG}{\text{unit vol.}}\,p(E),$$

where

$$\frac{dG}{\text{unit vol.}} = \frac{4\pi P^2}{h^3}\,dP.$$

The factor 2 accounts for the spin. Therefore,

$$dn = 2 \times \frac{4\pi P^2}{h^3}\,dP\,p(E). \tag{2.45}$$

In Eq. (2.45), $(4\pi P^2/h^3)\,dP$ is the density of states in momentum space lying between momentum P and $P + dP$. Taking the components of the momentum P_x, P_y, and P_z,

$$\frac{4\pi P^2}{h^3}\,dP = \frac{dP_x\,dP_y\,dP_z}{h^3},$$

where we are considering momentum component intervals between P_x and $P_x + dP_x$, P_y and $P_y + dP_y$, P_z and $P_z + dP_z$ instead of the absolute momentum interval P and $P + dP$.

Equation (2.45) can therefore be rewritten

$$dn = \frac{2}{h^3}\,p(E)\,dP_x\,dP_y\,dP_z. \tag{2.46}$$

THERMIONIC EMISSION

Substituting Eq. (2.46) in Eq. (2.44),

$$dJ_z = \frac{2e}{h^3} v_z p(E) \, dP_x \, dP_y \, dP_z. \tag{2.47}$$

Replacing v_z by P_z/m,

$$dJ_z = \frac{2e}{mh^3} p(E) P_z \, dP_x \, dP_y \, dP_z. \tag{2.48}$$

The total emission current density in the z direction J_z is found by integrating Eq. (2.48) for all possible values of momenta in the x and y directions (from $-\infty$ to ∞) and the value of momentum in the z direction from a minimum positive value P'_z to the maximum value ∞. The minimum momentum in the z direction P'_z corresponds to the surface barrier potential. We shall correlate P'_z with the work function later.

$$J_z = \int dJ_z = \int_{P'_z}^{\infty} \int_{-\infty}^{\infty} \int_{-\infty}^{\infty} \frac{2e}{mh^3} P_z p(E) \, dP_x \, dP_y \, dP_z,$$

where $p(E) = \left[\exp\left(\frac{E - E_f}{kT}\right) + 1\right]^{-1}$, according to Fermi-Dirac statistics.

Therefore,

$$J_z = \int_{P'_z}^{\infty} \int_{-\infty}^{\infty} \int_{-\infty}^{\infty} \frac{2e}{mh^3} \left[\exp\left(\frac{E - E_f}{kT}\right) + 1\right]^{-1} P_z \, dP_x \, dP_y \, dP_z. \tag{2.49}$$

The energies involved in thermionic emission are much greater than E_f, and, therefore, $E - E_f \gg kT$. Hence

$$\exp\left(\frac{E - E_f}{kT}\right) + 1 \simeq \exp\left(\frac{E - E_f}{kT}\right).$$

Therefore Eq. (2.49) can be simplified as

$$J_z = \int_{P'_z}^{\infty} \int_{-\infty}^{\infty} \int_{-\infty}^{\infty} \frac{2e}{mh^3} e^{E_f/kT} P_z e^{-E/kT} \, dP_x \, dP_y \, dP_z.$$

Setting the energy E in terms of momentum components,

$$E = \frac{P_x^2 + P_y^2 + P_z^2}{2m},$$

$$J_z = \int_{P'_z}^{\infty} \int_{-\infty}^{\infty} \int_{-\infty}^{\infty} \left(\frac{2e}{mh^3} e^{E_f/kT}\right) P_z \exp\left(-\frac{P_x^2 + P_y^2 + P_z^2}{2mkT}\right) dP_x \, dP_y \, dP_z. \tag{2.50}$$

This equation is relatively simple to integrate because the three components of momentum are independent of each other:

$$\int_{-\infty}^{\infty} \exp\left(-\frac{P_x^2}{2mkT}\right) dP_x = (2\pi mkT)^{1/2}$$

$$\int_{-\infty}^{\infty} \exp\left(-\frac{P_y^2}{2mkT}\right) dP_y = (2\pi mkT)^{1/2}$$

$$\int_{P_z'}^{\infty} P_z \exp\left(-\frac{P_z^2}{2mkT}\right) dP_z = +mkT \exp\left(-\frac{P_z'^2}{2mkT}\right).$$

Hence

$$J_z = \frac{2e}{mh^3} e^{E_f/kT} 2\pi(mkT)^2 \exp\left(-\frac{P_z'^2}{2mkT}\right).$$

Simplifying,

$$J_z = \left(\frac{4\pi mek^2}{h^3}\right) T^2 e^{E_f/kT} \exp\left(-\frac{P_z'^2}{2mkT}\right).$$

Let the minimum momentum of the electron correspond to zero energy and let this level be E_a below the vacuum level. As such, an electron leaving the metal with the minimum energy has an energy

$$E_a = \frac{P_z'^2}{2m},$$

$$J_z = \left(\frac{4\pi mek^2}{h^3}\right) T^2 \exp\left(\frac{E_f - E_a}{kT}\right). \tag{2.51}$$

Referring to Fig. 2.6, $E_a - E_f$ is the energy required to bring an electron from inside the material lying at the Fermi energy level to the vacuum level outside the metal. We have seen in Sec. 2.4 that vacant states exist only near the Fermi level, or the highest occupied state is close to the Fermi

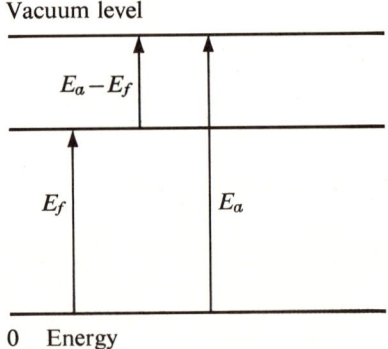

Fig. 2.6
Energy-level diagram for thermionic emission

level. It is these electrons that gain energy and are emitted. If $E_a - E_f = e\Phi$, where e is the electronic charge, Φ is called the *surface work function*.

Equation (2.51) for the thermionic current density can be rewritten in terms of surface work function as

$$J_z = \left(\frac{4\pi mek^2}{h^3}\right)T^2 e^{-e\Phi/kT}. \tag{2.52}$$

Putting $4\pi mek^2/h^3 = A$,

$$J_z = AT^2 e^{-e\Phi/kT}. \tag{2.53}$$

Equation (2.53) is the famous *Richardson-Dushman equation* for thermionically emitted current density. The current density increases with decreasing work function and increasing temperature. The value of A calculated from constants m, e, k, and h is 120.4×10^4 A/m²/(°K)². The value obtained experimentally does not agree with the calculated value. The discrepancy may be attributed to the wave nature of the electrons. The constant A has been calculated on the basis of the energy of the electron without considering the fact that the electron as a wave may be reflected back from the surface contamination, such as absorbed gas layers. If γ is the reflection coefficient,

$$A_{\text{eff}} = A(1 - \gamma),$$

where A_{eff} is the effective value of A.

Energy of thermally emitted electrons. In the following the average energy of the thermally emitted electrons will be investigated. This can be determined by investigating the velocity distribution of the emitted electrons. This investigation makes use of the principles of statistics and should be a good concluding exercise in the chapter.

The differential current density in the z direction is given by Eq. (2.48),

$$dJ_z = \frac{2e}{mh^3} p(E) P_z \, dP_x \, dP_y \, dP_z. \tag{2.48}$$

Substituting the value of $p(E)$, assuming $E - E_f \gg kT$, and integrating over the momentum ranges in the x and y directions,

$$dJ_z = \frac{2e}{mh^3} (2\pi mkT) e^{E_f/kT} P_z \exp\left(-\frac{P_z^2}{2mkT}\right) dP_z$$

or

$$dJ_z = \frac{4\pi ekT}{h^3} e^{E_f/kT} \exp\left(-\frac{P_z^2}{2mkT}\right) P_z \, dP_z,$$

$$E_a - e\Phi = E_f, \qquad \frac{4\pi ekT}{h^3} = \frac{4\pi mek^2}{h^3} \cdot \frac{T}{mk} = \frac{AT}{mk},$$

$$dJ_z = \frac{AT}{mk} \exp\left(\frac{E_a - e\Phi}{kT}\right) P_z \exp\left(-\frac{P_z^2}{2mkT}\right) dP_z. \tag{2.54}$$

Since dJ_z is the differential current density of the emitted electrons, their minimum energy is E_a.

Let $P_a^2/2m = E_x + E_z$, where E_z is the energy of the free electron in the z direction. Therefore,

$$\frac{P_z\, dP_z}{m} = dE_z,$$

$$dJ_z = \frac{AT}{mk}\, me^{-e\Phi/kT} e^{-E_z/kT}\, dE_z. \tag{2.55}$$

Simplifying Eq. (2.55),

$$dJ_z = \frac{1}{kT} AT^2 e^{-e\Phi/kT} e^{-E_z/kT}\, dE_z. \tag{2.56}$$

From Eq. (2.53),

$$J_z = AT^2 e^{-e\Phi/kT}. \tag{2.53}$$

Dividing Eq. (2.53) by Eq. (2.56),

$$\frac{dJ_z}{J_z} = \frac{1}{kT} e^{-E_z/kT}\, dE_z. \tag{2.57}$$

The above equation states that the fraction of emitted electrons having energies between E_z and $E_z + dE_z$ is a modified one-dimensional energy distribution function,

$$\frac{dJ_z}{J_z} = \frac{d(nev_z')}{nev_z'} = \frac{dn}{n} = \Delta P.$$

dn/n is by definition the probability of an emitted electron ΔP to have an energy in the direction of emission between E_z and $E_z + dE_z$. v_z' is the velocity of the electron after emission in the z direction. Therefore,

$$dP = \frac{dn}{n} = \frac{1}{kT} e^{-E_z/kT}\, dE_z = f(v_z')\, dv_z',$$

where $f(v_z')$ is the velocity distribution function:

$$E_z = \tfrac{1}{2} mv_z'^2,$$

$$dE_z = mv_z'\, dv_z'.$$

Therefore,

$$\frac{dn}{n} = f(v_z')\, dv_z' = \frac{1}{kT} mv_z' dv_z' \exp\left(-\frac{1}{2}\frac{mv_z'^2}{kT}\right).$$

Therefore,

$$f(v_z') dv_z' = \frac{mv_z'}{kT} \exp\left(-\frac{1}{2}\frac{mv_z'^2}{kT}\right) dv_z'$$

The mean-square velocity is

$$\overline{v_m^2} = \int_0^\infty v_z'^2 f(v_z')\, dv_z' = \int_0^\infty \frac{mv_z'^3}{kT} \exp\left(-\frac{1}{2}\frac{mv_z'^2}{kT}\right) dv_z'$$

With the help of gamma functions,

$$\overline{v_m^2} = 2kT/m.$$

The total energy of n electrons per unit volume is

$$W = n(\tfrac{1}{2}m\overline{v_m^2}) = nkT,$$

where W is the energy per unit volume.

The results obtained are interesting because they show that the average energy resulting from the z component of velocity of the emitted electron is twice that given by the equipartition law for particles obeying Boltzmann statistics.

Problem 1

Consider a chamber containing monoatomic potential-free molecules of a gas. Determine the root-mean-square velocity of the molecules if the gas is at room temperature (27°C). The mass of the molecule may be taken as the rest mass of a proton. Determine the pressure P exerted on the walls of the container if the number of molecules per cubic meter is 10^{23}.   *Ans.* 2.74×10^3 m/sec; 414 N/m²

Problem 2

The distribution function of velocity v of the gas molecules of mass m at a temperature T may be written

$$\frac{\Delta n}{n} = f(v)\,\Delta v = C \exp\left(-\frac{1}{2}\frac{mv^2}{kT}\right) v^2\,\Delta v,$$

where v can assume all values from zero to infinity.
a. Normalize the distribution function.
b. Calculate the average value of $\bar{v}$.
c. Calculate the average value of $\bar{v}^n$, where n is an integer.
d. Find the v_0 for which $f(v)$ is a maximum.

Ans. **a.** $c = \left(\dfrac{2}{\pi}\right)^{1/2}\left(\dfrac{m}{kT}\right)^{3/2}$ **b.** $\bar{v} = 2^{3/2}\left(\dfrac{kT}{\pi m}\right)^{1/2}$

c. $\bar{v}^n = \dfrac{n+1}{(\pi^{1/2})}\left(\dfrac{2kT}{m}\right)^{n/2}\left|\dfrac{n+1}{2}\right.$ **d.** $v_0 = \left(\dfrac{2kT}{m}\right)^{1/2}$

Problem 3

A gas possesses a Maxwellian velocity distribution function. Show that the fraction of molecules in a given volume that possesses a velocity v_x in one direction only and whose magnitude is greater than some selected value v_0 is

$$\int_{v_0}^{\infty} f(v_x)\,dv_x = \tfrac{1}{2} - \pi^{-1/2} \int_0^{[\frac{1}{2}(mv_0^2/kT)]^{1/2}} \exp\left(-\frac{1}{2}\frac{mv_0^2}{kT}\right) d\left(\frac{1}{2}\frac{mv_0^2}{kT}\right)^{1/2}.$$

Problem 4

Using a Maxwellian distribution function:

a. Find the mean-square fluctuation of the velocity; i. e.,

$$\overline{\Delta v^2} = \overline{(v - \bar{v})^2} = \overline{v^2} - \bar{v}^2.$$

b. Find the mean and mean-square value of the kinetic energy of an atom. Also determine the mean-square fluctuation of the kinetic energy of an atom.

Ans.   a. $\dfrac{kT}{m}\left(3 - \dfrac{8}{\pi}\right)$ b. $\dfrac{3kT}{2}, \dfrac{15}{4}(kT)^2, \dfrac{3}{2}(kT)^2$

Problem 5

Find the probability of an electronic state being occupied at room temperature, if the energy of the state lies 0.2 eV above the Fermi level. Do the same for a state that lies 0.2 eV below the Fermi level. Repeat the computations at 100 and 1000°K
Ans.   At room temperature, $4 \times 10^{-4}, \simeq 1$; at 100°K, $0, \simeq 1$; at 1000°K, 0.0834 0.524.

Problem 6

Show that the kinetic energy of a free electron gas at 0°K is $\tfrac{3}{5}NE_{f0}$, where N is the number of electrons and E_{f0} is the Fermi level at 0°K.

Problem 7

For free electrons the density of states between energy interval E and $E + \Delta E$ is given by $A E^{1/2} \Delta E$, where A is a constant. If n is the number of electrons per unit volume and all states from energy value zero to E_f are occupied, what would A be in terms of known quantities?

Problem 8

Show that the wavelength of an electron having an energy equal to the Fermi energy is given by

$$\lambda = 2\left(\frac{\pi}{3n}\right)^{1/3},$$

where n is the number of electrons per unit volume.

Problem 9

a. Show that the derivative of $p(E) = [\exp(E - E_f/kT) + 1]^{-1}$ is symmetrical about E_f and that $\int_{-\infty}^{\infty} (\partial p/\partial E) \, dE = -1$.

b. Show that $\partial p/\partial E$ has an appreciable value in an energy region of the order of kT on either side of the Fermi level.

Problem 10

Compute the Fermi level in copper at absolute zero and at 500°C. There are four atoms per unit cell and the unit cell is a cube with side 3.61 Å.
Ans. $E_{f0} = 7.1$ eV, $E_f(500°C) = 7.1$ eV.

Problem 11

a. Energy is needed to bring an electron from the inside of a thermionic emitter to the outside. The loss of energy would cool the emitter. Show that the temperature of the emitter can be kept constant if an extra power, $P = I_s[\Phi + (2kT/e)]$, is supplied to it after it starts emitting I_s, the thermionic current.

b. An impregnated thermionic emitter consisting of a monoatomic layer of barium on oxygen has a current density of 100 A/cm² at 1320°C. The emitter has an area of 0.2 cm². If the work function of the emitter is 1.6 eV, compute the extra heater power needed to maintain the temperature of the emitter, if a continuous current is drawn from the emitter.   *Ans.* 37.4 W.

3

Thermal Properties of Solids

Specific heat and thermal conductivity are two quantities that describe the thermal properties of solids. The specific heat of a material is defined as the amount of thermal energy absorbed by a unit mass when its temperature is raised by one degree.

Specific heat at constant volume. The following analysis, although specifically for gases, can be extended to solids under the appropriate conditions. The first law of thermodynamics states that the amount of heat added to a system dQ must equal the increase in energy dU of the system plus the amount of work done by the system dW:

$$dQ = dU + dW.$$

If the work done is of a mechanical nature,

$$dW = p\,dV,$$

where p is the pressure and dV is the change in volume. U is determined uniquely by temperature and volume on $U(V, T)$. Therefore,

$$dU = \left(\frac{\partial U}{\partial T}\right)_V dT + \left(\frac{\partial U}{\partial V}\right)_T dV.$$

Hence

$$dQ = \left(\frac{\partial U}{\partial T}\right)_V dT + \left(\frac{\partial U}{\partial V}\right)_T dV + p\,dV$$

or

$$dQ = \left(\frac{\partial U}{\partial T}\right)_V dT + \left[\left(\frac{\partial U}{\partial V}\right)_T + P\right] dV. \tag{3.1}$$

If the volume is constant, the second term on the right side of Eq. (3.1) is zero and

$$\left(\frac{dQ}{dT}\right)_V = \left(\frac{\partial U}{\partial T}\right)_V = C_V, \qquad (3.2)$$

where C_V is the specific heat at constant volume and is normally defined as the amount of heat energy required to change the temperature of a gram molecule of the solid by 1 degree.

Contributions to specific heat of solids. As the temperature of a solid is increased, the energy input in the solid can have a twofold contribution.

1. An increase in temperature results in a more rigorous motion of the atoms, called a *lattice vibration*.
2. In metals and semiconductors there are conduction electrons, and an increase in temperature results in an increased energy of the motion of electrons.

In the first five sections we shall be concerned with the specific heat contribution due to lattice vibrations.

3.1 Classical theory of specific heats

Suppose it is possible to fix the position of all atoms in a crystal such that they are all in equilibrium position. If one of the atoms is now displaced over a small distance much smaller than the interatomic distance and set free, the displaced atom would carry out harmonic vibrations about its equilibrium position and its energy of vibration would be the same as that of three one-dimensional harmonic oscillators, one for each direction of motion. Therefore, the vibrational energy of a crystal containing N atoms is equivalent to the energy of a system of $3N$ harmonic oscillators, each one-dimensional.

The energy of a one-dimensional harmonic oscillator is

$$E = \frac{P^2}{2m} + \frac{fx^2}{2}.$$

Because $f/m = \omega^2$,

$$E = \frac{P^2}{2m} + \frac{1}{2}m\omega^2 x^2. \qquad (3.3)$$

The first term on the right side is the kinetic energy and the second term is the potential energy.

THERMAL PROPERTIES OF SOLIDS

According to statistical mechanics, the average energy of a harmonic oscillator at any temperature T is

$$\bar{E} = \frac{\int_0^\infty E e^{-E/kT}\, dE}{\int_0^\infty e^{-E/kT}\, dE},$$

because, according to (2.26), the probability of an atom or harmonic oscillator having an energy E is

$$P(E) = A e^{-E/kT} \quad \text{and} \quad \int_0^\infty P(E)\, dE = 1$$

or

$$A = \left(\int_0^\infty e^{-E/kT}\, dE\right)^{-1}.$$

By solving the expression for $\bar{E}$, one obtains $\bar{E} = kT$. Although the average total energy is kT, one still must calculate the average kinetic energy and the average potential energy. In the equation of the energy of an oscillator given by Eq. (3.3), the energy is instantaneous. The average energy would depend upon the average value of the square of the momentum $\overline{P^2}$ and on the average value of the square of the displacement of the atom or nucleus from its equilibrium position $\overline{x^2}$. Therefore,

$$\bar{E} = \frac{\overline{P^2}}{2m} + \frac{m\omega^2}{2} \overline{x^2},$$

and applying statistical mechanics,

$$\bar{E} = \frac{1}{2m} \frac{\int_{-\infty}^\infty P^2 \exp\left(-\frac{P^2}{2mkT}\right) dP}{\int_{-\infty}^\infty \exp\left(-\frac{P^2}{2mkT}\right) dP} + \frac{m\omega^2}{2} \frac{\int_{-\infty}^\infty x^2 \exp\left(-\frac{m\omega^2}{2kT} x^2\right) dx}{\int_{-\infty}^\infty \exp\left(-\frac{m\omega^2 x^2}{2kT}\right) dx}.$$

Solving the integrals by gamma functions,

$$\frac{1}{2m} \frac{\int_{-\infty}^\infty P^2 \exp\left(-\frac{P^2}{2mkT}\right) dP}{\int_{-\infty}^\infty \exp\left(-\frac{P^2}{2mkT}\right) dP} = \frac{kT}{2},$$

$$\frac{m\omega^2}{2} \frac{\int_{-\infty}^\infty x^2 \exp\left(-\frac{m\omega^2}{2kT} x^2\right) dx}{\int_{-\infty}^\infty \exp\left(-\frac{m\omega^2 x^2}{2kT}\right) dx} = \frac{kT}{2}.$$

Hence
$$\overline{E} = \frac{kT}{2} + \frac{kT}{2} = kT.$$

The average energy of a one-dimensional oscillator is kT. If there are N atoms, the energy of the vibration of the atoms corresponds to the total energy of vibration of $3N$ one-dimensional harmonic oscillators:
$$U = 3N\overline{E} = 3NkT.$$

If one considers a gram molecule, the number of atoms is N_0 in a gram molecule and the total vibrational energy of the atoms is
$$U = 3N_0 kT.$$

Therefore, the specific heat at constant volume is
$$C_V = \partial U/\partial T = 3N_0 k = 3R, \tag{3.4}$$

where R is the gas constant.

The above analysis shows that the specific heat at constant volume is a constant and equals $3R$ or 6 cal/deg/gram molecule. This is the so-called *Dulong-Petit value* of the specific heat at constant volume. It is independent of temperature and, therefore, according to the classical theory of specific heat, the lattice-vibration contribution to the specific heat is constant and equals $3R$. The results of the classical theory are in agreement with the experimental results of the specific heat of many solids at high temperatures. The experimental results do not agree with the theory at low temperatures. The specific heat falls as the temperature is lowered. For nonmetallic crystals the specific heat varies as T^3 as the temperature T approaches zero, and for metallic crystals the specific heat varies as T as the temperature T approaches zero. This is in disagreement with the results of Dulong and Petit and hence against the results of the classical theory of lattice vibrations.

3.2 Einstein's theory of specific heat

To account for the discrepancy between the classical theory and the experimental results of the specific heat, Einstein in 1905 proposed a theory for the specific heat of nonmetallic solids. He assumed that a solid containing N atoms could be represented by $3N$ harmonic one-dimensional oscillators, each with a common frequency ν. In his theory the atoms vibrate independently of each other with the same frequency of vibration ν because of their identical surroundings. Einstein assumed Planck's quantum hypothesis — that all possible values of the oscillator energy do not exist and that the energy is quantized. An atomic oscillator can assume energies given by

$$E_n = nh\nu,$$

where $n = 0, 1, 2, 3, \ldots$ and h is Planck's constant.

According to classical statistics the average energy of the atomic oscillator is

$$\bar{E} = \frac{\sum_0^\infty nh\nu e^{-nh\nu/kT}}{\sum_0^\infty e^{-nh\nu/kT}}. \tag{3.5}$$

The integration sign is replaced by sigma, because n is quantized. To evaluate the right side of Eq. (3.5) consider the denominator D,

$$D = \sum_0^\infty e^{-nh\nu/kT} = 1 + e^{-h\nu/kT} + e^{-2h\nu/kT} + \cdots = (1 - e^{-h\nu/kT})^{-1}.$$

Differentiating the denominator D partially with respect to $1/kT$,

$$\frac{\partial D}{\partial (kT)^{-1}} = \frac{\partial}{\partial (kT)^{-1}}\left(\sum_0^\infty e^{-nh\nu/kT}\right) = \frac{\partial}{\partial (kT)^{-1}}(1 - e^{-h\nu/kT})^{-1}$$

or

$$\sum_0^\infty - nh\nu e^{-nh\nu/kT} = -\frac{h\nu e^{-h\nu/kT}}{(1 - e^{-h\nu/kT})^2}. \tag{3.6}$$

Equation (3.6) suggests that the numerator Nu of the right side of Eq. (3.5) is

$$Nu = \sum_0^\infty nh\nu e^{-nh\nu/kT} = \frac{h\nu e^{-h\nu/kT}}{(1 - e^{-h\nu/kT})^2}.$$

Hence

$$\bar{E} = \frac{h\nu e^{-h\nu/kT}}{1 - e^{-h\nu/kT}}$$

or

$$\bar{E} = \frac{h\nu}{\exp(h\nu/kT) - 1}. \tag{3.7}$$

Equation (3.7) represents the average energy of an atomic oscillator with Einstein's assumptions. The total vibrational energy in a system of N atoms corresponds to the energy of $3N$ one-dimensional oscillators and, therefore,

$$U = 3N \frac{h\nu}{\exp(h\nu/kT) - 1}. \tag{3.8}$$

For a gram molecule of a solid containing N_0 atoms,

$$U = 3N_0 \frac{h\nu}{\exp(h\nu/kT) - 1}. \tag{3.9}$$

Therefore,

$$C_V = \frac{\partial U}{\partial T} = 3N_0 k \left(\frac{h\nu}{kT}\right)^2 \frac{e^{h\nu/kT}}{\exp(h\nu/kT) - 1|^2}. \tag{3.10}$$

Before discussing the results of Eq. (3.10) it may be mentioned that Einstein assumed only Planck's hypothesis regarding the quantization of energy and did not use the results of a quantum-mechanical treatment of an oscillator. If we take the atomic oscillator to be a quantum oscillator, we concluded in Sec. 1.5 by Eq. (1.49) that the possible values of energy of a quantum oscillator are

$$E_n = (n + \tfrac{1}{2})h\nu.$$

As such, the average energy of an atomic oscillator should be

$$\bar{E} = \frac{h\nu}{2} + \frac{h\nu}{\exp(h\nu/kT) - 1}. \tag{3.11}$$

Equation (3.11) can be obtained by following a procedure similar to the one adopted in arriving at Eq. (3.7). The total vibrational energy of a system of N atoms is

$$U = 3N \left\{ \frac{h\nu}{2} + \frac{h\nu}{\exp(h\nu/kT) - 1} \right\}, \tag{3.12}$$

and for a gram molecule

$$C_V = 3N_0 k \left(\frac{h\nu}{kT}\right)^2 \frac{e^{h\nu/kT}}{[\exp(h\nu/kT) - 1]^2}. \tag{3.13}$$

Equations (3.13) and (3.10) are similar, showing that the specific heat at constant volume is the same in either of the two cases — assuming Planck's hypothesis for the quantization of energies or taking the possible values of energies of a quantum oscillator.

In Eq. (3.10), if $kT \gg h\nu$,

$$C_V = 3N_0 k \left(\frac{h\nu}{kT}\right)^2 \left[\left(1 + \frac{h\nu}{kT} + \cdots - 1\right)^2\right]^{-1} = 3N_0 k = 3R.$$

At low temperature $T \to 0$,

$$h\nu \gg kT \quad \text{and} \quad e^{h\nu/kT} \gg 1.$$

Therefore Eq. (3.10) reduces to

$$C_V = 3R \left(\frac{h\nu}{kT}\right)^2 e^{-h\nu/kT}. \tag{3.14}$$

The specific heat at low temperatures varies as $e^{-h\nu/kT} T^{-2}$; however, at very low temperatures the exponential term is dominant, and so the low-temperature variation of the specific heat in Einstein's model is said to vary according to a negative exponential function. This makes the specific heat fall off more rapidly than experimental results indicate. In Sec. 3.3 we shall discuss Debye's approximations resolving this difficulty. It is convenient to express Eq. (3.10) in terms of a characteristic temperature θ_E, where

$$h\nu = k\theta_E,$$

and hence Eq. (3.10) can be rewritten

$$C_V = 3R\left(\frac{\theta_E}{T}\right)^2 \frac{e^{\theta_E/T}}{(e^{\theta_E/T} - 1)^2}. \tag{3.15}$$

3.3 Vibration modes of a continuous medium; Debye's approximations

The shortcomings in Einstein's model may be attributed to the oversimplification of his model. In 1912 Debye pointed out that the vibrational spectrum of a solid contained a broad range of frequencies; in other words, it is possible to propagate waves through solids over a broad wavelength spectrum. Einstein's theory assumed the atomic vibrations to be independent of each other and each at the same frequency, whereas Debye's approximations consider the vibrational modes of a continuous medium.

Vibrational modes of a continuous medium. To determine the vibrational modes of a three-dimensional crystal having fixed boundaries or sides, we would first study the case of a one-dimensional crystal or a string tightened at both ends. If, in Fig. 3.1, y is the deflection of the string,

$$y = 0 \begin{cases} \text{at } x = 0, \\ \text{at } x = L \end{cases} \quad \text{because the string is tightened.}$$

If $y(x, t)$ represents the deflection of the string at a distance x and time t, the motion of the string may be described by a one-dimensional wave equation,

$$\frac{\partial^2 y}{\partial x^2} = \frac{1}{v^2}\frac{\partial^2 y}{\partial t^2}, \tag{3.16}$$

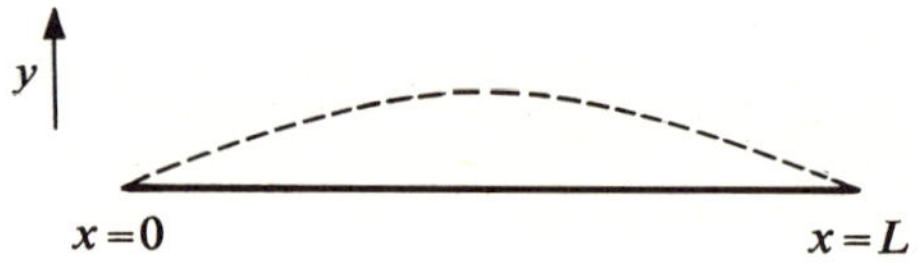

Fig. 3.1
String with both ends fixed

where v is the velocity of propagation of the waves. Equation (3.16) can be easily derived for the string by writing down stress-and-strain relationships for a length Δx of the string at a distance x:

$$v = \nu\lambda = \frac{2\pi\nu}{2\pi/\lambda} = \frac{\omega}{K},$$

where ω is the frequency (angular) of the wave and K is $2\pi/\lambda$, λ being the wavelength of the wave. Equation (3.16) may be rewritten

$$\frac{\partial^2 y}{\partial x^2} = \frac{K^2}{\omega^2}\frac{\partial^2 y}{\partial t^2}. \tag{3.17}$$

Using the method of separation of variables and assuming that y can be represented as

$$y = X(x)T(t),$$

$$\frac{1}{T}\frac{d^2T}{dt^2} = \frac{\omega^2}{K^2}\frac{1}{X}\frac{d^2X}{dx^2}. \tag{3.18}$$

In Eq. (3.18) the left side is a function of t only and the right side a function of x only. They can only be equal if both are equal to a constant, say $-\omega^2$. Therefore,

$$\frac{1}{T}\frac{d^2T}{dt^2} = \frac{\omega^2}{K^2}\frac{1}{X}\frac{d^2X}{dx^2} = -\omega^2.$$

The negative sign for the constant is needed to make the equations periodic Hence

$$\frac{d^2X}{dx^2} + K^2 X = 0, \tag{3.19}$$

$$\frac{d^2T}{dt^2} + \omega^2 T = 0. \tag{3.20}$$

Solving Eqs. (3.19) and (3.20),

$$X = A \sin Kx + B \cos Kx,$$

$$T = C \sin \omega t + D \cos \omega t,$$

where A, B, C, and D are arbitrary constants. The string is tightened at both ends, so

$$X = 0 \quad \text{at} \quad x = 0.$$

Therefore, $B = 0$ and

$$y = XT = (AC \sin \omega t + AD \cos \omega t) \sin Kx.$$

Also, $y = 0$ at $x = L$ at all values of t. Therefore, $\sin KL = 0$ or $KL = n\pi$, where n is an integer excluding zero. If $n = 0$, the deflection is always zero at all values of x and t and the string would not be vibrating, a condition we would not be interested in:

$$K = n\pi/L.$$

Each value of n corresponds to a different value K, hence a different value of λ, and hence a different mode of vibration:

$$y = (AC \sin \omega t + AD \cos \omega t) \sin(n\pi x/L). \tag{3.21}$$

Because $K = 2\pi/\lambda = n\pi/L$

$$\lambda = 2L/n.$$

Depending upon the value of n, we have different wavelengths of the vibration of the string, the wavelengths are discrete, and are $2L$, L, $2L/3$, $2L/4$, ... For a given velocity of propagation of the wave v,

$$\nu = v/\lambda = vn/2L,$$

which results in a discrete frequency spectrum depending upon the value of n. Hence one can write

$$\lambda_n = 2L/n \quad \text{and} \quad \nu_n = vn/2L. \tag{3.22}$$

The number of possible modes of vibration dn in a frequency interval $d\nu$ are obtained by differentiating,

$$dn = d\nu \frac{2L}{v}. \tag{3.23}$$

Although n is still an integer, we have differentiated. This is only possible by assuming that n is very large, and the frequency spectrum may be considered nearly continuous.

Because ω is a function of n, ω may be replaced by ω_n in Eq. (3.21),

$$y = (AC \sin \omega_n t + AD \cos \omega_n t) \sin(n\pi x/L). \tag{3.24}$$

Equation (3.24) can be simplified to

$$y = G \cos \omega_n t \sin(n\pi x/L), \tag{3.25}$$

where G is a new constant. The solution corresponds to standing waves. In fact, if a string is fixed at both ends, the solutions are those corresponding to standing waves.

To obtain the vibrational modes of a three-dimensional solid, we would have to write an equation for the wave motion of the continuous solid in three dimensions. Extending Eq. (3.16) to three dimensions and calling the deflection U, where $U(x,y,z,t)$,

$$\nabla^2 U = \frac{1}{v^2} \frac{\partial^2 U}{\partial t^2}, \tag{3.26}$$

where v is the velocity of propagation of the waves assumed to be the same in all directions. Equation (3.26) can be rewritten

$$\frac{\partial^2 U}{\partial x^2} + \frac{\partial^2 U}{\partial y^2} + \frac{\partial^2 U}{\partial z^2} = \frac{1}{v^2} \frac{\partial^2 U}{\partial t^2}. \tag{3.27}$$

Assuming the three-dimensional solid to be the shape of a cube of side L and assuming the faces of the solid to be fixed, the possible solutions of U are standing waves:

$$U(x,y,z,t) = G \sin \frac{n_x \pi x}{L} \sin \frac{n_y \pi y}{L} \sin \frac{n_z \pi z}{L} \cos \omega t, \tag{3.28}$$

where n_x, n_y, and n_z are nonzero integers. Substituting the solution (3.28) in the three-dimensional wave equation (3.27), we obtain

$$\frac{\pi^2}{L^2} (n_x^2 + n_y^2 + n_z^2) = \frac{4\pi^2 \nu^2}{v^2}, \tag{3.29}$$

and because $v = \nu\lambda$,

$$\frac{\pi^2}{L^2} (n_x^2 + n_y^2 + n_z^2) = \frac{4\pi^2}{\lambda^2}. \tag{3.30}$$

The possible frequencies of vibrations ν's [Eq. (3.29)] or the possible wavelengths [Eq. (3.30)] are determined by three quantum number numbers n_x, n_y, and n_z.

The numbers n_x, n_y, and n_z are integers, and therefore truly speaking, the possible frequencies of vibration are discrete. For larger values of the numbers n_x, n_y, and n_z, the differences between frequencies are so small that the possible frequency spectrum may be treated as continuous. Defining the possible modes of vibration between frequency ν and $\nu + d\nu$, calling $Z(\nu) \, d\nu$, we would determine it in the following lines. Drawing or assuming a coordinate system with n_x, n_y, and n_z, the frequency ν of vibration for a given value of n_x, n_y, and n_z depends upon the distance R of the n_x, n_y, and n_z values from the origin of the coordinate system:

$$R^2 = n_x^2 + n_y^2 + n_z^2.$$

Therefore,

$$R^2 = \frac{4L^2 \nu^2}{v^2}. \tag{3.31}$$

Any change in the values of n_x, n_y, and n_z changes R and hence ν. The problem of finding the modes of vibration between frequency ν and $\nu + d\nu$ is therefore equivalent to the problem of finding the number of vibrational modes in a shell of thickness dR at a distance R. Because n_x, n_y, and

n_z are only positive, the number of modes corresponds to the octant of thickness dR:

$$Z(\nu)\,d\nu = \tfrac{1}{8} 4\pi R^2\,dR = \frac{\pi R^2}{2}\,dR.$$

From Eq. (3.31),

$$dR = \frac{2L}{v}\,d\nu.$$

Therefore,

$$Z(\nu)\,d\nu = \frac{4\pi L^3}{v^3}\nu^2\,d\nu$$

or

$$Z(\nu)\,d\nu = \frac{4\pi V}{v^3}\nu^2\,d\nu, \tag{3.32}$$

where V is the volume of the solid.

Equation (3.32) shows that the modes of vibration of a continuous three dimensional solid are proportional to ν^2 and that there is no limitation to the magnitude of the frequency ν. The area enclosed by the curve $Z(\nu)$ and ν with the abscissa gives us the total modes of vibration in a desired frequency interval. The total modes of vibration from 0 to ν would be (Fig. 3.2)

$$\int_0^\nu Z(\nu)\,d\nu = \frac{4\pi V}{v^3}\int_0^\nu \nu^2\,d\nu = \frac{4\pi V}{3v^3}\nu^3.$$

In the above treatment only one possible mode of vibration has been considered. Actually three independent modes of vibration must be considered. For each atom there may be two transverse and one longitudinal

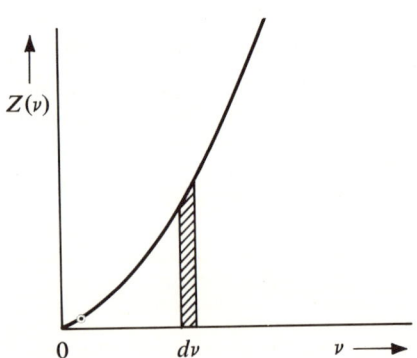

Fig. 3.2
Frequency spectrum for a three-dimensional continuum

modes and if the velocity of propagation in the transverse mode is v_t and if the velocity of propagation in the longitudinal mode is v_l,

$$Z(\nu)\,d\nu = 4\pi\left(\frac{2}{v_t^3} + \frac{1}{v_l^3}\right)V\nu^2\,d\nu. \tag{3.33}$$

Debye's approximation. Debye assumed that the continuum model could be employed to explain the variation of specific heat of solid at low temperatures. It has been seen that in the continuum model all frequencies are possible, but a solid containing N atoms can only have $3N$ modes of vibration. Debye assumed a cutoff frequency in the modes of vibration so as to comply with a total of $3N$ modes.

If ν_D is the cutoff frequency, called the *Debye frequency*,

$$\int_0^{\nu_D} Z(\nu)\,d\nu = 3N. \tag{3.34}$$

Substituting Eq. (3.33) in Eq. (3.34),

$$4\pi V\left(\frac{2}{v_t^3} + \frac{1}{v_l^3}\right)\int_0^{\nu_D} \nu^2\,d\nu = 3N$$

or

$$\nu_D^3 = \frac{9N}{4\pi V}\left(\frac{2}{v_t^3} + \frac{1}{v_l^3}\right)^{-1}. \tag{3.35}$$

An approximate idea of the cutoff frequency ν_D can be obtained as follows. Assuming v_t and $v_l \sim 10^3$ m/sec is the velocity of sound,

$$N/V \simeq 10^{28}\,m^{-3}, \qquad \nu_D \simeq 10^{13}\,\text{sec}^{-1}.$$

The maximum frequency of vibration of the solid is 10^{13} HZ, and therefore the minimum wavelength corresponding to maximum frequency is 10^{-10} or 1 Å. Since the wavelengths are greater than 1 Å, in other words, greater than interatomic distances, the crystal looks like a continuum from the point of view of waves and hence Debye's assumption may be justified.

Associating with each vibrational mode a harmonic oscillator of the same frequency,

$$U = \int_0^{\nu_D} Z(\nu)\,d\nu \left\{\frac{h\nu}{\exp(h\nu/kT) - 1}\right\}. \tag{3.36}$$

Substituting Eq. (3.33) in Eq. (3.36),

$$U = 4\pi V\left(\frac{2}{v_t^3} + \frac{1}{v_l^3}\right)\int_0^{\nu_D} \frac{h\nu^3\,d\nu}{\exp(h\nu/kT) - 1}. \tag{3.37}$$

According to Eq. (3.35),
$$4\pi V\left(\frac{2}{v_t^3} + \frac{1}{v_l^3}\right) = \frac{9N}{v_D^3},$$
and, therefore,
$$U = \frac{9N}{v_D^3} \int_0^{v_D} \frac{h\nu^3\, d\nu}{\exp(h\nu/kT) - 1}. \tag{3.38}$$

Introducing $x = h\nu/kT$ and $x_m = h\nu_D/kT$,
$$U = 9N\left(\frac{kT}{h\nu_D}\right)^3 kT \int_0^{x_m} \frac{x^3\, dx}{e^x - 1}. \tag{3.39}$$

Calling Debye's characteristic temperature θ_D and defining $\theta_D = h\nu_D/k$,
$$U = 9N\left(\frac{T}{\theta_D}\right)^3 kT \int_0^{\theta_D/T} \frac{x^3\, dx}{e^x - 1}. \tag{3.40}$$

Equation (3.40) represents the total vibrational energy of a solid containing N atoms. If we consider *high temperatures*,
$$T \gg \theta_D \quad \text{and} \quad x \ll 1.$$

Equation (3.40) may be modified as
$$U = 9N\left(\frac{T}{\theta_D}\right)^3 kT \int_0^{\theta_D/T} x^2\, dx = 3NkT.$$

For a gram molecule
$$U = 3RT,$$
and the specific heat at constant volume is
$$C_V = \frac{\partial U}{\partial T} = 3R \text{ or } 6 \text{ cal/gram molecule.}$$

The result at high temperatures is in conformity with experiment and classical theory.

If we consider *low temperatures*,
$$T \ll \theta_D,$$
x is large and in the limit approaches infinity. Equation (3.40) may be modified as
$$U = 9N\left(\frac{T}{\theta_D}\right)^3 kT \int_0^\infty \frac{x^3\, dx}{e^x - 1}, \quad \int_0^\infty \frac{x^3\, dx}{e^x - 1} = \frac{\pi^4}{15},$$
and therefore
$$U = \tfrac{3}{5}\pi^4 NkT\left(\frac{T}{\theta_D}\right)^3.$$

For a gram molecule

$$C_V = \frac{\partial U}{\partial T} = \frac{12}{5}\pi^4 R\left(\frac{T}{\theta_D}\right)^3. \tag{3.41}$$

Equation (3.40) shows that the specific heat at constant volume of a solid at low temperatures varies as T^3. Accurate measurements show that there are various deviations in the low-temperature region from Debye's theory. The discrepancy may be accounted for by the maximum frequency assumed by Debye, ν_D. It has been shown that minimum wavelength λ_D, corresponding to ν_D, is of the order of 1 Å and hence of the order of interatomic distances. Therefore, the solid may not be an exact continuum at these low wavelengths. Various other cutoff procedures are postulated but we shall not go into the details of these.

3.4 Equivalence of the elastic waves in an infinite one-dimensional array of identical atoms and a low-pass filter

Let the atoms be considered as mass points in an infinite array, each having a mass m and separated from each other by a distance a. Assuming nearest-neighbor interaction and assuming Hooke's law is obeyed, the modes of vibrations of the atoms would be calculated and it would be seen that this one-dimensional array of atoms is equivalent in its modes of vibration to an electrical low-pass filter with electrical element inductances and capacitances. Let the displacement of the nth atom from its equilibrium position be x_n. Therefore, the displacements $x_1, x_2, \ldots, x_{n-1}, x_n, x_{n+1}, \ldots$ refer to deflections of $1, 2, \ldots, n-1, n, n+1, \ldots$ atoms from their equilibrium positions. The equation of motion for the nth atom would be

$$m\frac{d^2x_n}{dt^2} = -f(x_n - x_{n-1}) - f(x_n - x_{n+1}), \tag{3.42}$$

where f is the force constant describing the nearest-neighbor action. (See Fig. 3.3) Equation (3.42) may be rewritten

$$m\frac{d^2x_n}{dt^2} = f(x_{n-1} + x_{n+1} - 2x_n). \tag{3.43}$$

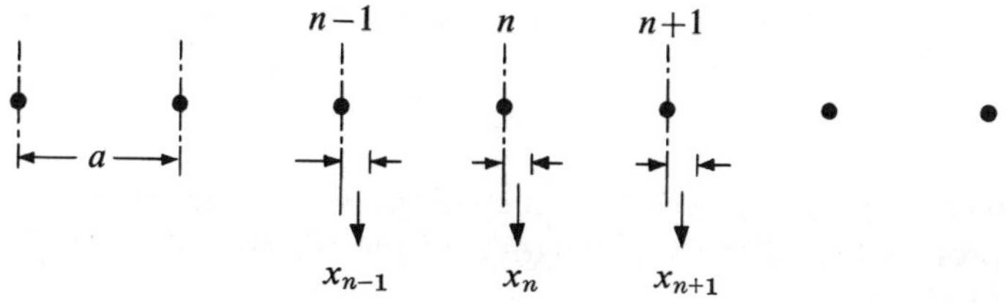

Fig. 3.3
Linear chain of identical mass points

The general solution of Eq. (3.43) may be written as the sum of two remaining waves, one propagating to the right and the other propagating to the left:

$$x_n(t) = (A_1 e^{ikna} + A_2 e^{-ikna})e^{-i\omega t}, \tag{3.44}$$

where A_1 and A_2 are constants. $K = 2\pi/\lambda$, $\omega = 2\pi\nu$, λ is the wavelength of the propagating wave, and ν is the frequency of the propagating wave.

If Eq. (3.44) is a solution of Eq. (3.43), we would like to determine the conditions that must be satisfied so that it is a solution:

$$m\ddot{x}_n(t) = -m\omega^2 e^{-i\omega t}(A_1 e^{ikna} + A_2 e^{-ikna}), \tag{3.45}$$

$$x_{n-1} = [A_1 e^{iK(n-1)a} + A_2 e^{-i(n-1)Ka}]e^{-i\omega t},$$

$$x_{n+1} = [A_1 e^{iK(n+1)a} + A_2 e^{-i(n+1)Ka}]e^{-i\omega t},$$

$$2x_n = 2e^{-i\omega t}(A_1 e^{iKna} + A_2 e^{-iKna}),$$

or the right side of Eq. (3.43) is

$$A_1 f[e^{iK(n-1)a} + e^{iK(n+1)a} - 2e^{iKna}]e^{-i\omega t} + A_2 f[e^{-iK(n-1)a} + e^{-iK(n+1)a} - 2e^{-iKna}]e^{-i\omega t},$$

or the right side of Eq. (3.43) is

$$f(e^{iKa} + e^{-iKa} - 2)(A_1 e^{iKna} + A_2 e^{-iKna})e^{-i\omega t}.$$

Equating the above equation to (3.45),

$$-m\omega^2 e^{-i\omega t}(A_1 e^{iKna} + A_2 e^{-iKna}) = f(e^{iKa} + e^{-iKa} - 2)$$
$$\times (A_1 e^{iKna} + A_2 e^{-iKna})e^{-i\omega t},$$

$$m\omega^2 = -f(e^{iKa} + e^{-iKa} - 2),$$

$$m\omega^2 = -f(e^{iKa/2} - e^{-iKa/2})^2,$$

or

$$\omega^2 = \frac{4f}{m}\sin^2\frac{Ka}{2}.$$

The frequency of vibration ω^2 has a maximum with a value $4f/m$,

$$\omega^2 = \omega_{max}^2 \sin^2\frac{Ka}{2}. \tag{3.46}$$

Equation (3.46) expresses the frequency of the vibrational waves in terms of K, a representative of the wavelength λ of the vibrational wave (Fig. 3.4). In Eq. (3.46), if $Ka \ll 1$,

$$\omega = \omega_{max}\frac{Ka}{2} \quad \text{or} \quad \omega/K = \text{const.}$$

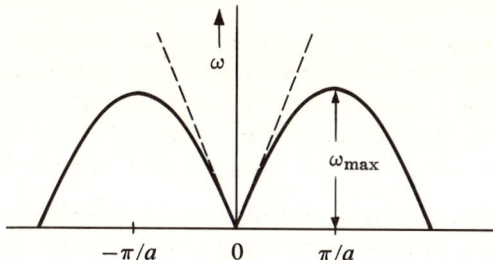

Fig. 3.4
Frequency of vibrational waves as a function of the wave vector K in a monoatomic linear lattice

The ratio ω/K is the velocity of propagation of a wave in a continuous string. From the above analysis it can be concluded that a continuous string and an array of mass points are identical in vibration only if

$$Ka \ll 1,$$
$$K \ll 1/a,$$
$$2\pi/\lambda \ll 1/a,$$

or
$$\lambda \gg a,$$

i.e., when the wavelength of the propagating wave is large compared to interatomic distances.

Equation (3.46) further shows that the ω/K, the velocity of propagation, changes with the K value or the wavelength. The condition

$$Ka/2 = \pi/2$$

corresponds to, or results in, $\omega = \omega_{max}$:

$$K = \pi/a \quad \text{or} \quad \lambda = 2a.$$

The maximum vibrational frequency of the chain occurs for $\lambda = 2a$. Because $a \sim 10^{-10}$ m (interatomic distance), λ (corresponding to maximum frequency) or $\lambda_{min} \simeq 10^{-10}$ m.

The maximum angular frequency is

$$\omega_{max} = \left(\frac{4f}{m}\right)^{1/2},$$

$$2\pi\nu_{max} = \left(\frac{4f}{m}\right)^{1/2},$$

$$\nu_{max} = \frac{1}{\pi}\left(\frac{f}{m}\right)^{1/2},$$

The above analysis shows that the array of atoms may be considered to work as low-pass filters for all vibrational frequencies, $\nu \leq \nu_m$.

Because $\omega = \omega_{max} \sin(Ka/2)$, the frequency of vibration remains the same if K is replaced by $K + (2\pi m/a)$ with $m = \pm 1, \pm 2, \pm 3$. To obtain a unique relationship between the frequency of vibration of the lattice and K, the values of K must be confined to a range of values $2\pi/a$. The range of K values from $-\pi/a$ to π/a is called the *first Brillouin zone*.

A low-pass electric filter. Figure 3.5 shows an electric low-pass filter. L is the inductance of each element and C is the shunt capacitance. The

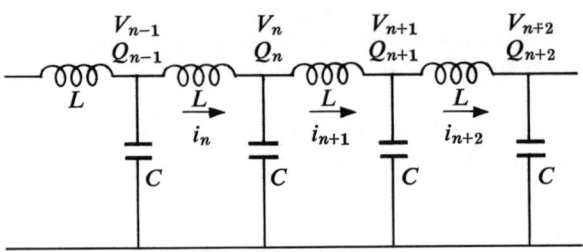

Fig. 3.5
Electric low-pass filter

inductances are connected in series and the capacitances in shunt. The capacitances shunt out the high frequencies and the low frequencies are allowed to pass. Let V_n and Q_n be the potential and charge on the nth condenser. Calling i_n the current flowing between condensers $n-1$ and n, the equation of the line can be obtained as follows:

$$L\frac{di_n}{dt} = V_{n-1} - V_n = \frac{Q_{n-1}}{c} - \frac{Q_n}{c}, \qquad (3.47)$$

$$i_n - i_{n+1} = \frac{dQ_n}{dt}$$

because $V_n = Q_n/c$.
Differentiating Eq. (3.47),

$$L\frac{d^2i_n}{dt^2} = \frac{1}{c}\left(\frac{dQ_{n-1}}{dt} - \frac{dQ_n}{dt}\right).$$

Substituting the values of dQ_{n-1}/dt and dQ_n/dt in terms of currents,

$$L\frac{d^2i_n}{dt^2} = \frac{1}{C}(i_{n-1} + i_{n+1} - 2i_n). \qquad (3.48)$$

Equation (3.48) results in a value of i_n, the current flowing in the line. The potential differences and charges on the capacitors can be determined. Equation (3.48) is identical with the equation of motion of a one-dimensional array of atoms given by Eq. (3.43). In the equation f/m is replaced by $1/LC$ and i_n replaces x_n. The results of the solution of Eq. (3.43) can be applied to Eq. (3.48). The cutoff frequency of the low-pass filter is

$$\nu_{max} = \frac{1}{\pi(LC)^{1/2}}. \tag{3.49}$$

The low-pass electric filter contains no resistance. We thus conclude that both the low-pass electric filter and an array of identical atoms are identical in their response to electrical and vibrational frequencies. The elastic forces act as coupling forces in the lattice and the inductances as coupling forces in lines. The masses and condensers can be thought of as supplying internal forces to the systems.

3.5 Diatomic linear lattice

A diatomic linear lattice possesses modes of vibration that are of considerable interest. We shall analyze these possible modes of vibration. In a diatomic lattice (Fig. 3.6) let the nearest-neighbor distance be a. The

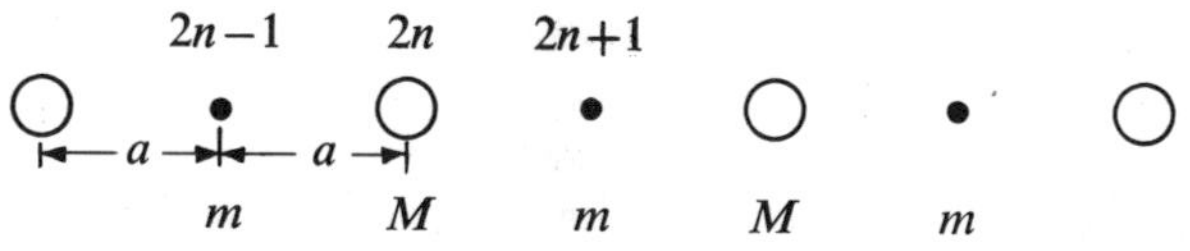

Fig. 3.6
Linear chain of equidistance mass points M and m ($M > m$)

particles in the linear lattice are numbered in such a way that even-numbered particles have mass M and odd-numbered m. Assuming nearest-neighbor interaction would mean that the small masses m interact only with the nearest large mass and that the large masses M interact only with the nearest small mass.

The equations of motion of the two types of particles are different because of their different masses. If x_{2n}, x_{2n+1}, and x_{2n-1} are the deflections of $2n$th, $(2n+1)$th, and $(2n-1)$th particles, respectively, from their equilibrium positions, the equations

$$M\frac{d^2x_n}{dt^2} = f(x_{2n-1} + x_{2n+1} - 2x_{2n}), \tag{3.50}$$

$$m\frac{d^2x_{2n+1}}{dt^2} = f(x_{2n} + x_{2n+2} - 2x_{2n+1}), \tag{3.51}$$

describe the motion of $2n$th and $(2n+1)$th particles, respectively.

Let us assume a wave solution to these equations of the following form:

$$x_{2n}(t) = Ae^{-i(\omega t - 2nKa)}, \quad (3.52)$$

$$x_{2n+1}(t) = Be^{-i(\omega t - \overline{2n+1}\, Ka)}, \quad (3.53)$$

where $K = 2\pi/\lambda$ and A and B are amplitudes for the motion of masses M and m, respectively. Equation (3.52) represents a wave propagating only through particles of mass M; Eq. (3.53) represents a wave propagating only through mass m. The wavelength and frequencies for a given disturbance must be equal. The amplitudes of the two waves are not necessarily equal. They may differ in magnitude as well as in phase.

In order that Eqs. (3.52) and (3.53) are solutions of (3.50) and (3.51), respectively, certain relations must be imposed on the constants. The substitution of the solutions in the equations of motion results in

$$-M\omega^2 A e^{-i(\omega t - 2nKa)} = f[Be^{-i(\omega t - \overline{2n+1}\, Ka)} + Be^{-i(\omega t - \overline{2n-1}\, Ka)} - 2Ae^{-i(\omega t - 2nKa)}],$$

$$(-M\omega^2 + 2f)A - fB(e^{iKa} + e^{-iKa}) = 0,$$

or

$$(M\omega^2 - 2f)A + 2fB \cos Ka = 0. \quad (3.54)$$

Similarly, from Eqs. (3.52), (3.53), and (3.51),

$$(m\omega^2 - 2f)B + 2fA \cos Ka = 0. \quad (3.55)$$

Equations (3.54) and (3.55) are two linear equations in A and B. The condition that these equations give nonvanishing solutions for A and B is that the determinant of the coefficients of A and B vanishes:

$$\begin{vmatrix} M\omega^2 - 2f & 2f \cos Ka \\ 2f \cos Ka & m\omega^2 - 2f \end{vmatrix} = 0,$$

$$(M\omega^2 - 2f)(m\omega^2 - 2f) - 4f^2 \cos^2 Ka = 0,$$

$$Mm\omega^4 - 2f(M + m)\omega^2 + 4f^2 - 4f^2 \cos^2 Ka = 0,$$

or

$$\omega^4 - 2f\left(\frac{1}{m} + \frac{1}{M}\right)\omega^2 + \frac{4f^2}{Mm} \sin^2 Ka = 0. \quad (3.56)$$

Equation (3.56) possesses two solutions for ω^2 and hence two solutions for ω, because the frequency is always taken to be positive. This means that for each value of K there will be two values of the frequency. In contrast to a monoatomic linear lattice, there are now two angular frequencies ω_+ and ω_-, corresponding to a single value of K:

$$\omega^2 = f\left(\frac{1}{m} + \frac{1}{M}\right) \pm f\left[\left(\frac{1}{m} + \frac{1}{M}\right)^2 - \frac{4 \sin^2 Ka}{Mm}\right]^{1/2}. \quad (3.57)$$

Equation (3.57) shows that the angular frequency is a periodic function of K.

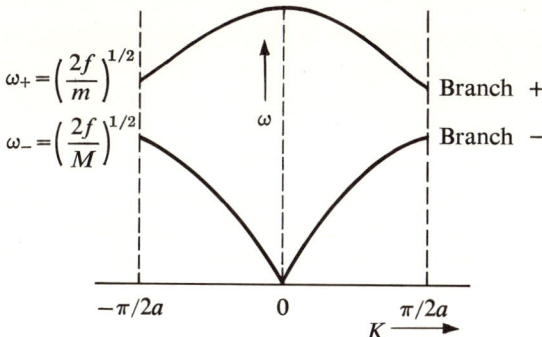

Fig. 3.7
ω_+ and ω_- as functions of K for the case $M > m$

Assuming $M > m$, the mass of the even-numbered atoms is greater than odd-numbered atoms and $K = 0$:

$$\omega^2 = f\left(\frac{1}{m} + \frac{1}{M}\right) \pm f\left(\frac{1}{m} + \frac{1}{M}\right)$$

or

$$\omega_+ = 2f\left(\frac{1}{m} + \frac{1}{M}\right), \quad \omega_- = 0.$$

The value of the frequency for the modes of vibration for $K = \pm\pi/2a$ is

$$\omega_+ = 2f/m, \quad \omega_- = 2f/M.$$

Figure 3.7 shows the values of ω_+ and ω_- plotted as functions of K for the case $M > m$. The larger the ratio M/m, the wider the frequency gap between the two branches. With increasing values of K, it can be easily shown that the variations of ω_+ and ω_- are as shown in Fig. 3.7.

Upper branch. For $K = 0$,

$$\omega_+ = (2f)^{1/2}\left(\frac{1}{m} + \frac{1}{M}\right)^{1/2}. \tag{3.58}$$

Substituting Eq. (3.58) in Eqs. (3.54) or (3.55), a relationship between A and B can be found:

$$(M\omega^2 - 2f)A + 2fB = 0,$$

$$MA\left[2f\left(\frac{1}{m} + \frac{1}{M}\right)\right] - 2fA + 2fB = 0,$$

or

$$\frac{M}{m}A = -B \quad \text{since } f \neq 0 \quad \text{or} \quad MA = -mB.$$

This shows that for $K = 0$ the waves for the upper branch are out of phase; i.e., the displacement of particles of mass M is opposite that of the neighboring atom m. Evidently the center of mass of the two neighboring particles is stationary, but the restoring force center is such that the frequencies of waves are no longer zero. This means that for $K = 0$ or $\lambda \to \infty$, the particles or atoms oscillate in opposite directions, the lighter particles with larger amplitude. As K increases or λ decreases, the amplitude of heavy particles decreases and at $K = \pi/2a$,

$$\omega_+ = (2f/m)^{1/2},$$

$$\left(\frac{M}{m} 2f - 2f\right) A = 0,$$

or

$$A = 0,$$

so the heavy particles are at rest. The upper branch is frequently called the *optical branch* because of the fact that its frequencies are of the order of magnitude of infrared frequencies.

Lower branch. For $K = 0$,

$$\omega_- = 0.$$

Substituting this value of ω_- in Eq. (3.54),

$$-2fA + 2fB = 0 \quad \text{or} \quad A = B.$$

This shows that at $K = 0$, both the heavy and light particles are in phase and with amplitude for infinite wavelength. As the K value increases or λ decreases, the amplitude of light particles decreases, and at $K = \pi/2a$,

$$\omega_- = (2f/M)^{1/2},$$

$$\left(\frac{M}{M} 2f - 2f\right) B = 0,$$

or

$$B = 0,$$

so the light particles are at rest.

The lower branch is frequently called the *acoustical branch*. This name originates from the fact that the frequencies in this branch are of the same order as acoustical or supersonic vibrations.

Electrical analogue of the one-dimensional diatomic lattice. To construct an electrical line analogous to a diatomic linear lattice, we shall have to take two inductance elements corresponding to two different masses L_1 and L_2. To allow different coupling between two masses, two capaci-

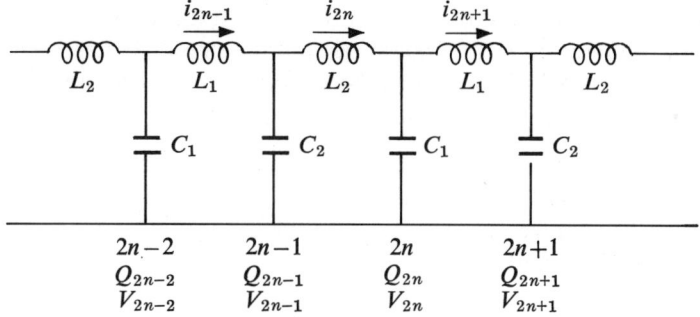

Fig. 3.8
Electrical analogue of the one-dimensional diatomic lattice

tances C_1 and C_2 are assumed. The arrangement of Fig. 3.8 can be shown to be analogous in frequency response to a diatomic linear lattice.

If i_{2n} represents the current flowing from condenser $2n - 1$ to condenser $2n$,

$$i_{2n} - i_{2n+1} = \frac{dQ_{2n}}{dt},$$

$$i_{2n+1} - i_{2n+2} = \frac{dQ_{2n+1}}{dt},$$

and

$$L_1 \frac{di_{2n+1}}{dt} = V_{2n} - V_{2n+1},$$

or

$$L_1 \frac{di_{2n+1}}{dt} = \frac{Q_{2n}}{C_1} - \frac{Q_{2n+1}}{C_2}. \tag{3.59}$$

Similarly,

$$L_2 \frac{di_{2n}}{dt} = \frac{Q_{2n-1}}{C_2} - \frac{Q_{2n}}{C_1}. \tag{3.60}$$

Differentiating Eqs. (3.59) and (3.60),

$$L_1 \frac{d^2 i_{2n+1}}{dt^2} = \frac{i_{2n} - i_{2n+1}}{C_1} - \frac{i_{2n+1} - i_{2n+1}}{C_2}, \tag{3.61}$$

$$L_2 \frac{d^2 i_{2n}}{dt^2} = \frac{i_{2n-1} - i_{2n}}{C_2} - \frac{i_{2n} - i_{2n+1}}{C_1}. \tag{3.62}$$

Equations (3.61) and (3.62) are similar to Eqs. (3.50) and (3.51) and are identical if $C_1 = C_2$ and we replace the capacitance by elastic constant and inductance by mass.

Phonons. In the case of vibration of solids it has been shown that the frequency of vibration ω is a function of K or ω_K or ν_K. It has been shown by Dekker that a vibrational mode is equivalent to a harmonic oscillator.

The average energy of an oscillator of frequency ν is

$$\overline{E} = \frac{h\nu}{\exp(h\nu/kT) - 1}. \tag{3.7}$$

Therefore, the average energy associated with a particular mode of vibration of frequency ν and certain K is

$$\overline{E} = \frac{h\nu_K}{\exp(h\nu_K/kT) - 1}.$$

The number of quanta n_K associated with the vibrational mode each of energy $h\nu_K$ is

$$n_K = \frac{\overline{E}}{h\nu_K} = (e^{h\nu_K/kT} - 1)^{-1}.$$

The quanta are referred to as *phonons*. The concept of phonons is convenient in the discussion of interaction of electrons and phonons because like a photon a phonon has both particle and wave effects.

3.6 Heat capacity of conduction electrons

In metals the valence electrons may be treated as conduction electrons. These electrons are available for conduction and in many cases may be treated as free electrons. Conduction means that the application of an electric field leads to a transfer of electrons and as such flow of current. The number of charge carriers that may be treated as free electrons in the case of semiconductors and insulators will be calculated in Chapter 4. If the conduction electrons are treated as free electrons such as gas molecules and we use Boltzmann statistics, gas laws are applicable to the electrons. If n is the number of electrons per unit volume, the total energy of the electrons due to thermal agitation at a temperature T per unit volume is given by

$$W = n\tfrac{3}{2}kT. \tag{3.63}$$

For a gram molecule, assuming one conduction electron per atom,

$$W = \tfrac{3}{2}RT. \tag{3.64}$$

If, however, there are z electrons (conduction) per atom,

$$W = \tfrac{3}{2}RzT, \tag{3.65}$$

and the specific heat at constant volume is

$$C_V = \left(\frac{\partial W}{\partial T}\right)_V = \frac{3}{2}Rz. \tag{3.66}$$

At high temperatures in solids, the specific heat at constant C volume due to lattice vibrations C_{V_L} is $3R$ (see Secs. 3.2 and 3.3). If C_{V_e} is the specific heat at constant volume due to electronic motion given by Eq. (3.66), the total specific heat at constant volume is

$$C_V = C_{V_L} + C_{V_e} = (3R + \tfrac{3}{2}Rz).$$

Even for $z = 1$,

$$C_V = 9R/2 \text{ per gram molecule.}$$

The above analysis would indicate that the specific heat at constant volume for metals at high temperature is $9R/2$. Experimental measurements indicate that even in the case of metals, the specific heat at constant volume at large temperatures is the same as in insulators and equals $3R$ for a gram molecule. No evidence shows such a high contribution of the conduction electrons toward the specific heat; the discrepancy may therefore be found in the classical statistics of the electrons assumed above.

The actual thermal behavior of the conduction electrons should take into account the Fermi-Dirac statistics, the density of states, and Pauli's exclusion principle.

If $p(E)$ is the probability of occupation to a state of energy E, $g(E)\,dE$ is the density of states in an energy interval E and $E + dE$ and N is the total number of electrons:

$$N = \int_0^\infty p(E)g(E)\,dE.$$

The total energy is

$$W = N\bar{E} = \int_0^\infty Ep(E)g(E)\,dE. \tag{3.67}$$

To solve the integral in Eq. (3.67), let us consider the integral

$$I = \int_0^\infty p(E)\frac{dF(E)}{dE}\,dE.$$

It was shown in Sec. 2.4 that

$$I = F(E_f) + \frac{\pi^2}{6}(kT)^2 F''(E_f).$$

If $F(E) = \int_0^E Eg(E)\,dE$,

$$I = \int_0^\infty Ep(E)g(E)\,dE = W.$$

Therefore,

$$W = F(E_f) + \frac{\pi^2}{6}(kT)^2 F''(E_f).$$

Substituting the value of $F(E)$,

$$W = \int_0^{E_f} Eg(E)\, dE + \frac{\pi^2}{6}(kT)^2 \left[\frac{d}{dE}[Eg(E)]\right]_{E_f}.$$

According to Eq. (2.27)

$$g(E)\, dE = 2\frac{4\pi}{h^3} m^{3/2} 2^{1/2} E^{1/2}\, dE. \qquad (2.27)$$

Therefore,

$$Eg(E)\, dE = \frac{1}{2\pi^2}\left(\frac{2m}{\hbar^2}\right)^{3/2} E^{3/2}\, dE$$

$$W = \int_0^{E_f} \frac{1}{2\pi^2}\left(\frac{2m}{\hbar^2}\right)^{3/2} E^{3/2}\, dE + \frac{\pi^2}{6}(kT)^2 \frac{1}{2\pi^2}\left(\frac{2m}{\hbar^2}\right)^{3/2}\frac{3}{2}E^{1/2}\bigg|_{E_f},$$

or

$$W = W_0 + (E_f - E_{f0})E_{f0}\, g(E_{f0}) + \frac{3}{2}\frac{\pi^2}{6}(kT)^2\frac{1}{2\pi^2}\left(\frac{2m}{\hbar^2}\right)^{3/2}E_f^{1/2}.$$

Substituting Eq. (2.43)

$$W = W_0 - \frac{\pi^2}{12}(kT)^2\frac{1}{2\pi^2}\left(\frac{2m}{\hbar^2}\right)^{3/2}E_f^{1/2} + \frac{3}{2}\frac{\pi^2}{6}(kT)^2\frac{1}{2\pi^2}\left(\frac{2m}{\hbar^2}\right)^{3/2}E_f^{1/2},$$

or

$$W = W_0 + \frac{\pi^2}{6}(kT)^2\frac{1}{2\pi^2}\left(\frac{2m}{\hbar^2}\right)^{3/2}E_f^{1/2}.$$

The first term on the right side represents the total energy of the electrons at absolute zero and is independent of a term T. If the Fermi level at absolute zero is E_{f0},

$$N = \int_0^{E_{f0}} g(E)\, dE = \int_0^{E_{f0}} \frac{1}{2\pi^2}\left(\frac{2m}{\hbar^2}\right)^{3/2} E^{1/2}\, dE = \frac{2}{3}\frac{1}{2\pi^2}\left(\frac{2m}{\hbar^2}\right)^{3/2} E_{f0}^{3/2}$$

or

$$\frac{3N}{2}E_{f0}^{-3/2} = \frac{1}{2\pi^2}\left(\frac{2m}{\hbar^2}\right)^{3/2}.$$

Therefore,

$$W = W(0) + \frac{\pi^2}{4}(kT)^2 N E_f^{1/2} E_{f0}^{-3/2}.$$

Because

$$E_f \simeq E_{f0},$$

$$W = W(0) + \frac{\pi^2}{4}(kT)^2 N E_{f0}^{-1}.$$

THERMAL CONDUCTIVITY OF SOLIDS

E_{f0}, the energy corresponding to the Fermi level at absolute zero, may be written in terms of the translational energy of the electron at a hypothetical temperature called the *Fermi temperature* T_f:

$$E_{f0} = kT_f.$$

Hence

$$W = W(0) + \frac{\pi^2}{4}(kT)^2 \frac{N}{kT_f}.$$

The specific heat at constant volume due to the conduction electron C_{V_e} is

$$C_{V_e} = \left(\frac{\partial W}{\partial T}\right)_V = \frac{\pi^2}{2}\frac{kT}{T_f} N.$$

Taking a gram molecule of the solid (metal) and assuming one conduction electron per atom,

$$C_{V_e} = \frac{\pi^2}{2} kN_0 \frac{T}{T_f} = \frac{\pi^2}{2}\frac{RT}{T_f}.$$

If, however, there are z electrons per atom available for conduction,

$$N = N_0 z$$

$$C_{V_e} = \frac{\pi^2}{2} Rz \frac{T}{T_f}. \tag{3.68}$$

Putting $(\pi^2/2)\frac{Rz}{T_f} = \gamma$,

$$C_{V_e} = \gamma T.$$

The specific heat contribution of conduction electrons is proportional to the temperature. The contribution depends upon γ, which is of the order of 10^{-4} in metals. The contribution at normal temperatures is negligible compared to the lattice contribution. At sufficiently low temperatures the electronic contribution to the specific heat becomes larger than the lattice contribution. This is due to the fact that C_{V_e} decreases as T, whereas C_{V_L} decreases as T^3 at low temperatures.

3.7 Thermal conductivity of solids

The thermal conductivity $\varkappa$ of a solid is most easily defined with respect to the steady-state flow of heat down a long bar within which there exists a temperature gradient

$$Q = -\varkappa \frac{\partial T}{\partial x}, \tag{3.69}$$

where Q is the thermal energy flux transmitted across a unit area per unit time.

$\varkappa$ is expressed in units of calories per meter-second-degree or watts per meter-degree. The form of Eq. (3.69) defining the thermal conductivity implies that the energy-transfer process may be a random process. By this we take it that the energy does not enter at one end of the specimen and proceed directly in a straight path to the other end. The energy diffuses through the specimen from the hot end to the cold end in the form of collisions. If the energy were propagated directly through the specimen, the thermal energy flux transmitted would not require a temperature gradient but would only depend upon the difference in temperature ΔT between the ends of the specimen regardless of the length of the specimen. The thermal conductivity may be due to conduction of heat flux by electrons or lattice vibrations.

Calculations of thermal conductivity of an electron gas. In metals the valence electrons may be treated as free electrons in a manner similar to gas molecules. In the following we develop an expression for $\varkappa$ for the electron gas $\varkappa_e$. Assume a temperature gradient $\partial T/\partial x$ along the x direction; the average of the electron would be a function of x. Let $E(0)$ be the average energy of an electron in a plane at $x = 0$. At a distance Δx from $x = 0$, the average energy of the electron is

$$E(\Delta x) = E(0) + \frac{\Delta E}{\Delta x} \Delta x.$$

Because the change in energy is due to the change in temperature or, E is dependent upon T,

$$\lim_{\Delta x \to 0} \frac{\Delta E}{\Delta x} = \frac{\partial E}{\partial x} = \frac{\partial E}{\partial T} \frac{\partial T}{\partial x}.$$

Therefore,

$$E(\Delta x) = E(0) + \frac{\partial E}{\partial T} \frac{\partial T}{\partial x} \Delta x.$$

Referring to Fig. 3.9, consider a plane of unit area at $x = 0$. We shall calculate the net transport of energy through this plane resulting from electrons that transport from $x > 0$ and $x < 0$. Assume that the energy of an electron crossing the plane at $x = 0$ is the energy an electron had at a distance Δx where it last collided. Let us consider an electron of velocity component in the x direction v_x. Let $-\Delta x$ represent the location of a plane where the electron last collided with the lattice. Considering a group of such electrons in number n_i per cubic meter, each one having collided last at $-\Delta x_i$, the energy of one electron crossing the plane at $x = 0$ is

$$E(-\Delta x_i) = \left[E(0) - \frac{\partial E}{\partial T} \frac{\partial T}{\partial x} \Delta x_i \right].$$

The total energy transferred across a unit plane at $x = 0$ per second due to the group of n_i electrons is

$$Q_i = \left[E(0) - \frac{\partial E}{\partial T} \frac{\partial T}{\partial x} \Delta x_i \right] v_{xi} n_i.$$

Consider all such groups of electrons,

$$Q = \sum_i Q_i = \sum_i \left[E(0) - \frac{\partial E}{\partial T} \frac{\partial T}{\partial x} \Delta x_i \right] v_{xi} n_i.$$

Because the average energy transferred by electrons at $x = 0$ is zero, the sum containing $E(0)$ vanishes:

$$Q = - \sum_i \frac{\partial E}{\partial T} \frac{\partial T}{\partial x} \Delta x_i\, v_{xi} n_i.$$

If n is the total number of electrons in a unit volume,

$$n \langle v_x \Delta x \rangle = \sum_i n_i \Delta x_i\, v_{xi},$$

where $\langle v_x \Delta x \rangle$ is the average value of the product. Hence

$$Q = -n \langle v_x \Delta x \rangle \frac{\partial E}{\partial T} \frac{\partial T}{\partial x}, \tag{3.70}$$

where E is the energy of an electron and n is the number of electrons per unit volume. Therefore, $n\, \partial E/\partial T = C_V$, the specific heat at constant volume per unit volume. Therefore,

$$Q = -C_V \langle v_x \Delta x \rangle \frac{\partial T}{\partial x}. \tag{3.71}$$

Comparing Eq. (3.71) with Eq. (3.69),

$$\varkappa = -C_V \langle v_x \Delta x \rangle . \tag{3.72}$$

It now remains to determine the average of the product of the x component of the velocity and the distance the electron collided last.

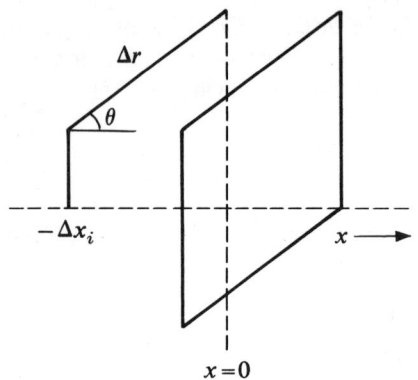
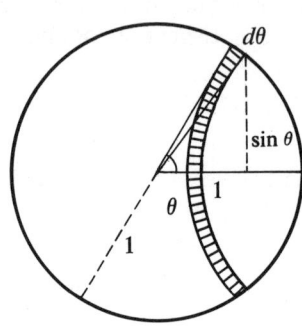

Fig. 3.9
Model for calculation of the net transport of thermal energy through a plane

Fig. 3.10
Electron moving in a direction θ with respect to the x axis

Referring to Fig. 3.9, consider an electron moving in a direction θ with respect to the x axis. If v is the velocity of the electron,

$$v_x = v \cos \theta.$$

If the electron covers a distance Δr until it crosses $x = 0$,

$$\Delta r = \Delta x / \cos \theta.$$

Therefore,

$$\langle v_x \Delta x \rangle = \langle v \Delta r \cos^2 \theta \rangle.$$

The average value of $\cos^2 \theta$ is shown below to be $\frac{1}{3}$ if $N(\theta) \, d\theta$ is the probability of finding an electron between θ and $\theta + d\theta$ from the position of its last collision while approaching the plane at $x = 0$. Considering a sphere of unit radius as shown in Fig. 3.10,

$N(\theta) \, d\theta \propto$ surface of the sphere between θ and $\theta + d\theta$,

$N(\theta) \, d\theta = A \, 2\pi \sin \theta \, d\theta.$

The constant A is determined by the normalization condition, which states that the probability of finding the electron over the entire space is unity:

$$\int_0^\pi N(\theta) \, d\theta = 1 = \int_0^\pi A \, 2\pi \sin \theta \, d\theta,$$

$$\langle \cos^2 \theta \rangle = \frac{\int_0^\pi \cos^2 \theta \, 2\pi \sin \theta \, d\theta}{\int_0^\pi 2\pi \sin \theta \, d\theta} = \frac{1}{3}.$$

Hence Eq. (3.72) may be rewritten

$$\varkappa = \tfrac{1}{3} C_V \langle v \, \Delta r \rangle.$$

$\langle v \, \Delta r \rangle$ in the case of electron statistics may be replaced by the product of the Fermi velocity v_f and λ, the mean free path of the electrons. The Fermi velocity is referred to as the velocity of an electron having energy corresponding to the Fermi energy level:

$$\tfrac{1}{2} m v_f^2 = E_f.$$

It will be shown in Chapter 4 that at ordinary temperature the average energy of the electrons is in the neighborhood of E_f, because electrons occupying states near the Fermi level are the only ones that can move. Hence

$$\varkappa = \tfrac{1}{3} C_V v_f \lambda. \tag{3.73}$$

If τ is the relaxation time, the average time between two consecutive collisions is

$$\lambda = v_f \tau,$$

and hence

$$\varkappa = \tfrac{1}{3} C_V v_f^2 \tau. \tag{3.73a}$$

It has been shown that in electronic heat capacity $C_V = C_{V_e}$ and is given by Eq. (3.68) for a gram molecule:

$$C_{V_e} = \frac{\pi^2}{2} Rz \frac{T}{T_f}. \tag{3.68}$$

Therefore,

$$\varkappa_e = \frac{\pi^2}{6} Rz \frac{T}{T_f} v_f^2 \tau, \tag{3.74}$$

where T_f is the Fermi temperature.

Now

$$E_f \simeq E_{f0} = kT_f = \tfrac{1}{2} m v_f^2.$$

Hence

$$\varkappa_e = \frac{\pi^2}{4} \frac{Rk\tau T}{m}. \tag{3.75}$$

The conductivity $\varkappa_e$ is only the contribution of conduction electrons

$$\varkappa_e = \frac{\pi^2}{3} \frac{Rk\tau}{m} T. \tag{3.76}$$

The electronic contribution is directly proportional to the temperature of the material and the relaxation time.

Thermal conductivity due to lattice vibrations. In both metals and insulators a part of the heat or the entire heat flux may be transferred from the hot end to the cold end by lattice waves. In metals the contribution of the heat conduction due to lattice waves is much smaller than the contribution of the heat conduction due to electrons. In insulators the reverse is true.

Debye in 1914 developed a theory for lattice-vibration contribution to thermal conductivity. If the forces between atoms were purely harmonic, there would be no coupling between the lattice waves and hence no mechanism for collision between phonons. The mutual scattering of waves is not possible for harmonic waves. Because solids expand upon heating, the waves must be anharmonic. The mean free path of a phonon λ_p, which is a measure of the distance traveled by a wave to alternate its intensity by a factor e, is infinite for purely harmonic waves. Anharmonic lattice vibrations limit the value of the mean free path.

The thermal-conductivity contribution due to lattice vibration $\varkappa_L$ may be written in a form similar to Eq. (3.73):

$$\varkappa_L = \tfrac{1}{3} C_V v \lambda_p,$$

where C_V is the specific heat at constant volume per unit volume, v the velocity of the wave propagation, and λ_p the mean free path of phonons. Both Debye in 1914 and Peierls in 1929 showed that λ_p is proportional to $1/T$ at high temperatures. This may be understood by considering the number of phonons at high temperatures. The number of phonons n_K results from Eq. (3.7):

$$n_k = \left[\exp\left(\frac{h\nu_K}{kT}\right) - 1 \right]^{-1}.$$

At high temperature $h\nu_K/kT < 1$, and as a first-order approximation,

$$n_k = kT/h\nu_K \quad \text{or} \quad n_k \propto T.$$

The number of phonons increases directly with temperature. The mean free path of phonons λ_p depends inversely on the number of phonons and hence

$$\lambda_p \propto 1/T.$$

The specific heat at high temperatures is constant and hence at high temperatures

$$\varkappa_L \propto 1/T.$$

At low temperatures λ_p is determined by the dimensions of the crystal and $C_V \propto T^3$. Therefore, at low temperatures

$$\varkappa_L \propto T^3$$

Problem 1

Find the average energy of an atom subjected to the following potentials using Boltzmann's statistics:

a. $-\boldsymbol{\mu}\cdot\mathbf{H}$ (magnetic moment in a magnetic field).
b. mgz (uniform gravitational potential).

Ans. a. $\bar{E} = \frac{5}{2}kT - \mu H \coth(\mu H/kT)$ b. $\bar{E} = \frac{5}{2}kT + \dfrac{mgL}{1 - e^{(mgL/kT)}}$

Problem 2

Using anharmonic potential energy,

$$V(x) = ax^2 - bx^3.$$

Show that the approximate heat capacity of a classical one-dimensional anharmonic oscillator is

$$C_V \cong k\left[1 + \left(\frac{15b^2}{8a^3}\right)kT\right].$$

Problem 3

N independent one-dimensional oscillators are in thermal equilibrium at a temperature T in a field of force described by the potential energy $V(x)$, where

$$V(x) = ax^2 + bx.$$

a. Determine the thermal energy of the system in a classical limit.
b. Find the specific heat of the system.

Ans. a. $\bar{E} = NkT - \dfrac{Nb^2}{4a}$ b. $C_V = Nk$

Problem 4

Find the specific heat of a one-dimensional lattice of identical atoms at low- and high-temperature limits:

a. $C_V = \dfrac{\pi}{3}\left(\dfrac{Nk^2}{\omega_m \hbar}\right)T.$

b. $C_V = (Nk)\left[1 - \dfrac{1}{24}\left(\dfrac{\omega_m \hbar}{kT}\right)^2\right].$

Problem 5

There are two branches of the $\omega(k)$ relation for the linear chain of atoms with two atoms of different masses per cell and assuming nearest-neighbor interaction. What kind of relation of the two atoms corresponds to the two values of ω at zero boundary?

Problem 6

Using Debye's theory, compute the lattice specific heat at constant volume for aluminum and copper at 300°K.   *Ans.* For Al, $C_V \cong 6$ cal/gram molecule; for Cu $\cong 5.76$ cal/gram molecule.

Problem 7

Estimate the electronic specific heat of Al and Cu at 300°K. Assume $E_{f0} = 11.7$ eV for Al, 7.1 eV for Cu. Compare the results of Problems 6 and 7 and comment.

4

Metals, Semiconductors, and Insulators

4.1 Band theory of solids; Kronig-Penny model

In Chapter 1 we discussed the motion of an electron in a potential box, potential well, and in a hydrogen atom. These have been rather simple cases for the calculation of electronic wave functions. The problems of calculating electronic wave functions and energy levels in a solid are very complicated, because we have to deal with a large number of interacting particles. To obtain a simplified picture of electronic motion in a solid, the following assumptions are made:

1. The nuclei in the crystalline state are at rest. This seems to be rather strange after spending a good part of Chapter 3 discussing atomic or lattice vibrations. We are interested in the motion of electrons, so the effect of nuclear motion on the behavior of electrons may be treated as a perturbation.

2. The field seen by a given electron is assumed to be that caused by fixed nuclei plus some average field produced by the charge distribution of all other electrons.

As a result of the above assumptions, the problem of the motion of an electron in a crystal reduces to the motion of an electron in a potential which has the periodicity of the crystal lattice. The analysis would result, among other things, in a natural distinction between metals, semiconductors, and insulators. The results would differ in certain respects from those obtained for the motion of electrons in a potential box $V = 0$ (free electrons). From Sec. 1.3 we remember that the energy of a free electron is a function of K and is given by

$$E = E_x + E_y + E_z = \hbar^2 K^2/2m, \tag{4.1}$$

where

$$K_x^2 + K_y^2 + K_z^2 = K^2 \qquad K = \frac{2\pi}{\lambda} = \frac{P}{\hbar}.$$

From Eq. (4.1) it can be concluded that the energy is a function of K and that there is no upper limit to energy.

We shall see in this section that this is not the case for the motion of an electron in a periodic potential. There exist allowed energy bands separated by forbidden energy gaps. The energy of the electron is a periodic function of K.

Before discussing the details of the analysis of the motion of electrons in a periodic potential we refer to an analogy that exists between electron motion in a periodic potential and elastic wave propagation in a periodic lattice, as discussed in Chapter 3. It has been shown in Sec. 3.5 that in a diatomic lattice with mass points separated by a distance a, two characteristic features describe the modes of vibrations of the lattice points.

1. The frequency of vibration is a function of K, the wave number $2\pi/\lambda$. It is not proportional to K but is a periodic function of K.
2. Not all frequencies of vibration are permitted. Bands of possible frequencies of vibration separated by forbidden frequency gaps exist.

After developing a theory of the electronic motion in a periodic structure we shall find that again two characteristic features describe the energy values of the electron.

1. The energy of the electron E is a function of K, the wave number $2\pi/\lambda$. It is not proportional to K but is a periodic function of K.
2. Not all energies of the electron are permitted. Bands of possible energies of the electron separated by forbidden energy gaps exist.

We therefore would expect to find an analogy in the two problems.

Physical basis for the existence of a forbidden energy gap. It was shown in Sec. 1.6 that the electronic wave function for an electron in the hydrogen atom in the ground state is given by

$$\varphi = A e^{-r/r_1} = (\pi r_1^3)^{-1/2} e^{-r/r_1}$$

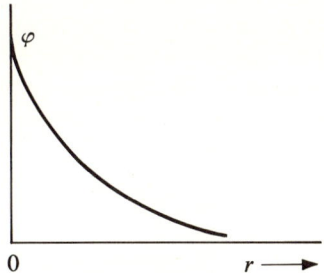

Fig. 4.1
Wave function against r for an electron in a hydrogen atom in the ground state

Figure 4.1 shows a plot of the wave function for values of r. It may be mentioned that the analysis, which resulted in the above equation of the wave function, assumed a single atom, or an atom uninfluenced by other charges or atoms. This means that the electron was considered to be in a potential field only due to the nucleus of the atom to which it belonged. Let us consider that two atoms are brought together; i.e., the separation between them is decreased. Let φ_A and φ_B be the wave functions of the electron in atoms A and B, respectively, when the two atoms are sufficiently far apart. As the atoms are brought together, their wave functions φ_A and φ_B overlap. The resultant wave function of an electron due to two nuclei could be either $\varphi_A + \varphi_B$ or $\varphi_A - \varphi_B$. Each combination shown in Fig. 4.2 preserves the equality of electron distribution between two protons, but an electron in the state $\varphi_A + \varphi_B$ would have slightly lower energy than the electron in the state $\varphi_A - \varphi_B$. In the case of the wave function $\varphi_A + \varphi_B$, the electron has a probability of being found midway between the two nuclei and hence spends part of its time in the region, although here the electron is under the influence of the binding force of both nuclei and hence the binding energy increases. In the state $\varphi_A - \varphi_B$ the probability density

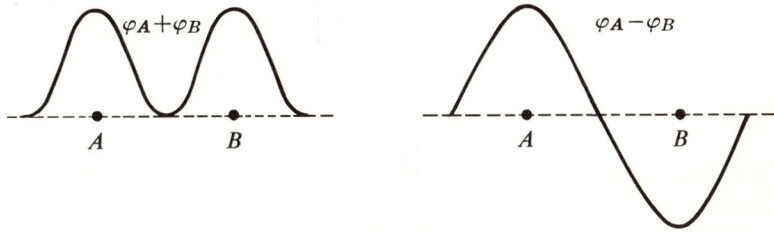

Fig. 4.2
Resultant wave function of an electron due to two nuclei

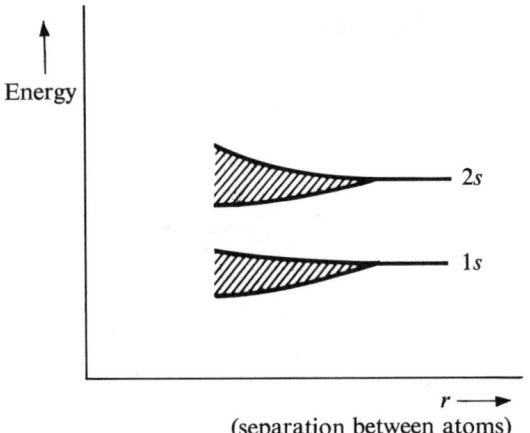

Fig. 4.3
Band formation for 1s and 2s states

vanishes between the two nuclei, and therefore the extra contribution of energy does not appear. Thus as the two atoms are brought together, each energy state splits into two separate energy states. If N atoms are brought together, each energy splits into N different energy states. Each energy level splits up into a sort of band of energy levels. Figure 4.3 shows band formation for 1s and 2s states. The width of a band depends upon the strength of interaction and the overlap between neighboring atoms.

Each of the $2l + 1$ energy states degenerate to form a band. Two or more energy bands may coincide in energy at special positions. This would lead to distinction between metals, semiconductors, and insulators.

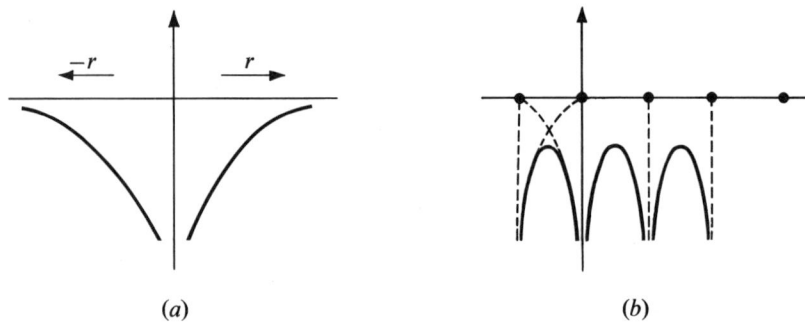

Fig. 4.4
Potential energy as a function of distance for a one-dimensional hydrogen crystal

BAND THEORY OF SOLIDS; KRONIG-PENNY MODEL

Potential energy of an electron. For a hydrogen atom the potential energy of the electron in the field of the nucleus is given by

$$V = -\frac{e^2}{4\pi\epsilon_0 r}$$

[see Fig. 4.4(a)]. In a crystal we have regular arrays of atoms. If it is possible to make a one-dimensional hydrogen crystal, the potential energy as a function of distance would be given by Fig. 4.4(b). For mathematical simplification, the potential-energy curve is idealized in a rectangular shape in Fig. 4.5.

The Kronig-Penny model. The motion of an electron in a periodic potential may be discussed in one dimension by referring to Fig. 4.5. Assume that the periodic potential is in the form of square wells. The period of the periodic potential is assumed to be $a + b$. From $x = 0$ to $x = -b$, the potential energy $V = V_0$. The region $x = 0$ to $x = -b$ is assumed to be the region of an atom. The Schrödinger equations for the two regions are

$$\frac{d^2\varphi}{dx^2} + \frac{2m}{\hbar^2} E\varphi = 0 \quad \text{from } x = 0 \text{ to } x = a, \quad (4.2a)$$

$$\frac{d^2\varphi}{dx^2} + \frac{2m}{\hbar^2}(-V_0 + E)\varphi = 0 \quad \text{from } x = 0 \text{ to } x = -b. \quad (4.2b)$$

Assuming that the total energy of the electron E is less than V_0 and defining

$$\alpha^2 = \frac{2mE}{\hbar^2}, \quad (4.3a)$$

$$\beta^2 = \frac{2m}{\hbar^2}(V_0 - E), \quad (4.3b)$$

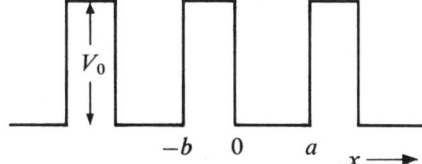

Fig. 4.5
One-dimensional Kronig-Penny potential

α and β are real and the Schrödinger equations for the two regions are

$$\frac{d^2\varphi}{dx^2} + \alpha^2\varphi = 0 \quad \text{from } x = 0 \text{ to } x = a, \tag{4.4a}$$

$$\frac{d^2\varphi}{dx^2} - \beta^2\varphi = 0 \quad \text{from } x = 0 \text{ to } x = -b. \tag{4.4b}$$

The potential is periodic, which means that

$$V(x) = V(x + a + b) = V(x + 2a + 2b).$$

As shown by Bloch, the solution of the Schrödinger equations has the form

$$\varphi(x) = u_K(x)e^{iKx}, \tag{4.5}$$

where $u_K(x)$ is a periodic function having the periodicity of the lattice; i.e.,

$$u_K(x) = u_K(x + a + b). \tag{4.6}$$

According to Bloch, the solution of the Schrödinger equation in this case is a plane wave modulated by the function $u_K(x)$, which has the same periodicity as the lattice. Substituting the solution (4.5) in Eqs. (4.4a) and (4.4b),

$$\frac{d\varphi(x)}{dx} = e^{iKx}\frac{du}{dx} + uiKe^{iKx},$$

$$\frac{d^2\varphi}{dx^2} = \frac{d^2u}{dx^2}e^{iKx} + iKe^{iKx}\frac{du}{dx} + iKe^{iKx}\frac{du}{dx} + u(iK)^2e^{iKx};$$

$$\left[\frac{d^2u}{dx^2} + 2iK\frac{du}{dx} + (\alpha^2 - K^2)u\right]e^{iKx} = 0, \tag{4.7a}$$

$$\left[\frac{d^2u}{dx^2} + 2iK\frac{du}{dx} - (\beta^2 + K^2)u\right]e^{iKx} = 0. \tag{4.7b}$$

Because $e^{iKx} \neq 0$,

$$\frac{d^2u}{dx^2} + 2iK\frac{du}{dx} + (\alpha^2 - K^2)u = 0, \tag{4.8a}$$

$$\frac{d^2u}{dx^2} + 2iK\frac{du}{dx} - (\beta^2 + K^2)u = 0. \tag{4.8b}$$

The simultaneous equations for $u(x)$ [Eqs. (4.8a) and (4.8b)] have general solutions for the two regions given by Eqs. (4.9a) and (4.9b):

$$u_1 = Ae^{i(\alpha-K)x} + Be^{-i(\alpha+K)x}, \tag{4.9a}$$

$$u_2 = Ce^{(\beta-iK)x} + De^{-(\beta+iK)x}. \tag{4.9b}$$

A, B, C, and D are constants and can be determined from the fact that the wave function φ and its normal derivative $\partial\varphi/\partial x$ are single-valued and continuous. u_1, we may recall, is for the region $x = 0$ to $x = a$ and u_2 is for the region $x = 0$ to $x = -b$. Hence the condition of single-valuedness and continuity would be satisfied by the following four boundary conditions:

$$u_1(0) = u_2(0), \tag{4.10}$$

$$u_1(a) = u_2(-b), \tag{4.11}$$

$$\left.\frac{du_1}{dx}\right|_{x=0} = \left.\frac{du_2}{dx}\right|_{x=0}, \tag{4.12}$$

$$\left.\frac{du_1}{dx}\right|_{x=a} = \left.\frac{du_2}{dx}\right|_{x=-b}. \tag{4.13}$$

Substituting Eq. (4.10) in Eqs. (4.9a) and (4.9b),

$$A + B = C + D. \tag{4.14}$$

Substituting Eq. (4.12) in Eqs. (4.9a) and (4.9b),

$$i(\alpha - K)A - i(\alpha + K)B = (\beta - iK)C - (\beta + iK)D. \tag{4.15}$$

Similarly, substituting Eqs. (4.11) and (4.13) in Eqs. (4.9a) and (4.9b),

$$Ae^{i(\alpha-K)a} + Be^{-i(\alpha+K)a} = Ce^{-(\beta-iK)b} + De^{(\beta+iK)b}, \tag{4.16}$$

$$i(\alpha - K)Ae^{i(\alpha-K)a} - i(\alpha + K)Be^{-i(\alpha+K)a} = (\beta - iK)Ce^{-(\beta-iK)b}$$
$$- (\beta + iK)De^{(\beta+iK)b}. \tag{4.17}$$

Equations (4.14), (4.15), (4.16), and (4.17) are homogeneous and so would have solutions provided the determinant of the coefficients A, B, C, and D vanishes. This would lead to involved equations. To obtain a handier solution we represent the periodic potential by a periodic delta function passing to a limit where $b \to 0$ and $V_0 \to \infty$ in such a way that $\beta^2 b$ remains constant and finite. Now

$$\alpha^2 = \frac{2mE}{\hbar^2},$$

$$\beta^2 = \frac{2m(V_0 - E)}{\hbar^2}.$$

Because $V_0 \gg E$,

$$\beta \gg \alpha,$$

and as α is in the region of $V = 0$, $\alpha = K$,

$$\beta \gg K.$$

By definition, in the region of a delta function

$$\frac{d^2u}{dx^2} \gg \frac{du}{dx}.$$

Now

$$\left.\frac{du_1}{dx}\right|_{x=a} = \left.\frac{du_2}{dx}\right|_{x=-b} \tag{4.13}$$

or

$$\left.\frac{du_1}{dx}\right|_{x=a} = \left.\frac{du_1}{dx}\right|_{x=0} - b\left.\frac{d^2u_2}{dx^2}\right|_{x=0}.$$

Referring to Eq. (4.8b), which represents the equation of u in the region of the delta function, and knowing that $d^2u/dx^2 \gg du/dx$,

$$\frac{d^2u}{dx^2} = (\beta^2 + K^2)u.$$

Further, $\beta \gg K$, and hence $d^2u/dx^2 = \beta^2 u$ in the region of the delta function. Therefore, the boundary condition corresponding to Eq. (4.13) reduces to

$$\left.\frac{du}{dx}\right|_{x=a} = \left.\frac{du}{dx}\right|_{x=0} - \beta^2 b u_{(x=0)}. \tag{4.18}$$

Defining,

$$\lim_{\substack{b \to 0 \\ \beta \to \infty}} \frac{\beta^2 ab}{2} = P.$$

Equation (4.18) reduces to

$$\left.\frac{du_1}{dx}\right|_{x=a} = \left.\frac{du_1}{dx}\right|_{x=0} - \frac{2P}{a} u_1\big|_{x=0}. \tag{4.19}$$

Equation (4.19) is a boundary condition involving only Eq. (4.9a), hence only constants A and B, but satisfying both (4.9a) and (4.9b). We need another boundary condition.

At $x = a$,

$$u_1 = Ae^{i(\alpha-k)a} + Be^{-i(\alpha+K)a} = u_2(x = b) = Ce^{-(\beta-iK)b} + De^{(\beta+iK)b},$$

and because $b \to 0$,

$$Ae^{i(\alpha-k)a} + Be^{-i(\alpha+K)a} = C + D.$$

The boundary condition $u_1(x = 0) = u_2(x = 0)$ results in

$$A + B = C + D.$$

Therefore,

$$Ae^{i(\alpha-K)a} + Be^{-i(\alpha+K)a} = A + B. \tag{4.20}$$

BAND THEORY OF SOLIDS; KRONIG-PENNY MODEL

Equations (4.19) and (4.20) would give us two equations with two unknowns A and B.

Equation (4.19), upon substitution of the values of u_1 and du_1/dx, would be

$$Ai(\alpha - K)e^{i(\alpha-K)a} - Bi(\alpha + K)e^{-i(\alpha+K)a} = Ai(\alpha - K) - Bi(\alpha + K)$$
$$- \frac{2P}{a}(A + B). \quad (4.21)$$

Writing Eqs. (4.20) and (4.21) in terms of coefficients of A and B,

$$[1 - e^{i(\alpha-K)a}]A + [1 - e^{-e(\alpha+K)a}]B = 0, \quad (4.22)$$

$$\left[i(\alpha - K)(1 - e^{-i(\alpha-K)a}) - \frac{2P}{a}\right]A + \left[-i(\alpha + K)(1 - e^{-i(\alpha+K)a}) - \frac{2P}{a}\right]B$$
$$= 0. \quad (4.23)$$

The introduction of the delta function reduces a set of four unknown constants to two, A and B; and Eqs. (4.22) and (4.23), which are homogeneous equations, allow us to determine A and B. This follows from the condition that the determinant given below be zero:

$$\begin{vmatrix} [1 - e^{i(\alpha-K)a}] & [1 - e^{-i(\alpha+K)a}] \\ i(\alpha - K)(1 - e^{-i(\alpha-K)a}) - \frac{2P}{a} & -i(\alpha + K)(1 - e^{-i(\alpha+K)a}) - \frac{2P}{a} \end{vmatrix} = 0.$$

Upon multiplication and simplification,

$$P\frac{\sin \alpha a}{\alpha a} + \cos \alpha a = \cos Ka. \quad (4.24)$$

The transcendental equation (4.24) must have a solution in α so that the electronic wave function has the form

$$\varphi(x) = e^{iKx}u_K(x).$$

It may be recalled that

$$P = \lim_{\substack{b \to 0 \\ \beta \to \infty}} \frac{\beta^2 ab}{2}.$$

Taking this definition of P we can confirm the results for two limiting cases as follows:

1. $P = 0$
   If $P = 0$, Eq. (4.24) reduces to
   $$\cos \alpha a = \cos Ka \quad \text{or} \quad \alpha = K.$$

Substituting the value of α,
$$2mE/\hbar^2 = K^2 \quad \text{or} \quad E = \hbar^2 K^2/2m,$$

which shows that the possible values of $P = 0$ correspond to an electron in a potential free region.

2. $P \to \infty$

If $P \to \infty$, because the right side of Eq. (4.24) is a cosine function the left side cannot be infinite and, therefore,

$$\sin \alpha a = 0,$$

$$\alpha a = n\pi,$$

or

$$E = \frac{\hbar^2 \pi^2 n^2}{2ma^2},$$

which shows that the possible values of energies are quantized and correspond to the case of an electron in a box (one-dimensional) surrounded by infinite walls [compare with Eq. (1.36)].

Equation (4.24) is expressed graphically in Fig. 4.6. The graphical representation, which is a plot of $[(P \sin \alpha a/\alpha a + \cos \alpha a]$ versus αa helps to get a physical interpretation of the equation. The left side of the equation is a function of α and hence the energy of the electron. The solid oscillating line in Fig. 4.6 represents a plot of the left side of Eq. (4.24). The right side of the equation in consideration sets a limit to the values the left side can have. Because cosine Ka can only assume values from -1 to $+1$, the physically meaningful solution of the left side must lie between the two horizontal dashed lines at -1 and $+1$. This limitation has a very important consequence — certain values of α or E do not lead to physically meaningful solutions. In other words, this means that an electron moving in a periodic potential can have energy values lying only in certain allowed

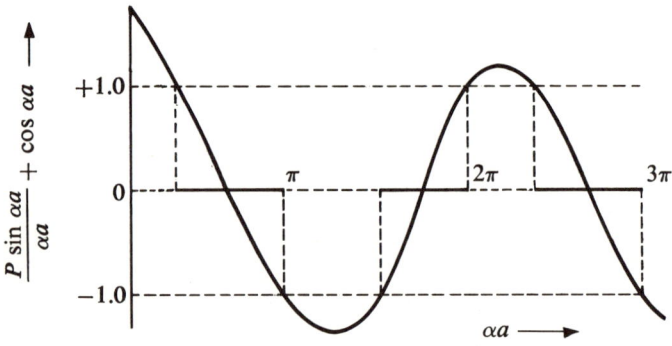

Fig. 4.6
Left side of Eq. (4.24) plotted as a function of α

BAND THEORY OF SOLIDS; KRONIG-PENNY MODEL

zones. The allowed energies in these zones are indicated by the heavy horizontal lines along the $a\alpha$ axis. The regions of αa that do not satisfy the condition of cos aK in the region -1 to $+1$ are forbidden zones of energy for such an electron. As the size of the barrier $V_0 b$ increases, P increases, the first term on the left side of Eq. (4.24) becomes predominant, and the allowed energy bands become narrower. On the other hand, as $V_0 b$ decreases the allowed bands become broader, until in the limiting case $V_0 = 0$, the solution degenerates to that of a free-electron gas.

With the help of Eq. (4.24) it is possible to plot E, the total energy of the electron as a function of the wave number K of the electron (Fig. 4.7).

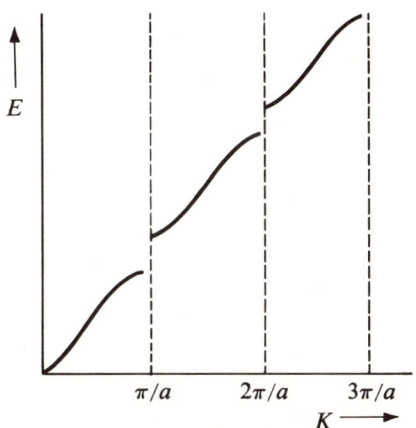

Fig. 4.7
Total energy of the electron as a function of the wave number K

The values of the right side ± 1 are obtained at $Ka = n\pi$, and hence the discontinuities in the E versus K plot occur at $n\pi/a$, where $n = \pm 1, \pm 2, \pm 3, \ldots$. Figure 4.7 shows the E versus K curve only for positive values of n. The zone extending from $K = -\pi/a$ to $K = \pi/a$ is called the *first Brillouin zone*. The *second Brillouin zone* extends from $-2\pi/a$ to $-\pi/a$ and from π/a to $2\pi/a$. In similar fashion we can define the high-order Brillouin zones. In Eq. (4.24) if K is replaced by $K + (2\pi n/a)$, where n is an integer, the right side of the equation remains the same. Therefore, K is not uniquely determined and it is frequently necessary to use a reduced wave vector in the region

$$-\frac{\pi}{a} \leq K \leq \frac{\pi}{a}$$

(Fig. 4.8). E_g is the forbidden energy gap at $K = \pm \pi/a$.

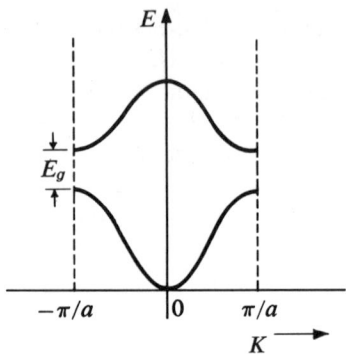

Fig. 4.8
E as a function of a reduced wave vector

Motion of electrons in a one-dimensional periodic potential. In this section we have established the fact that the possible values of energies of an electron in a one-dimensional periodic potential are bands of energy separated by a forbidden energy gap. It would be interesting to know the mass and velocity of the electron if it is occupying any given energy state.

1. *Velocity.* From the wave-mechanical theory it follows that the particle velocity equals the group velocity of the waves representing the electron:

$$v = d\omega/dK,$$

where ω is the angular frequency of de Broglie's waves and can be expressed in terms of energy by Planck's relationship, $E = \hbar\omega$. Therefore,

$$v = \frac{1}{\hbar}\frac{dE}{dK}. \tag{4.25}$$

Referring to Fig. 4.8, E is a function of K and therefore dE/dK is a function of K. If the electron has energies corresponding to $K = 0$ or $\pm\pi/a$, dE/dK is zero and therefore the velocity of the electron is zero.

In fact, as K increases from zero to π/a, the velocity increases from zero to a maximum value at the point of inflection ($d^2E/dK^2 = 0$) and then decreases to zero at $K = \pi/a$ (Fig. 4.9).

Expression (4.25) is a general equation for the velocity of electrons. In the case of free electrons,

$$E = \frac{\hbar^2 K^2}{2m},$$

$$v = \frac{1}{\hbar}\frac{dE}{dK} = \frac{\hbar K}{m},$$

BAND THEORY OF SOLIDS; KRONIG-PENNY MODEL

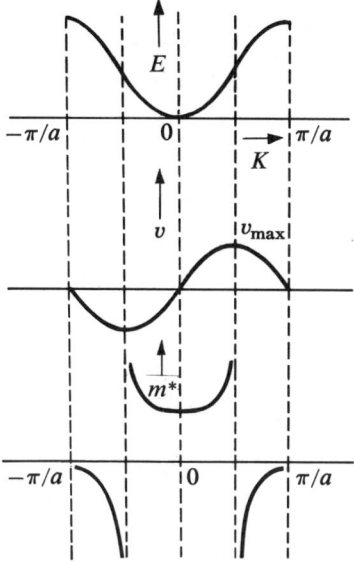

Fig. 4.9
Energy, velocity, effective mass, and f_k as a function of K

or the velocity of the free electrons is proportional to K. In the case of electrons in a one-dimensional periodic potential, E may not be proportional to K^2 [refer to Eq. (4.24)]. The velocity versus K curves for an electron in a one-dimensional periodic potential are plotted in Fig. 4.9. The constant of proportionality is referred to as an *effective mass*.

2. *Effective mass.* Effective mass can be defined by considering the motion of an electron with a given K in an externally applied electric field $\mathcal{E}$. The gain in energy of the electron in a time dt is

$$dE = e\mathcal{E}v\, dt,$$

where v is the velocity of the electron.

Substituting the value of v from Eq. (4.25),

$$dE = \frac{e\mathcal{E}}{\hbar} \frac{dE}{dK} dt.$$

The change in energy is always associated with a change in K and, therefore,

$$dE = \frac{dE}{dK} dK.$$

Therefore,
$$dK = \frac{e\mathcal{E}}{\hbar} dt$$
or
$$\frac{dK}{dt} = \frac{e\mathcal{E}}{\hbar}.$$

The acceleration of the particle or electron is
$$a = \frac{dv}{dt} = \frac{1}{\hbar}\frac{d^2E}{dK^2}\frac{dK}{dt},$$
and substituting the value of dK/dt,
$$a = \frac{1}{\hbar^2} e\mathcal{E} \frac{d^2E}{dK^2}. \tag{4.26}$$

In Eq. (4.26) a is the acceleration of the electron and $e\mathcal{E}$ is the force experienced by the electron due to an externally applied electric field. $e\mathcal{E}/a$ must be the effective mass of the electron:
$$m^* = \frac{e\mathcal{E}}{a} = \frac{\hbar^2}{d^2E/dK^2}. \tag{4.27}$$

Equation (4.27) is the equation of the effective mass and depends upon the form of the E versus K curve. It can be readily shown that for a free electron $m^* = m$, because
$$E = \hbar^2 K^2/2m \quad \text{and} \quad d^2E/dK^2 = \hbar^2/m.$$

One expects this result. However, for an electron moving in a periodic potential, the effective mass m^* is defined for various values of K by Eq. (4.24). Figure 4.9 shows a variation of m^* with K. m^* is positive in the lower half of the energy band and negative in the upper half. At the inflection point in the curve $E(K)$, m^* becomes infinite. Physically this means that if $K = 0$ and an electric field is applied to the electron, the wave number increases with time until the velocity reaches a maximum at $d^2E/dK^2 = 0$. A further gain in energy results in a decrease in velocity, which could be interpreted as the electron having a negative effective mass or behaving as a positively charged particle. In the upper half of the band the electron behaves as a positively charged particle referred to as a *hole*. (The concept of hole will be discussed in Sec. 4.5.) The degree of freedom of an electron is generally defined by a factor
$$f_K = \frac{m}{m^*} = \frac{m}{\hbar^2}\frac{d^2E}{dK^2}.$$

If f_K is small, the particle behaves as a heavy particle, and if f_K is large and approaches unity, the particle behaves as a free particle. According to Fig. 4.9, f_K is positive in the lower half of the band and negative in the upper half of the band.

Number of electrons per band. To determine the number of electrons per band we consider a one-dimensional crystal of N lattices. The length L of the lattice is $N(a+b)$. For a linear lattice of microscopic dimensions the boundary condition may be $\varphi(x) = \varphi(x+L)$, which is strictly true for a circular lattice.

According to Eq. (4.5),

$$\varphi(x) = u_K(x) e^{iKx},$$

$$\varphi(x+L) = u_K(x+L) e^{iK(x+L)}.$$

In accordance with the boundary condition,

$$e^{iKx} u_K(x) = u_K(x+L) e^{iK(x+L)}.$$

Because $L = N(a+b)$,

$$u_K(x) = u_K(x+L) \qquad \text{[refer to Eq. (4.6)]}.$$

Hence

$$e^{iKx} = e^{iK(x+L)}$$

or

$$K = 2\pi n/L,$$

with $n = \pm 1, \pm 2, \pm 3, \ldots$. The number of possible values of n or states in a wave-number interval dK is

$$dn = dK \frac{L}{2\pi}.$$

The total number of possible states in a band is

$$n = \int dn = \int_{-\pi/a}^{\pi/a} dK \frac{L}{2\pi} = \frac{L}{a}.$$

Because $b \to 0$, $a + b \simeq a$. $n = N$, where N is the number of unit cells. The total number of possible states in a band is the number of unit cells. Taking into account the spin of the electron and Pauli's exclusion principle, each state can be occupied at most by two electrons, and hence the total number of electrons in a band is $2N$. A band is filled if it has $2N$ electrons. We have worked out the case of a one-dimensional lattice but the same arguments hold good for a three-dimensional lattice; hence the total number of electrons in a band is $2N$, where N is the number of unit cells.

4.2 Distinction between metals, semiconductors, and insulators

This distinction is known to an electrical engineer from the point of view of electrical conductivity; the conductivity is a minimum for insulators and metals are good electrical conductors. It will be shown that the conductivity of a piece of material could be defined if electrons are treated as a free-electron gas and that the conductivity is directly proportional to the number of free electrons per unit volume taking part in the process of electrical conduction. It was established in Sec. 4.1 that f_K is the degree of freedom of an electron:

$$f_K = \frac{m}{\hbar^2} \frac{d^2E}{dK^2}.$$

d^2E/dK^2 depends upon K.

Knowing f_K and the total number of electrons in a band, the total number of effectively free electrons can be determined. The number of wave functions in an interval dK for a one-dimensional lattice of length L is $(L/2\pi)\,dK$. Two electrons can have the same function, so the number of electrons in an interval dK is $(L/\pi)\,dK$. The number of effectively free electrons dn_{eff} in the interval dK is

$$dn_{\text{eff}} = \frac{L}{\pi} dK f_K.$$

We substitute the value of f_K, and therefore

$$dn_{\text{eff}} = \frac{Lm}{\pi\hbar^2} \frac{d^2E}{dK^2} dK.$$

If the band is filled to $K = K_1$, momenta up to $\hbar K_1$ are possible. The momenta may be either positive or negative, and hence $K = K_1$ means that the band is filled from $K = -K_1$ to $K = K_1$. Hence

$$n_{\text{eff}} = \int_{-K_1}^{K_1} \frac{Lm}{\pi\hbar^2} \frac{d^2E}{dK^2} dK = \frac{2Lm}{\pi\hbar^2} \int_0^{K_1} \frac{d^2E}{dK^2} dK$$

or

$$n_{\text{eff}} = \frac{2mL}{\pi\hbar^2} \left(\frac{dE}{dK}\right)_{K=K_1}.$$

If a band is completely filled,

$$K = \pm \frac{n\pi}{a} \quad \text{where } n = 1, 2, 3, 4, \cdots,$$

$$\frac{dE}{dK} = 0 \quad \text{at } K = \pm\frac{n\pi}{a}.$$

In such a case

$$n_{\text{eff}} = 0.$$

The effective number of free electrons in a completely filled band is zero. Further, because dE/dK reaches a maximum at the point of inflection, the effective number of free electrons is a maximum if the band is filled to the point of inflection.

Insulators and intrinsic or pure semiconductors. Insulators and intrinsic or pure semiconductors are those materials in which at a temperature of absolute zero a certain number of energy bands are completely filled while others are completely empty. Hence the effective number of free electrons for current conduction is zero, and the electrical conductivity σ is zero. Figure 4.10 shows an energy diagram for an insulator and

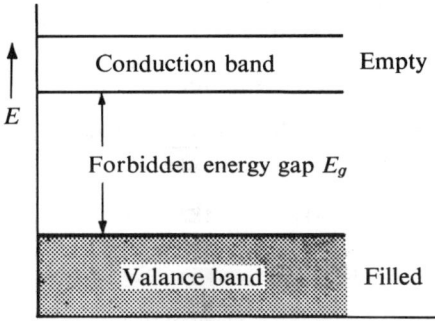

Fig. 4.10
Energy diagram for an insulator and pure semiconductor at absolute zero

pure semiconductor at absolute zero. The uppermost filled band is called the *valence band* and the next higher completely empty band is called the *conduction band*. The forbidden energy gap is denoted E_g. The only difference between an insulator and an intrinsic semiconductor is the order of magnitude of the energy gap E_g. If the forbidden energy gap is of the order of several electron volts, the solid is an insulator. In semiconductors the forbidden energy gap is either approximately 1 eV or less. Hence the difference between pure semiconductors and insulators is arbitrary.

Metals. If the highest filled band is incompletely filled, the number of effective free electrons is finite and the solid has a finite conductivity and is called a *metal*.

4.3 Free-electron model of metals

If a band is partially filled, in fact, if filled only in the neighborhood of $K = 0$, the effectiveness of an electron as a free electron is

$$f_K = \frac{m}{m^*} = \frac{m}{\hbar^2} \frac{d^2E}{dK^2} \to 1.$$

The electrons may be treated as free electrons. The model of a solid which assumes the highest-energy electrons as free electrons is called the *free-electron model*. According to this model the valence electrons in a metal are able to move about freely through the volume of the specimen. The interaction of the electrons with the ions of the atoms is considered negligibly small. The electrons are therefore treated as free particles such as gas molecules surrounded by the boundaries or walls of the solid. In earlier theories of the free electron, Boltzmann statistics or classical statistics was applied to the gas of electrons. This treatment assumed electrons to be rigid spheres. If they follow classical statistics the mean-square velocity of the electrons at a temperature T is

$$\overline{v^2} = 3kT/m. \qquad (2.3)$$

The electronic contribution to the specific heat per unit volume from Eq. (3.63) is

$$C_{V_e} = \frac{\partial W}{\partial T} = \frac{3}{2} nk$$

and the electronic thermal conductivity is

$$K_e = \tfrac{1}{3} C_{V_e}(\overline{v^2})^{1/2} \lambda,$$

where λ is the mean free path of the electrons. Hence

$$K_e = \frac{1}{3} \times \frac{3}{2} nk \left(\frac{3kT}{m}\right)^{1/2} \lambda.$$

It will be shown in Sec. 4.4 that the electrical conductivity can be defined as

$$\sigma = ne^2\tau/m,$$

where τ is the relaxation time. Therefore,

$$\lambda = (\overline{v^2})^{1/2}\tau,$$

$$\sigma = \frac{ne^2}{m} \frac{\lambda}{(\overline{v^2})^{1/2}} = \frac{ne^2}{m} \frac{\lambda(m^{1/2})}{(3kT)^{1/2}}.$$

The ratio of electronic thermal conductivity to electrical conductivity is

$$\frac{K_e}{\sigma} = \frac{1}{2} nk \frac{3kT}{m} \frac{m}{ne^2} = \frac{3}{2}\left(\frac{k}{e}\right)^2 T. \qquad (4.28)$$

A relationship of this type was first observed by Wiedmann and Franz and is called the *Wiedemann-Franz ratio*. One of the few successes of the classical model of the free-electron model is the semiquantitative agreement of the Wiedemann-Franz law. Many other discrepancies between the results of the classical theory and the experimental results could not be explained and the quantum-mechanical model of the electron gas was adopted. This model made use of Pauli's exclusion principle. In metals the outermost electrons may be treated as free electrons. The density of states for these electrons corresponds to the density of states of particles or electrons in a potential box given by Eq. (1.39), which incorporates Pauli's exclusion principle. The probability of occupation of a state of energy E is given by Eq. (2.33). Therefore, the free-electron model using quantum-mechanical treatment is applicable to metals, semiconductors, and insulators, the only difference being that in semiconductors and insulators the effective mass m^* may not be equal to the free-electron mass.

4.4 Effect of electric field on metals; impurities in metals

It is known from basic electrical engineering that if a voltage of magnitude V is applied to a specimen of resistance R, the current I flowing in the specimen is given by Ohm's law:

$$V = IR.$$

If the specimen has length l and A is the area of cross section,

$$R = \rho l / A,$$

where ρ is the resistivity of the material:

$$V = I \frac{\rho l}{A} \quad \text{or} \quad \frac{V}{l} = \frac{I}{A} \rho,$$

where V/l is the applied electric field $\mathcal{E}$ and I/A is the current density J. Therefore,

$$J = \sigma \mathcal{E}, \tag{4.29}$$

where σ is $1/\rho$, the conductivity of the specimen.

Equation (4.29) is Ohm's law in terms of current density, conductivity, and applied electric field. To determine the effect of the applied electric field in terms of microscopic quantities, we shall first consider the classical free-electron gas model. The quantum statistics in the free-electron model will be discussed after the classical treatment.

Classical free-electron gas. Suppose there are n electrons per unit volume. Imagine that the electrons are moving randomly with a velocity distribution decided by the temperature of the material and hence of the free-electron gas. In the absence of an applied electric field, the average velocity in any direction, say x, is

$$\langle v_x \rangle = \frac{1}{n} \sum_{i=1}^{n} v_{xi} = 0,$$

where v_{x_i} is the x component of velocity of the ith electron. The average velocity $\langle v_x \rangle = 0$, because in equilibrium there are just as many electrons moving in one direction as in other directions.

Consider a single electron under the influence of an applied electric field in the x direction $\mathcal{E}_x$. The force experienced by the electron is $e\mathcal{E}_x$. As a result of this force, the electron with mass m accelerates. The equation of motion of the electron is

$$m\frac{d\langle v_x \rangle}{dt} = e\mathcal{E}_x. \tag{4.30}$$

Integrating,

$$\langle v_x \rangle = \frac{e\mathcal{E}_x}{m} t + \text{const.}$$

If the field is applied at $t = 0$, $\langle v_x \rangle = 0$ at $t = 0$:

$$\langle v_x \rangle = \frac{e\mathcal{E}_x}{m} t,$$

which means that the velocity increases linearly with time. If all the electrons in a unit volume were assumed to be noninteracting, noncolliding with each other, the current density in the x direction would be

$$J_x = ne\langle v_x \rangle = \frac{(ne^2\mathcal{E}_x \cdot t)}{m},$$

which means that as a result of the applied electric field the current would increase linearly with time, thus violating the basic Ohm's law. We are, therefore, not justified in treating the electrons as noncolliding.

During the acceleration of the electron as a result of the Lorentz force $e\mathcal{E}_x$, it may collide with an ion and as a result lose its entire momentum and energy. If τ is the relaxation time, the average time between two consecutive collisions, the frictional force of the electron due to collisions, may be expressed $m \langle v_x \rangle / \tau$. Hence the modified equation of motion is

$$m\frac{d\langle v_x \rangle}{dt} + \frac{m\langle v_x \rangle}{\tau} = e\mathcal{E}_x. \tag{4.31}$$

The solution of Eq. (4.31) is

$$\langle v_x \rangle = \langle v_x \rangle_\infty (1 - e^{-t/\tau}),$$

where $\langle v_x \rangle_\infty$ is the final average velocity in the x direction referred to as drift velocity:

$$\langle v_x \rangle_\infty = \frac{e\tau}{m} \mathcal{E}_x,$$

$$\langle v_x \rangle = \frac{e\tau}{m} \mathcal{E}_x (1 - e^{-t/\tau}),$$

and the current density is

$$J_x(t) = ne\langle v_x \rangle = \frac{ne^2\tau}{m} \mathcal{E}_x (1 - e^{-t/\tau}). \tag{4.32}$$

The current density rises exponentially with time, the time constant of the exponential rise being the relaxation time. τ is the order of 10^{-14} sec. This is the reason we do not note an exponential rise of the current, even on very fast oscilloscopes.

The steady-state current density is

$$J_x = \frac{ne^2\tau}{m}\mathcal{E}_x. \tag{4.33}$$

Comparing Eq. (4.33) with Eq. (4.29),

$$\sigma = \frac{ne^2\tau}{m}. \tag{4.34}$$

The electrical conductivity is related to the number of free electrons per unit volume, the charge of the electron, the mass of the electron, and the relaxation time.

The results of the derivation of the expression for conductivity can be easily understood. The charge transported is proportional to ne; e/m enters because of acceleration in a given field. The time τ describes the duration of the field. Mobility is defined as the steady-state drift velocity $\langle v_x \rangle_\infty$ per unit of electric field strength. Hence the mobility is

$$\mu = \frac{\langle v_x \rangle_\infty}{\mathcal{E}_x} = \frac{e\tau}{m}, \tag{4.35}$$

and therefore

$$\sigma = ne\mu. \tag{4.36}$$

Mobility is very useful in dealing with semiconductors.

Quantum-mechanical free-electron gas. It can be shown that the introduction of Fermi-Dirac statistics in place of Maxwell-Boltzmann statistics has very little influence on electrical conductivity. Using the classical free-electron gas model, we found

$$\sigma = \frac{ne^2\tau}{m}. \tag{4.34}$$

Before taking up the quantum-mechanical free-electron gas model, we shall discuss the derivation of Eq. (4.34) from another point of view. Taking a one-dimensional classical electron gas, the number of free electrons having velocities between v_x and $v_x + dv_x$ is

$$f(v_x)\, dv_x = A \exp\left(-\frac{1}{2}\frac{mv_x^2}{kT}\right) dv_x$$

(refer to Sec. 2.1). The distribution function $f(v_x)$ is plotted against v_x in Fig. 4.11 as a solid line. The distribution function is shifted to the right by an amount $(e\tau/m)E_x$ as a result of the application of an electric field $\mathcal{E}_x$. The

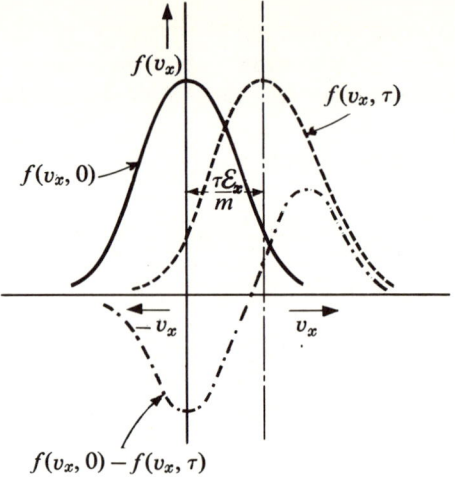

Fig. 4.11
Distribution function $f(v_x)$ plotted against v_x without electric field (solid line) and with electric field (dashed line)

new distribution function is shown as a dashed line in Fig. 4.11 and represents the steady-state distribution function in the presence of a field. The difference in the distribution function as a result of the applied electric field $\mathcal{E}_x$ is given approximately by a first-order Taylor's expansion:

$$f(v_x, \tau) \simeq f(v_x, 0) - \frac{e\mathcal{E}_x\tau}{m}\left[\frac{df(v_x)}{dv_x}\right]_{t=0},$$

where $t = 0$ is the instant of application of the field. The steady state in the presence of an electric field is assumed to have been reached after time τ:

$$f(v_x, 0) - f(v_x, \tau) \simeq \frac{e\mathcal{E}_x\tau}{m}\left[\frac{df(v_x)}{dv_x}\right]_{t=0}.$$

The difference distribution function is plotted in Fig. 4.11 by a dashed-and-dotted line. The average velocity without the field $\mathcal{E}_x$ is

$$\overline{v_x} = \int_{-\infty}^{\infty} v_x f(v_x)\, dv_x = 0,$$

because

$$\int_{-\infty}^{\infty} \exp\left(-\frac{1}{2}\frac{mv_x^2}{kt}\right) dv_x = \int_{-\infty}^{\infty} f(v_x)\, dv_x = 1$$

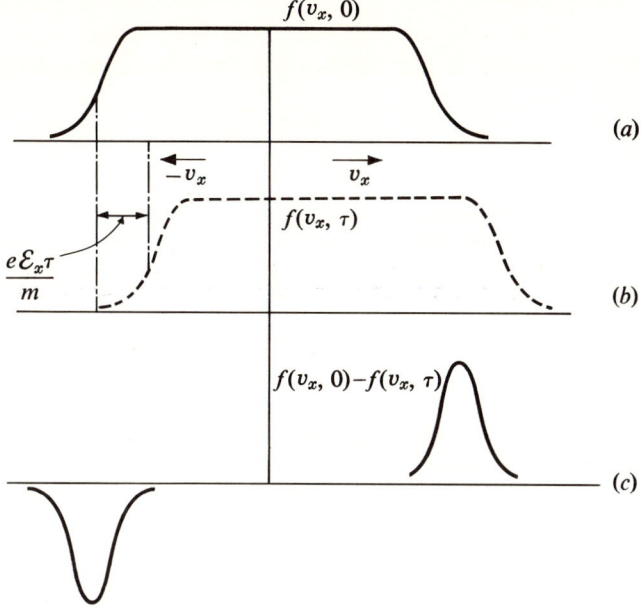

Fig. 4.12
Fermi-Dirac velocity distribution in one dimension. (a) Equilibrium velocity distribution (without electric field). (b) Distribution shifted by electric field. (c) Difference between steady-state distribution and equilibrium distribution

for normalization. However, as a result of the applied field, the average velocity or drift velocity v_D is

$$v_D = \langle v_x \rangle = \int_{-\infty}^{\infty} v_x \frac{e\mathcal{E}_x \tau}{m} \left| \frac{df(v_x)}{dv_x} \right|_{t=0} dv_x$$

or

$$v_D = \frac{e\mathcal{E}_x \tau}{m} \left\{ \left[f(v_x)_{t=0}\, v_x \right]_{-\infty}^{\infty} - \int_{-\infty}^{\infty} f(v_x)_{t=0}\, dv_x \right\};$$

$$\int_{-\infty}^{\infty} f(v_x)_{t=0}\, dv_x = 1 \quad \text{and} \quad \left[f(v_x)_{t=0} v_x \right]_{-\infty}^{\infty} = 0,$$

and, therefore,

$$v_D = \frac{e\mathcal{E}_x \tau}{m},$$

$$\sigma = \frac{ne^2 \tau}{m}.$$

It is evident that the distribution function undergoes changes in the neighborhood of the Fermi level, so that relaxation times τ_f for electrons having energy corresponding to E_f are the ones that come in question, and, as such,

$$\sigma = \frac{ne^2\tau_f}{m}. \tag{4.36}$$

Although the treatment given here is based on the free-electron approximation, a similar treatment may be given for an electron moving in a periodic potential. The number of effectively free electrons is given by n_{eff} and, as such, in general,

$$\sigma = \frac{n_{\text{eff}}e^2\tau_f}{m}. \tag{4.37}$$

Wiedemann-Franz ratio. The ratio has been defined in Sec. 4.3 and is given by Eqs. (3.76) and (4.34):

$$\frac{K_e}{\sigma} = \text{Wiedmann-Franz ratio},$$

$$K_e = \frac{\pi^2}{3} R \frac{kT}{m} T, \tag{3.76}$$

and for a unit volume,

$$K_e = \frac{\pi^2}{3} n \frac{k^2 T}{m} T,$$

$$\sigma = \frac{ne^2\tau}{m}. \tag{4.34}$$

Therefore,

$$\frac{K_e}{\sigma} = \frac{\pi^2}{3} \frac{k^2}{e^2} T. \tag{4.38}$$

Equation (4.38) is similar to Eq. (4.28), differing only by a constant.

Effect of time-varying fields. In this section we have so far discussed only the effect of static fields on electron gases in metals. We shall now investigate the effect of an alternating field on metals.

Let us assume that the applied field is sinusoidal and can be represented by $\mathcal{E}_{x_0} \cos \omega t$. The equation of motion of the electrons in the electron cloud in terms of the average velocity in the x direction $\langle v_x \rangle$ or the drift velocity v_D is given by

$$m \frac{dv_D}{dt} + \frac{mv_D}{\tau} = -e\mathcal{E}_{x_0} \cos \omega t. \tag{4.39}$$

Assuming a solution of the above equation given by
$$v_D = \text{Re}(A^* e^{j\omega t}) \quad \text{in the steady state,}$$
$$m\,\text{Re}(A^* j\omega e^{j\omega t}) + m\,\text{Re}\left(\frac{A^*}{\tau} e^{j\omega t}\right) = -e\mathcal{E}_{x0}\,\text{Re}(e^{j\omega t})$$
or
$$\text{Re}[A^*(1 + j\tau\omega)e^{j\omega t}] = -\frac{e\mathcal{E}_{x0}\tau}{m}\,\text{Re}(e^{j\omega t}),$$
which shows that
$$A^* = -\frac{e\tau}{m}\mathcal{E}_{x0}\frac{1}{1 + j\tau\omega}.$$
Hence
$$v_D = \text{Re}\left(-\frac{e\tau\mathcal{E}_{x0}}{m}\frac{1}{1 + j\tau\omega}e^{j\omega t}\right).$$
Simplifying,
$$v_D = -\frac{e\tau}{m}\mathcal{E}_{x0}\left(\frac{\cos\omega t}{1 + \tau^2\omega^2} + \frac{\tau\omega}{1 + \tau^2\omega^2}\sin\omega t\right). \qquad (4.40)$$

The current density is
$$J_x = nev_D.$$
Substituting Eq. (4.40) in the expression for current density,
$$J_x = \frac{ne^2\tau}{m}\mathcal{E}_{x0}\left(\frac{\cos\omega t}{1 + \tau^2\omega^2} + \frac{\tau\omega}{1 + \tau^2\omega^2}\sin\omega t\right). \qquad (4.41)$$
Calling $ne^2\tau/m = \sigma_{\text{static}}$,
$$J_x = \sigma_{\text{static}}\mathcal{E}_{x0}\left(\frac{1}{1 + \tau^2\omega^2}\cos\omega t + \frac{\tau\omega}{1 + \tau^2\omega^2}\sin\omega t\right). \qquad (4.42)$$

The current density J_x, according to Eq. (4.42), depends upon σ_{static}, the relaxation time τ, and the angular frequency of the applied field ω. Equation (4.42) further shows that the current density consists of two parts, one in phase with the applied field and the other shifted $\pi/2$ from the applied field. Equation (4.42) further shows that as long as $\tau\omega \ll 1$,
$$J_x = \sigma_{\text{static}}\mathcal{E}_{x0}\cos\omega t,$$
and the metal behaves as a pure resistance with conductivity given by the static value $\sigma_{\text{static}} = ne^2\tau/m$. Figure 4.13 shows the variation of σ with the angular frequency of the applied field. In the neighborhood of $\tau\omega \simeq 1$, the conductivity decreases sharply and goes to zero. If $\tau\omega = 1$,
$$\sigma = \frac{\sigma_{\text{static}}}{2}.$$

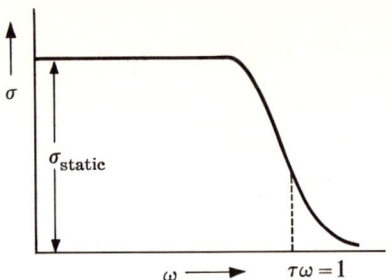

Fig. 4.13
Variation of σ with the angular frequency of the applied field

This occurs at approximately a frequency of 10^{14} Hz, because $\tau \simeq 10^{-14}$ sec. For the conductivity to decrease sharply, the frequency of the applied field should be in the region of ultraviolet light. The loss of conductivity may be explained by the fact that the electrons cannot move fast enough with the field and hence lag behind, resulting in no resultant drift velocity.

The second part of the current density in Eq. (4.42) is a maximum at $\tau\omega \simeq 1$.

Qualitative discussion of resistivity of metals; Impurities. Electrons have a dual nature, so the electronic wave number remains unchanged while moving through a periodic potential. Because electrons are also waves, they can pass through perfect crystals without resistance. Therefore, if we have a perfect crystal and all the nuclei are at rest, the mean free path of the electron between two successive collisions of a particle would be infinite, because in the language of particles there would be virtually no collisions. Hence the resistivity or conductivity of materials is based on deviations from periodicity of the perfect crystal structure. These deviations may be:

1. Thermal vibrations of the ions or lattices.
2. Imperfection in lattice, vacancies, dislocations, etc.
3. Impurities or impurity atoms.
4. Boundaries.

Variation of resistivity with temperature. To study the variation of resistivity with temperature, we shall first take up the classical model and then discuss the lattice vibrations and the wave nature of the electrons. In the classical model the atoms are assumed to be rigid spheres and electron scattering takes place at these rigid spheres. We further assume that the electrons obey Maxwell-Boltzmann statistics. It has been shown that the electrical conductivity is

$$\sigma = \frac{ne^2\tau}{m}.$$

The relaxation time τ is related to the mean free path λ by the relationship

$$\lambda = \langle v_x \rangle \tau,$$

where $\langle v_x \rangle$ is the average velocity in the x direction as a result of the applied electric field. Therefore,

$$\sigma = \frac{ne^2\lambda}{m\langle v_x \rangle}.$$

$\langle v_x \rangle$ can be determined with the help of Maxwell-Boltzmann statistics. The number of electrons having a velocity component lying between v_x and $v_x + dv_x$ is

$$f(v_x)\,dv_x = A \exp\left(-\frac{1}{2}\frac{mv_x^2}{kT}\right)dv_x,$$

and, therefore,

$$\langle v_x \rangle = \frac{\int_0^\infty v_x \exp\left(-\frac{1}{2}\frac{mv_x^2}{kT}\right)dv_x}{\int_0^\infty \exp\left(-\frac{1}{2}\frac{mv_x^2}{kT}\right)dv_x}.$$

Computing the right side,

$$\langle v_x \rangle = \left(\frac{2kT}{m\pi}\right)^{1/2};$$

hence

$$\sigma = \frac{ne^2\lambda}{m\langle v_x \rangle} = \frac{ne^2\lambda}{m}\left(\frac{m\pi}{2k}\right)T^{-1/2}. \tag{4.43}$$

The electrical conductivity according to the classical theory is proportional to $T^{-1/2}$. The results of the classical theory are not in conformity with the experimental results. Experimentally the electrical conductivity of metals is found to vary inversely with temperature at high temperatures. The discrepancy can be explained if the lattice vibrations and the wave nature of electrons is taken into account. Assuming that the effect of lattice defects, foreign impurity atoms, and boundaries is independent of temperature, the influence of lattice waves on the motion of electrons alone would affect the variation of resistance as a function of temperature. The investigation of the influence of lattice waves on the motion of electrons is involved and complicated. Simplifying assumptions are made to calculate the variation of resistance as a function of temperature. Because in the quantum theory

$$\sigma = \frac{ne^2\lambda_f}{mv_f},$$

we are interested in the scattering of electrons with Fermi energy. Einstein's model of lattice vibrations is employed, and the interaction between the electrons and the atomic vibrations is studied only qualitatively.

If M is the mass of the atom and ν is the vibrational frequency of the atom in the Einstein model, the equation of motion of the atom would be

$$M\frac{d^2x}{dt^2} + 4\pi^2\nu^2 Mx = 0.$$

The total energy of a one-dimensional simple harmonic oscillator is composed of potential and kinetic energy. The total average energy is

$$\bar{E} = \frac{h\nu}{\exp(h\nu/kT) - 1}. \tag{3.7}$$

If $h\nu/k = \theta_E$ and $\theta_E \ll T$,

$$\bar{E} = kT.$$

The energy $\bar{E}$ consists of the average kinetic energy and average potential energy, which are equal.

The average P.E. $= kT/2$. For a simple harmonic oscillator,

$$\text{P.E.} = 2\pi^2\nu^2 Mx^2.$$

The average potential energy is

$$\text{P.E.} = 2\pi^2\nu^2 M\langle x^2 \rangle = kT/2. \tag{4.44}$$

If Q_f is the scattering cross section of an atom and is referred to its capability of scattering an electron with Fermi energy, λ_f is defined as the mean free path of an electron with Fermi energy and as such has a unit probability of being scattered if it travels a distance λ_f. Q_f may be defined as

$$\lambda_f N Q_f = 1,$$

where N is the number of atoms per unit volume. If the displacement x is zero, the atoms are at rest and there is no scattering. Hence $\langle x^2 \rangle$, having the dimensions of area, must be proportional to Q_F, the scattering area of cross section.

$$\lambda_f = \frac{1}{NQ_f} \propto \frac{1}{\langle x^2 \rangle} \propto \frac{M\theta_E^2}{T}$$

or

$$\lambda_f = \text{const}\,\frac{M\theta_E^2}{T}.$$

Because $\sigma = ne^2\lambda_f/mv_f$,

$$\sigma = \text{const}\,\frac{M\theta_E^2}{T}. \tag{4.45}$$

Equation (4.45) shows that $\sigma \propto T^{-1}$, which agrees with experimental results at high temperatures. At temperatures such that $T \ll \theta_E$,

$$e^{h\nu/kT} \gg 1,$$

$$\overline{E} \simeq h\nu e^{-h\nu/kT},$$

$$\langle x^2 \rangle \propto e^{-h\nu/kT}.$$

The conductivity at low temperatures according to Einstein's model is proportional to $e^{\theta_E/T}$. This does not confirm with experimental results. At low temperature the contribution of impurities and imperfections to electrical resistivity is predominant. The presence of impurities results in a field near the impurity atom different from the parent atom, so the impurities produce deviations from the periodicity of the potential and therefore are scattering centers.

If τ_i is the relaxation time associated with impurities atoms and τ_{th} is the relaxation time associated with the thermal vibrations of the atoms, the resultant relaxation time is

$$\frac{1}{\tau} = \frac{1}{\tau_i} + \frac{1}{\tau_{\text{th}}}.$$

The probabilities of scattering are inversely proportional to the relaxation times, and the probability of scattering is enhanced if both impurities and thermal vibrations of atoms are taken into account.

Matthiessen rule. The *Matthiessen rule* states that the resistivity of a metal varies with temperature according to

$$\rho = \rho_i + \alpha T, \tag{4.46}$$

where ρ_i is the resistivity contribution of impurities and α is the coefficient of increase of resistance.

4.5 Intrinsic semiconductors and insulators

It was discussed in Sec. 4.2 that the difference between pure semiconductors and insulators is the magnitude of the forbidden energy gap. Pure semiconductors are referred to as *intrinsic semiconductors*. It was also shown that in intrinsic semiconductors and insulators at $T = 0$, the valence band is completely filled and the conduction band is completely empty. If the temperature is other than absolute zero, a certain number of electrons from the top of the valence band may be thermally excited into the conduction band. Those electrons that reach the conduction band partially fill the conduction band and thermally excited electrons would be available for conduction in an applied electric field. The valence band which is completely filled at $T = 0$ becomes partially filled, there being vacancies at the top of the band. The presence of such empty states in the valence band leads to conductivity, because electrons in the valence band can transfer to these states and hence acquire a drift velocity due to the applied field. The empty states are termed *holes*.

Let us consider a single hole in a valence band, corresponding electrons having been excited to the conduction band. Assuming an applied electric field, let us consider the behavior of electrons in this band. The current density due to electrons in the valence band of a completely filled valence band is

$$J_v = ne\langle v_x \rangle,$$

where $n \langle v_x \rangle$ is the sum of the velocity of all the electrons in this band:

$$n\langle v_x \rangle = \sum_{i=1}^{i=n} v_{x_i} = 0,$$

because the band is filled. If we now consider a vacant state and call it the jth state,

$$n\langle v_x \rangle = \left(\sum_{\substack{i=1 \\ i \neq j}}^{i=n} v_{x_i} \right) + v_j.$$

Therefore,

$$\sum_{i=1}^{i=n} v_{x_i} = 0 = \left(\sum_{\substack{i=1 \\ i \neq j}}^{i=n} v_{x_i} \right) + v_j$$

and hence

$$\sum_{\substack{i=1 \\ i \neq j}}^{i=n} v_{x_i} = -v_j. \tag{4.47}$$

The left side of Eq. (4.47) represents the resultant velocity of all the electrons in the valence band except one and is equal to the velocity of the missing electron but in the opposite direction. The vacant state therefore behaves as a mobile opposite charge, the hole. Hence the hole current density due to a single hole is

$$J_h = -ev_h,$$

where the subscript h has been inserted for j to represent the hole. If p is the number of holes per unit volume and $\langle v_h \rangle$ is the average drift velocity of vacancies,

$$J_h = -pe\langle v_h \rangle.$$

Because e is the electronic charge and is negative,

$$J_h = pe\langle v_h \rangle. \tag{4.48}$$

The average velocity of holes $\langle v_h \rangle$ is related to the applied electric field by the mobility equation (4.35),

$$\langle v_h \rangle = \mu_h \mathcal{E},$$

where μ_h is the mobility of holes. Therefore,

$$J_h = pe\mu_h \mathcal{E}.$$

The current density due to conduction band electrons is

$$J_e = ne\langle v_e \rangle, \tag{4.49}$$

where n is the number of electrons per unit volume and $\langle v_e \rangle$ is the average drift velocity of electrons. The average drift velocity of electrons $\langle v_e \rangle$ is related to the applied electric field by mobility equation (4.35). $\mu_e \mathcal{E} = \langle v_e \rangle$, where μ_e is the mobility of electrons. Therefore,

$$J_e = ne\mu_e \mathcal{E}. \tag{4.50}$$

The total current density in a semiconductor or an insulator is the sum of the current density due to holes and the current density due to electrons. The electronic current is in a direction opposite to the hole current. The total current density J is obtained by adding Eqs. (4.48) and (4.50) (see Fig. 4.14):

$$J = (J_h + J_e) = (pe\mu_h + ne\mu_e)\mathcal{E}. \tag{4.51}$$

Comparing Eq. (4.51) with Ohm's law, represented by (4.29),

$$J = \sigma \mathcal{E} \tag{4.29}$$

Therefore,

$$\sigma = (pe\mu_h + ne\mu_e). \tag{4.52}$$

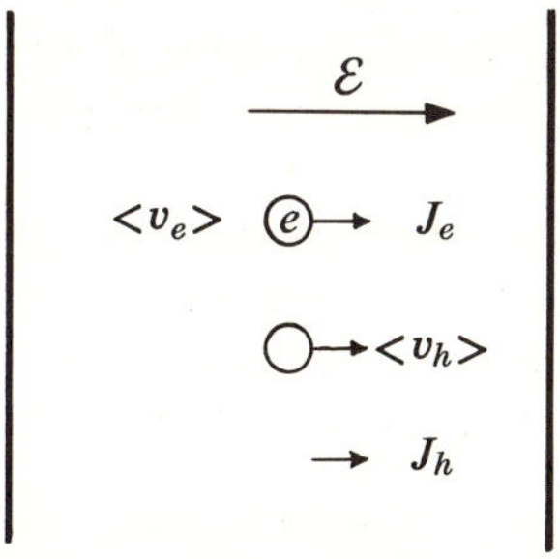

Fig. 4.14
Motion of electrons and holes in an electric field $\mathcal{E}$

Equation (4.52) is the total electrical conductivity of a semiconductor or insulator.

Electron distribution in insulators and intrinsic semiconductors. Equation (4.52) shows that the electric conductivity of insulators and intrinsic semiconductors, in fact, all semiconductors, depends on the number of holes per unit volume p, the number of electrons n, and the mobility of holes and electrons μ_h and μ_e, respectively. It will be shown later that the electrical properties are determined essentially by these quantities. We shall now calculate the number of electrons n in the conduction band. Before we discuss the problem of calculating the number of electrons in the conduction band statistically, we shall assume a simplified model. This model will help us estimate the location of the Fermi level and will give us an approximate idea of the number of electrons in the conduction band.

Simplified model. Referring to Fig. 4.15, we assume that the widths of the conduction band and valence band are small compared to the for-

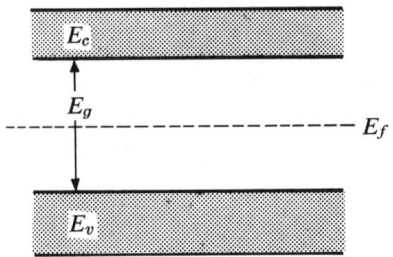

Fig. 4.15
Simplified model of an insulator or intrinsic semiconductor

bidden energy gap. We further assume that all the electrons in the conduction band have the same energy E_c and all electrons in the valence band have the same energy E_v. Let n be the number of electrons in the valence band at $T = 0$. At any temperature other than absolute zero, some of the electrons from the valence band are thermally excited into the conduction band. Let n_c be the number of electrons left behind in the conduction band and n_v be the number of electrons left behind in the valence band:

$$n = n_c + n_v.$$

The number of electrons in the conduction band is

$$n_c = np(E_c),$$

INTRINSIC SEMICONDUCTORS AND INSULATORS

where $p(E_c)$ is the probability of occupation of an electron state of energy E_c in the conduction band.

According to Fermi-Dirac statistics, Eq. (2.33),

$$p(E_c) = \left[\exp\left(\frac{E_c - E_f}{kT}\right) + 1\right]^{-1}.$$

Therefore,

$$n_c = \frac{n}{\{\exp[(E_c - E_f)/kT] + 1\}}. \tag{4.53}$$

The number of electrons in the valence band is

$$n_v = np(E_v),$$

where $p(E_v)$ is the probability of an electron occupying the valence band

$$n_v = \frac{n}{\{\exp[(E_v - E_f)/kT] + 1\}} \tag{4.54}$$

Now $n = n_c + n_v$ and, therefore, from Eqs. (4.53) and (4.54),

$$n = \frac{n}{\{\exp[(E_c - E_f)/kT] + 1\}} + \frac{n}{\{\exp[(E_v - E_f)/kT] + 1\}}$$

or

$$1 = \left[\exp\left(\frac{E_c - E_f}{kT}\right) + 1\right]^{-1} + \left[\exp\left(\frac{E_v - E_f}{kT}\right) + 1\right]^{-1}.$$

Simplifying,

$$\exp\left(\frac{E_c + E_v - 2E_f}{kT}\right) + \exp\left(\frac{E_c - E_f}{kT}\right) + \exp\left(\frac{E_v - E_f}{kT}\right) + 1$$

$$= \exp\left(\frac{E_v - E_f}{kT}\right) + 1 + 1 + \exp\left(\frac{E_c - E_f}{kT}\right) \cdot \exp\left(\frac{E_c + E_v - 2E_f}{kT}\right),$$

or

$$\frac{E_c + E_v}{2} = E_f. \tag{4.55}$$

Equation (4.55) shows that in this simplified model the Fermi level E_f is located halfway between the valence and conduction bands. The position is independent of temperature. The number of electrons in the conduction band is

$$n_c = \frac{n}{\{\exp[(E_c - E_f)/kT] + 1\}} \tag{4.53}$$

The Fermi level according to Eq. (4.55) lies lower than the conduction band by half the energy gap $E_g/2$. In intrinsic semiconductors and insulators E_g may vary from 0.1 to 10 eV, and hence at room temperature, where $kT \simeq \frac{1}{40}$ eV, $E_c - E_f \gg kT$ and

$$\exp\left(\frac{E_c - E_f}{kT}\right) \gg 1.$$

Therefore,

$$n_c = n \exp\left(\frac{E_f - E_c}{kT}\right). \tag{4.56}$$

Substituting the value of E_f from Eq. (4.55) in Eq. (4.56),

$$n_c = n \exp\left(\frac{E_v - E_c}{2kT}\right) = n e^{-E_g/2kT}. \tag{4.57}$$

The number of electrons in the conduction band is proportional to $e^{-E_g/2kT}$, and hence the carrier concentration increases with an increase in temperature. Further, the concentration decreases with an increase in energy gap E_g at a temperature T. This shows that in insulators there are fewer electrons in the conduction band than in semiconductors.

Having gotten an approximate idea of the Fermi-level location and the dependence of carrier concentration, we proceed to calculate the number of electrons and holes in a more exact way.

Improved model. It was discussed in Chapter 2 that the number of electrons available between energy interval E and $E + dE$ is given by

$$dn = g(E)p(E)\,dE, \tag{2.29}$$

where $g(E)\,dE$ is the density of states in the energy interval E and $E + dE$ and $p(E)$ is the probability that a state of energy E is occupied. The conduction band does not have all states of the same energy. The energy is E_c at the bottom of the conduction band and extends to higher values of energy as we consider the top of the conduction band (Fig. 4.16). The

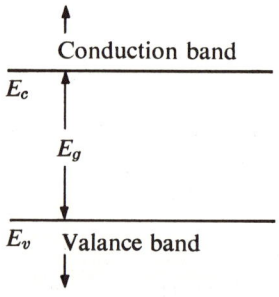

Fig. 4.16
Improved model of an insulator

INTRINSIC SEMICONDUCTORS AND INSULATORS

number of electrons in the conduction band should therefore be calculated by integrating Eq. (2.29) from E_c to the energy corresponding to the top of the conduction band:

$$n_c = \int_{E_c}^{top} g(E)p(E)\,dE$$

According to Eq. (2.27), the density of states for a potential free electron is

$$g(E)\,dE = 2 \times \frac{4\pi}{h^3} m^{3/2}(2E)^{1/2}\,dE. \tag{2.27}$$

The electron in the conduction band is not a free electron, but earlier in this section we showed that if the mass of the electron is replaced by the effective mass $m^* = \hbar^2/(d^2E/dK^2)$, an electron moving in a periodic potential could be treated as a free electron. Therefore, the density of states of electrons in the conduction band is

$$g(E)\,dE = 2\left(\frac{4\pi}{h^3}\right)2^{1/2}m_e^{*3/2}E^{1/2}dE. \tag{4.58}$$

If we consider the minimum energy of the electrons to be E_c and call m_e^* the effective mass of electrons in the conduction band,

$$g(E)\,dE = 2\left(\frac{4\pi}{h^3}\right)2^{1/2}(m_e^*)^{3/2}(E-E_c)^{1/2}\,dE. \tag{4.59}$$

Therefore,

$$dn = \frac{4\pi}{h^3}(2m_e^*)^{3/2}\frac{(E-E_c)^{1/2}\,dE}{\{\exp[(E-E_f)/kT]+1\}}, \tag{4.60}$$

and integrating,

$$n_c = \frac{4\pi}{h^3}(2m_e^*)^{3/2}\int_{E_c}^{\infty}\frac{(E-E_c)^{1/2}\,dE}{\{\exp[(E-E_f)/kT]+1\}}. \tag{4.61}$$

It should be noted that in Eq. (4.61) the upper limit of integration has been replaced by an infinite energy. This has been done to simplify the integration. It can be shown that the results obtained with an upper limit of infinite energy (calculated simply) hardly differ from those of actual upper limit (calculated in a complicated way). Assume

$$E_c - E_f \geq 4kT \approx 0.1 \text{ eV at room temperature}$$

i.e., assume the Fermi level is lower than the bottom of the conduction band by more than $4kT$. We have already found that in a simplified model $E_f = (E_c + E_v)/2$, and because the energy gaps are 0.1 to 10 eV for semiconductors and insulators, we can safely say that at room temperature the assumption is valid.

In Eq. (4.61), $E - E_f$ at the lower limit is $E_c - E_f$, and hence for the entire integration limit $E - E_f \geq 4kT$ and $\exp[(E - E_f)/kT] \gg 1$. Hence

$$n_c = \frac{4\pi}{h^3}(2m_e^*)^{3/2} \int_{E_c}^{\infty} (E - E_c)^{1/2} \exp\left(\frac{E_f - E}{kT}\right) dE. \quad (4.62)$$

To solve the integral (4.62), put

$$E = E_c + x,$$

$$n_c = \frac{4\pi}{h^3}(2m_e^*)^{3/2} e^{E_f/kT} \int_0^{\infty} x^{1/2} e^{-E_c/kT} e^{-x/kT} dx,$$

or

$$n_c = \frac{4\pi}{h^3}(2m_e^*)^{3/2} \exp\left(\frac{E_f - E_c}{kT}\right) \int_0^{\infty} x^{1/2} e^{-x/kT} dx. \quad (4.63)$$

It can be easily shown with the help of gamma functions that

$$\int_0^{\infty} x^{1/2} e^{-x/kT} dx = (kT)^{3/2} \frac{\pi^{1/2}}{2},$$

and, therefore,

$$n_c = 2\left(\frac{2\pi m_e^* kT}{h^2}\right)^{3/2} \exp\left(\frac{E_f - E_c}{kT}\right). \quad (4.64)$$

In an intrinsic semiconductor or insulator the number of electrons in the conduction band must equal the number of vacancies or holes in the valence band. To determine the position of the Fermi level according to this improved model, we shall calculate the number of holes statistically and shall equate it with the number of electrons in the conduction band, n. Therefore,

$$n_c = n = 2\left(\frac{2\pi m_e^* kT}{h^2}\right)^{3/2} \exp\left(\frac{E_f - E_c}{kT}\right). \quad (4.65)$$

The probability that a state of energy E is unoccupied in the conduction band is

$$1 - p(E) = 1 - \left[\exp\left(\frac{E - E_f}{kT}\right) + 1\right]^{-1}$$

or

$$1 - p(E) = \frac{\exp(E - E_f/kT)}{1 + \exp(E - E_f/kT)}.$$

Assuming the Fermi level lies at an energy distance $\geq 4kT$ from E_v, the top of the valence band is

$$\exp\left(\frac{E - E_f}{kT}\right) \ll 1$$

and hence

$$1 - p(E) = \exp\left(\frac{E - E_f}{kT}\right).$$

INTRINSIC SEMICONDUCTORS AND INSULATORS

The number of holes or vacancies in the energy interval E and $E + dE$ in the valence band is

$$dp = g(E)[1 - p(E)]\, dE$$

Taking the effective mass of holes m_h^*,

$$dp = \frac{4\pi}{h^3}(2m)_h^{*3/2} E^{1/2}\, dE \exp\left(\frac{E - E_f}{kT}\right).$$

The highest vacant state can have energy corresponding to the top of the valence band. The lowest energy for the sake of mathematical simplification may be taken as $-\infty$. Counting the energy values from E_v,

$$p = \int_{-\infty}^{E_v} \frac{4\pi}{h^3}(2m_h^*)^{3/2}(E_v - E)^{1/2} \exp\left(\frac{E - E_f}{kT}\right) dE. \qquad (4.66)$$

Equation (4.66) can be solved with the help of gamma functions, and

$$p = 2\left(\frac{2\pi m_h^* kT}{h^2}\right)^{3/2} \exp\left(\frac{E_v - E_f}{kT}\right). \qquad (4.67)$$

Equation (4.67) gives the number of holes or vacancies in the valence band. In intrinsic semiconductors and insulators $n = p$; therefore, equating the right sides of Eqs. (4.65) and (4.67),

$$2\left(\frac{2\pi m_e^* kT}{h^2}\right)^{3/2} \exp\left(\frac{E_f - E_c}{kT}\right) = 2\left(\frac{2\pi m_h^* kT}{h^2}\right)^{1/2} \exp\left(\frac{E_v - E_f}{kT}\right). \qquad (4.68)$$

Simplifying Eq. (4.68),

$$E_f = \frac{E_c + E_v}{2} + \frac{3}{4} kT \ln\left(\frac{m_h^*}{m_e^*}\right), \qquad (4.69)$$

which shows that if $m_h^* = m_e^*$,

$$E_f = \frac{E_c + E_v}{2}.$$

From equation (4.69) it can be calculated that

1. The simplified model was not far from reality. Although we assumed a single energy in the bands, it turns out that the effect of the effective mass is the only difference.
2. If $m_h^* \neq m_e^*$, the Fermi level is a function of temperature. In general $m_h^* > m_e^*$, so the Fermi level is raised slightly with temperature.

From Eqs. (4.65) and (4.67),

$$np = 4\left(\frac{2\pi kT}{h^2}\right)^3 (m_e^* m_h^*)^{3/2} \exp\left(\frac{E_v - E_c}{kT}\right).$$

Because $n = p$ in our case, we assume

$$n = p = n_i,$$

where n_i is called the *intrinsic carrier concentration*.

$$n_i^2 = 4\left(\frac{2\pi kT}{h^2}\right)^3 (m_e^* m_h^*)^{3/2} e^{-E_g/kT}.$$

Therefore,

$$n_i = 2\left(\frac{2\pi kT}{h^2}\right)^{3/2} (m_e^* m_h^*)^{3/4} e^{-E_g/2kT}, \qquad (4.70)$$

where E_g is the energy gap.

The electrical conductivity is

$$\sigma = (ne\mu_e + pe\mu_h). \qquad (4.52)$$

In the case of an intrinsic semiconductor,

$$\sigma = n_i e(\mu_e + \mu_h),$$

and substituting the value of n_i from Eq. (4.70),

$$\sigma_i = 2e\left(\frac{2\pi kT}{h^2}\right)^{3/2} (m_e^* m_h^*)^{3/4} e^{-E_g/2kT} (\mu_e + \mu_h). \qquad (4.71)$$

Equation (4.71) shows that the electric conductivity of an intrinsic semiconductor or insulator at a given temperature depends upon the negative exponential of the forbidden energy gap between the valence and conduction band. The room-temperature energy gap of some important materials is given below:

Material	E_g(eV)	Material	E_g(eV)
Diamond	6		
CdS	2.42	Ge	0.67
GaAs	1.4	InAs	0.33
InP	1.25	InSb	0.18
Si	1.1	PbTe	0.30

The intrinsic carrier concentration at room temperature is much smaller in diamond, with energy gap 6 eV, than in InSb, with energy gap 0.18 eV.

Equation (4.71) further shows that besides the energy gap the intrinsic electrical conductivity of a semiconducting material depends upon the mobility of both holes and electrons. The mobility in a pure semiconductor or insulator is determined by the interaction of the electron with lattice waves or phonons. According to Seitz, Kittel, and others, in such a case

INTRINSIC SEMICONDUCTORS AND INSULATORS

μ_e and μ_h are both proportional to $T^{-3/2}$. Incorporating the temperature dependence of mobility in Eq. (4.71),

$$\sigma_i = A e^{-E_g/2kT}, \tag{4.72}$$

where A is a constant.

Taking logarithms of both sides in Eq. (4.72),

$$\ln \sigma_i = \ln A - \frac{E_g}{2kT}. \tag{4.73}$$

Equation (4.73) suggests to us a method of determining the energy gap of an intrinsic material. A measurement of the intrinsic conductivity of the material at various temperatures helps us plot the results on a semilogarithmic paper. The slope of the curve is $E_g/2k$, and hence the energy gap can be determined (see Fig. 4.17).

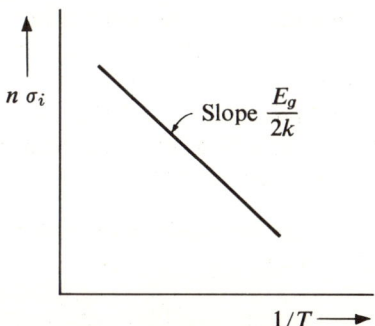

Fig. 4.17
Logarithm of the intrinsic conductivity conductivity versus $1/T$

Referring back to Eqs. (4.65) and (4.67), it is customary and conventional to define

$$N_C = 2\left(\frac{2\pi m_e^* kT}{h^2}\right)^{3/2},$$

$$N_V = 2\left(\frac{2\pi m_h^* kT}{h^2}\right)^{3/2}.$$

The Fermi level for an intrinsic semiconductor is denoted E_I. Hence for an intrinsic material,

$$n = N_C \exp\left(\frac{E_I - E_c}{kT}\right), \tag{4.74}$$

$$p = N_V \exp\left(\frac{E_V - E_I}{kT}\right). \tag{4.75}$$

The intrinsic carrier density n_i may be expressed

$$n_i^2 = np = N_c N_V \exp\left(\frac{E_V - E_c}{kT}\right), \qquad (4.76)$$

$$n_i = (N_c N_V)^{1/2} e^{-E_g/2kT}. \qquad (4.77)$$

The intrinsic Fermi-level equation (4.69) may be rewritten

$$E_I = \frac{E_c + E_V}{2} + kT\left[\ln\left(\frac{N_V}{N_C}\right)^{1/2}\right]. \qquad (4.78)$$

Equations (4.74) and (4.78) are very useful and are used later to correlate the Fermi levels of intrinsic and extrinsic semiconductors.

4.6 Extrinsic semiconductors

In intrinsic semiconductors the carrier concentration of both holes and electrons at normal temperatures such as room temperature is very low. The electric conductivity is very low, and hence for an appreciable current density through the semiconductor, the applied electric field must be very large. The following example shows the carrier concentrations in the case of germanium at room temperature, 300°K.

Carrier concentration in intrinsic germanium at room temperature. According to Eq. (4.74),

$$n = N_C \exp\left(\frac{E_I - E_c}{kT}\right),$$

$$N_C = 2\left(\frac{2\pi m_e^* kT}{h^2}\right)^{3/2}.$$

Assume $m_e^* = m_h^* = m = 9.1 \times 10^{-31}$ Kg. This assumption would not affect the results, because the order of magnitude of n is decided by the exponential factor in Eq. (4.74). m is the rest mass of the electron, $k = 1.38 \times 10^{-23}$ J/°K, and $h = 6.625 \times 10^{-34}$ J sec. Therefore,

$$N_C = 2\left(\frac{2\pi \times 9.1 \times 10^{-31} \times 1.38 \times 10^{-23}}{6.625 \times 6.625 \times 10^{-68}}\right)^{3/2} T^{3/2},$$

or

$$N_C = 2(2.4) \times 10^{21} T^{3/2}/\text{m}^3 = 4.8 \times 10^{21} T^{3/2}/\text{m}^3.$$

At 300°K,

$$N_C = 2.5 \times 10^{25} \text{m}^{-3}.$$

Because $m_e^* = m_h^* = m$,

$$N_C = N_V = 2.5 \times 10^{25} \text{m}^{-3}.$$

To determine n or p or n_i, a knowledge of E_I is necessary. Because $m_h^* = m_e^*$,

$$E_I = \frac{E_c + E_V}{2}.$$

Assuming $E_V = 0$,

$$E_g = 0.68 \text{ eV}, \quad E_c = 0.68 \text{ eV}, \quad E_I = 0.34 \text{ eV}$$

and

$$n = 2.5 \times 10^{25} \exp\left(-\frac{0.34}{1/40}\right), \quad kT(\text{at } 300°K) = \frac{1}{40} \text{ eV}.$$

Therefore,

$$n = \begin{cases} 2.5 \times 10^{25} e^{-13.6} = 5.6 \times 10^{19} \text{m}^{-3} \\ p = n_i = 5.6 \times 10^{19} \text{m}^{-3} \end{cases}$$

The mobility values are

$$\mu_e = 0.38 \text{ m}^2/\text{V sec}, \quad \mu_h = 0.18 \text{ m}^2/\text{V sec}.$$

Therefore,

$$\sigma_i = n_i e(\mu_h + \mu_e) = 5.6 \times 10^{19} \times 1.6 \times 10^{19}(0.56) = 5(\Omega \text{ m})^{-1},$$

$$\rho_i = 20(\Omega \text{ cm}) = \tfrac{1}{5}\Omega \text{ m}.$$

To obtain a current density of 10 A/cm² or 10^5/m²,

$$J = \sigma_i E,$$

$$E = \frac{10^5}{5} = 2 \times 10^4 \text{ V/m or } 200 \text{ V/cm}.$$

The above calculations show that in the case of intrinsic germanium a voltage of 200 V is required to pass a current of 10 A, when the specimen is 1 cm² in cross section and 1 cm in length. In the following it will be shown that addition of impurities in intrinsic semiconductors enhances the electrical conductivity and makes them more useful devices. Addition of boron to intrinsic or pure silicon in the proportion of 1 boron atom to 10^5 silicon atoms increases the conductivity of pure silicon by a factor of 10^3 at room temperature. The reason for this increase in conductivity lies mainly in the increase in carrier concentrations.

N-type semiconductors. Let us consider single crystals of semiconductor such as silicon and germanium. Each semiconductor atom has four valence electrons, with which it forms four electron pair bonds with four nearest neighbors. The electron pair bonding is called *covalent bonding*.

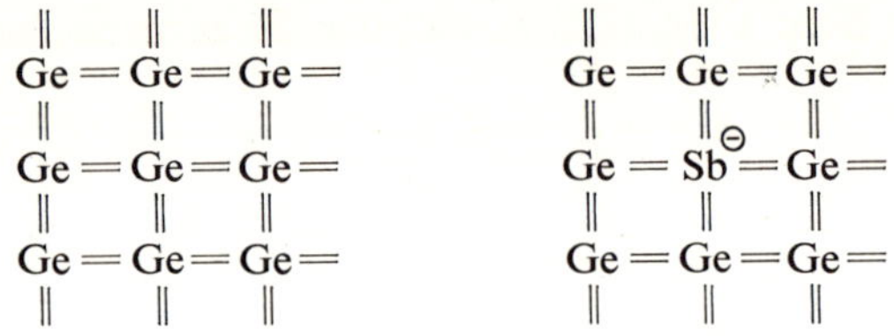

140 METALS, SEMICONDUCTORS, AND INSULATORS

```
 ||    ||    ||              ||    ||    ||
Ge = Ge = Ge =              Ge = Ge = Ge =
 ||    ||    ||              ||    ||⊖   ||
Ge = Ge = Ge =              Ge = Sb = Ge =
 ||    ||    ||              ||    ||    ||
Ge = Ge = Ge =              Ge = Ge = Ge =
 ||    ||    ||              ||    ||    ||

      (a)                         (b)
```

Fig. 4.18
(a) Covalent bonding in pure germanium. (b) Charges associated with impurity atoms in germanium

Figure 4.18(a) shows the covalent bonding in pure germanium. Now if an impurity atom of five valence electrons such as antimony, arsenic, or phosphorus is substituted in the lattice for the parent atom, after the electrons used in covalent bonding one valence electron from the impurity atom will be left over [Fig. 4.18(b)]. X rays and other studies indicate that if impurities are diffused or alloyed, the impurity atoms enter the lattice in place of a parent atom rather than occupying an interstitial position. The extra electron in the antimony or, for that matter, in any five valence atoms can be set free thermally with a rather low energy. In that case the antimony atom becomes positively ionized and an electron becomes available for conduction purposes. It may be mentioned here that the extra electron in the antimony atom is necessary for the neutrality of the atom but unnecessary for the covalent bonding. Impurity atoms that may be ionized to give up an electron are called *donors*. Semiconducting materials that have an excess of these types of impurities are called *N*-type materials. The excess electron in the pentavalent impurity atom is bound to the nucleus of the atom at low temperatures. The energy necessary to ionize the electron from the impurity atom can be calculated in a simple way by the Bohr model. According to this model the electron is assumed to be circling around the positive charge such that the centrifugal force equals the Coulomb force of attraction. Therefore,

$$\frac{m^*v^2}{r} = \frac{e^2}{4\pi\epsilon_0\epsilon_r r^2}, \qquad (4.79)$$

where m^* is the effective mass of the electron, v is the velocity of the electron, and ϵ_r is the dielectric constant of material (semiconductor), because the electron moves in the atmosphere of surrounding atoms of semiconductors.

According to Bohr's hypothesis, the angular momentum is quantized, and, therefore,

$$m^*vr = \frac{nh}{2\pi} = n\hbar. \quad (4.80)$$

The total energy of the electron is

$$E = \text{P.E.} + \text{K.E.},$$

$$\text{P.E.} = \int_\infty^r \frac{e^2}{4\pi\epsilon_0\epsilon_r r^2} dr = -\frac{e^2}{4\pi\epsilon_0\epsilon_r r},$$

$$\text{K.E.} = \tfrac{1}{2}m^*v^2 = \frac{e^2}{8\pi\epsilon_0\epsilon_r r}.$$

Eliminating v from Eqs. (4.79) and (4.80),

$$\frac{n^2\hbar^2}{m^*r^3} = \frac{e^2}{4\pi\epsilon_0\epsilon_r r^2}$$

or

$$r = \frac{4\pi\epsilon_0\epsilon_r n^2\hbar^2}{m^*e^2}. \quad (4.81)$$

Now

$$E = \frac{e^2}{8\pi\epsilon_0\epsilon_r r} - \frac{e^2}{4\pi\epsilon_0\epsilon_r r} = -\frac{e^2}{8\pi\epsilon_0\epsilon_r r}. \quad (4.82)$$

Substituting the value of r from Eq. (4.81) in Eq. (4.82),

$$E = -\frac{m^*e^4}{8\epsilon_0^2\epsilon_r^2 n^2 h^2}. \quad (4.83)$$

The relative dielectric constants for germanium and silicon are 15.8 and 11.7, respectively. If the effective mass of the excess electron is assumed to be 0.2, the rest mass of an electron and $n = 1$ the ground state of the electron, the ionization energy of the electron for the atom is

$$E_{\text{ionization}}(\text{germanium}) = +\frac{13.6 \times 0.2}{(15.8)^2} = 0.0105 \text{ eV},$$

$$E_{\text{ionization}}(\text{silicon}) = \frac{13.6 \times 0.2}{(11.7)^2} = 0.0191 \text{ eV}.$$

The ionization energies for the pentavalent excess electrons are 0.0105 and 0.0191 eV in the case of germanium and silicon semiconductors, respectively. The thermal energies required to ionize the electrons from pentavalent impurities are much smaller than the forbidden energy gap of the semiconductors. Even at low temperatures it is expected that most of the impurity atoms will contribute to electrons in the conduction band, and hence the electrical conductivity of an *N*-type material is more than the intrinsic material. The increase in conductivity depends upon the concentration of impurity atoms.

P-type semiconductors. Let us once again consider single crystals of semiconductors such as silicon and germanium. Now if an impurity atom of three valence electrons is substituted in the lattice in place of the parent atom, it is an electron short for covalent bonding. Typical examples of such trivalent atoms are boron, aluminum, gallium, and indium. It can be easily shown that the energy necessary for an electron to fill in the covalent bonding is very small. An electron from the parent atom with a very low energy might be accepted by the impurity atom. The impurity atoms are called *acceptors* (Fig. 4.19). When electrons are accepted by

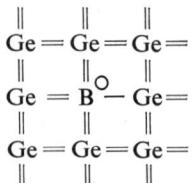

Fig. 4.19
Charges associated with impurity atoms in germanium, positive holes

acceptors, they become negatively ionized atoms, because they would have a surplus electron after accepting the electron from the parent atom. In the parent atom a hole is left behind. Because the energy required by electrons to be accepted by acceptors is low, even at low temperatures it is to be expected that most of the trivalent impurity atoms would accept electrons and leave behind an equal number of holes in the valence band. Holes contribute to conduction of electrical current in the presence of applied electric fields. Hence the electrical conductivity of a *P*-type material is more than intrinsic material. As in *N*-type material, the increase in conductivity depends upon the concentration of acceptor atoms.

Energy diagram of extrinsic semiconductors. At $T = 0$ in an intrinsic semiconductor the conduction band is completely filled and the valence band is completely empty. Pentavalent impurities are introduced in the semiconductors. The quantum state of the electron in a donor atom

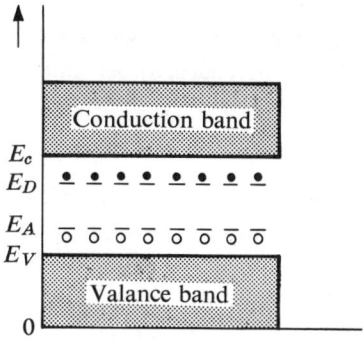

Fig. 4.20
Energy diagram for impurity semiconductor

is located a little below the conduction, because only a small amount of energy is required to free them or, in other words, put them in the conduction band. Such a level is called a *donor level*. Because the donor level has a specific energy as against bands, the donor states are discrete states. At $T = 0$ all electrons in the donor levels are occupying their positions with the impurity atoms. The energy state exists only in the vicinity of the impurity atom, so it is called a *localized state*. Figure 4.20 shows these states. At $T > 0$ some of the electrons leave these states and go over to the conduction band. E_D is the energy of a donor level.

In a similar way, in a semiconductor containing acceptors we would have acceptor levels E_A lying near the valence band. At $T = 0$ no acceptor level is occupied by electrons from valence bands. At $T > 0$ some of the electrons leave the valence band and go over to the acceptor levels. E_A is the energy of an acceptor level.

Statistics of an extrinsic semiconductor. Consider a semiconductor with both donor and acceptor levels. Referring to Fig. 4.20, let the bottom of the valence band be the reference energy. As the temperature of the semiconductor is increased, electrons are first excited from the donor levels to conduction bands and from the valence bands to acceptor levels. The electrons leaving the donor levels leave behind positively ionized donors and, similarly, electrons leaving the valence band make the acceptor atoms negatively charged atoms and leave behind holes in the valence band. As the temperature is further raised, electrons are able to go over to the conduction band directly from the valence band.

Let n be the number of electrons in the conduction band per unit volume, p the number of holes in the valence band per unit volume, n_A the number of ionized acceptors (negatively charged) per unit volume, and p_D the number of ionized donors (positively charged) per unit volume. Because of the charge neutrality at any temperature T,

$$n + n_A = p + p_D. \tag{4.84}$$

In Eq. (4.84) the quantities of interest for electrical conduction are n and p. We shall proceed to calculate the two quantities n and p statistically. If N_A is the number of acceptor atoms per unit volume and N_D the number of donor atoms per unit volume,

$$n_A = N_A p(E_A),$$

where $p(E_A)$ is the probability of occupation of an electron, the state of energy E_A.

From Eq. (2.33),

$$p(E_A) = \left[\exp\left(\frac{E_A - E_f}{kT}\right) + 1\right]^{-1}, \tag{2.33}$$

and therefore

$$n_A = N_A \left[\exp\left(\frac{E_A - E_f}{kT}\right) + 1\right]. \tag{4.85}$$

Similarly,

$$p_D = N_D[1 - p(E_D)],$$

where $[1 - p(E_D)]$ is the probability that an electron has left the energy state E_D.

Applying Eq. (2.33),

$$1 - p(E_D) = 1 - \left[\exp\left(\frac{E_D - E_f}{kT}\right) + 1\right]^{-1}$$

or

$$1 - p(E_D) = \frac{\exp(E_D - E_f/kT)}{1 + \exp(E_D - E_f/kT)} = \left[\exp\left(\frac{E_f - E_D}{kT}\right) + 1\right]^{-1};$$

hence

$$p_D = \frac{N_D}{[\exp(E_f - E_D/kT) + 1]}. \tag{4.86}$$

In Eqs. (4.85) and (4.86) E_f is the Fermi level.

Calculation of n. The number of electrons in the conduction band can be carried out in a manner similar to intrinsic semiconductors. The number of electrons in an energy interval between E and $E + dE$ in the conduction band is

$$dn = g(E)p(E)\,dE,$$

$$g(E)\,dE = \frac{4\pi}{h^3}(2m_e^*)^{3/2}(E - E_c)^{1/2}\,dE.$$

$(E - E_c)^{1/2}$ has been introduced in place of $E^{1/2}$ in the above equation because the minimum energy of an electron in the conduction band is E_c:

$$p(E) = \left[\exp\left(\frac{E - E_f}{kT}\right) + 1\right]^{-1}$$

and therefore

$$dn = \frac{4\pi}{h^3}(2m_e^*)^{3/2}(E - E_c)^{1/2}\left[\exp\left(\frac{E - E_f}{kT}\right) + 1\right]^{-1} dE.$$

If we integrate the above expression over the entire band using ∞ as the upper limit, we obtain n:

$$n = \int_{E_c}^{\infty} \frac{4\pi}{h^3}(2m_e^*)^{3/2}(E - E_c)^{1/2}\left[\exp\left(\frac{E - E_f}{kT}\right) + 1\right]^{-1} dE. \quad (4.87)$$

We shall assume that $E - E_f \geq 2kT$, bearing in mind that we shall discuss the assumption in the next subsection.

Using the approximation

$$\exp\left(\frac{E - E_f}{kT}\right) + 1 \simeq \exp\left(\frac{E - E_f}{kT}\right)$$

in the above integral one gets

$$n = 2\left(\frac{2\pi m_e^* kT}{h^2}\right)^{3/2} \exp\left(\frac{E_f - E_c}{kT}\right) \quad (4.88)$$

Calculation of p. The number of holes in the valence band can also be carried out in a manner similar to intrinsic semiconductors. The number of holes in an energy interval between E and $E + dE$ in the valence band is

$$dp = g(E)[1 - p(E)] dE,$$

where $1 - p(E)$ is the probability that there is a vacancy or hole of energy E and

$$g(E)\, dE = \frac{4\pi}{h^3}(2m_h^*)^{3/2}(E - E_v)^{1/2} dE.$$

$(E - E_v)^{1/2}$ has been introduced in place of $E^{1/2}$ in the above equation because the maximum energy of a vacancy or hole is E_v.
m_h^* is the effective mass of the hole. Therefore,

$$dp = \frac{4\pi}{h^3}(2m_h^*)^{3/2}(E - E_v)^{1/2}\left[1 + \exp\left(\frac{E_f - E}{kT}\right)\right]^{-1}.$$

We shall assume that $E_f - E \geq 2kT$ for the entire integration limit from

$-\infty$ to E_v, bearing in mind that we shall discuss the assumption in the next subsection. Hence

$$p = \int_{-\infty}^{E_v} \frac{4\pi}{h^3} (2m_h^*)^{3/2}(E - E_v)^{1/2} \exp\left(-\frac{E_f - E}{kT}\right) dE$$

Therefore,

$$p = 2\left(\frac{2\pi m_h^* kT}{h^2}\right)^{3/2} \exp\left(\frac{E_v - E_f}{kT}\right). \tag{4.89}$$

It may be shown from Eqs. (4.85), (4.86), (4.88), and (4.89) that for charge neutrality Eq. (4.84) may be rewritten as Eq. (4.90),

$$2\left(\frac{2\pi m_h^* kT}{h^2}\right)^{3/2} \exp\left(\frac{E_f - E_c}{kT}\right) + \frac{N_A}{\{\exp[(E_A - E_f)/kT] + 1\}}$$

$$= \frac{N_D}{\{\exp[(E_f - E_D)/kT] + 1\}} + 2\left(\frac{2\pi m_h^* kT}{h^2}\right)^{3/2} \times \exp\left(\frac{E_v - E_f}{kT}\right). \tag{4.90}$$

In Eq. (4.90) the only unknown term is E_f, the Fermi level. A solution of the equation for E_f will be taken up below.

Fermi level and its variation in semiconductors. The discussion of Fermi level in Sec. 2.4 cannot be carried over in the case of semiconductors because of energy-band formation there. However, it is of general nature that the probability of occupation of a state of energy E by an electron in a semiconductor or metal or for that matter any material is

$$p(E) = \left[\exp\left(\frac{E - E_f}{kT}\right) + 1\right]^{-1}. \tag{2.33}$$

In the case of semiconductors we have already discussed that in intrinsic semiconductors the Fermi level E_I is given by the expression (4.69),

$$E_I = \frac{E_c + E_v}{2} + \frac{3}{4} kT \ln\left(\frac{m_h^*}{m_e^*}\right). \tag{4.69}$$

In the expression (4.69), because m_h^* is not very much different than m_e^*, the intrinsic Fermi level increases with temperature, but the increase is rather little. On the other hand, in extrinsic semiconductors the Fermi level depends strongly on the concentration and temperature. Before discussion of a general expression for the Fermi level we shall take up some special cases.

N-type material at very low temperature. Let us consider an N-type material, which means that either there are no acceptor atoms present in the material, or if there are some, their number is much smaller than the number of donor atoms. At $T = 0$ the valence band and the donor levels are all filled and the conduction band is empty. At $T > 0$ but still low, some of the electrons from the donor states are thermally excited in the conduction band, leaving behind ionized donors. We are assuming very low temperatures, so it would be justified that practically no electron gets sufficient energy to be excited from the valence band into the conduction band. For simplicity assume that all electrons in the conduction band have the same energy E_c. The number of electrons in the conduction band per unit volume is

$$n = N_D p(E_c),$$

where N_D is the number of donor atoms per unit volume. $p(E_c)$ is the probability of finding an electron in the conduction band:

$$p(E_c) = \left[\exp\left(\frac{E_c - E_f}{kT}\right) + 1\right]^{-1},$$

where E_f is the location of the Fermi level in this N-type material at very low temperatures. Hence

$$n = \frac{N_D}{\{\exp\left[(E_c - E_f)/kT\right] + 1\}}$$

E_c is greater than E_f because of the definition of E_f and because the temperature is very low:

$$\exp\left(\frac{E_c - E_f}{kT}\right) \gg 1$$

hence

$$n = N_D \exp\left(\frac{E_f - E_c}{kT}\right). \tag{4.91}$$

The number of ionized donors per unit volume p_D is

$$p_D = N_D[1 - p(E_D)],$$

where $1 - p(E_D)$ means that an electron is missing from an energy state E_D:

$$1 - p(E_D) = 1 - \left[\exp\left(\frac{E_D - E_f}{kT}\right) + 1\right]^{-1} = \frac{\exp\left[(E_D - E_f)/kT\right]}{1 + \exp\left[(E_D - E_f)/kT\right]}$$

$$= \left[\exp\left(\frac{E_f - E_D}{kT}\right) + 1\right]^{-1}.$$

Because $E_f > E_D$ and the temperature is low,

$$1 - p(E_D) = \exp\left(\frac{E_D - E_f}{kT}\right).$$

Hence

$$p_D = N_D \exp\left(\frac{E_D - E_f}{kT}\right). \tag{4.92}$$

The number of electrons in the conduction band must be equal to the number of ionized donors in this case. Hence

$$n = p_D,$$

or from Eqs. (4.91) and (4.92),

$$N_D \exp\left(\frac{E_f - E_c}{kT}\right) = N_D \exp\left(\frac{E_D - E_f}{kT}\right)$$

or

$$E_f = \frac{E_c + E_D}{2}. \tag{4.93}$$

The Fermi level at very low temperatures in an N-type material lies halfway between the conduction-band lower edge and the donor energy levels. It may be mentioned that the calculation has been carried out with all energy states in the conduction band E_c of the same energy. It has been shown in a parallel case of the calculation of the Fermi level of an intrinsic semiconduction that a finite width of the conduction band hardly affects the position of the Fermi level.

N-type material at very high temperature. As the temperature is gradually increased from a low temperature, the contribution of electrons in the conduction band from the valence band increases and at very high temperatures (500°K in Ge) far exceeds the donor contribution to the conduction band even if the number of donor atoms per m³ is 10^{21}. With

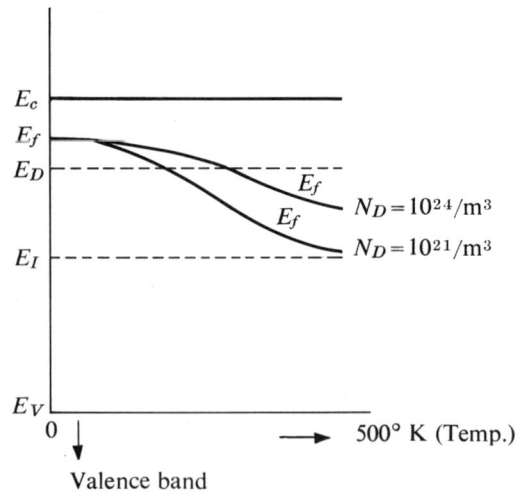

Fig. 4.21
Fermi level as a function of temperature for an N-type semiconductor

higher donor concentrations the intrinsic behavior predominates at higher temperatures. Figure 4.21 shows this behavior.

Variation of Fermi level with temperature and concentration of impurity atoms in N-type material. The Fermi level at any temperature T and donor concentration N_D in an N-type material can be determined by solving Eq. (4.90) for the case of an N-type material. Rewriting Eq. (4.90),

$$2\left(\frac{2\pi m_e^* kT}{h^2}\right)^{3/2} \exp\left(\frac{E_f - E_c}{kT}\right) + \frac{N_A}{\{\exp[(E_A - E_f)/kT] + 1\}}$$
$$= \frac{N_D}{\{\exp[(E_f - E_D)/kT] + 1\}} + 2\left(\frac{2\pi m_h^* kT}{h^2}\right)^{3/2} \times \exp\left(\frac{E_v - E_f}{kT}\right). \quad (4.90)$$

In an N-type material, assuming $N_A = 0$, i.e., no acceptors are present,

$$2\left(\frac{2\pi m_e^* kT}{h^2}\right)^{3/2} = N_c,$$

$$2\left(\frac{2\pi m_h^* kT}{h^2}\right)^{3/2} = N_v,$$

and hence

$$N_c \exp\left(\frac{E_f - E_c}{kT}\right) = \frac{N_D}{\{\exp[(E_f - E_D)/kT] + 1\}} + N_v \exp\left(\frac{E_v - E_f}{kT}\right). \quad (4.94)$$

A graphical solution of Eq. (4.94) for E_f would give us the Fermi level at any temperature T and donor concentration N_D. It would be simple to solve Eq. (4.94) for a special case. We assume that the temperature is not very high, and hence in an N-type material $n \gg p$. It will be shown that for all semiconductors

$$np = n_i^2,$$

and hence if n is very large, p is very small.

$$p = N_v \exp\left(\frac{E_v - E_f}{kT}\right) \to 0 \text{ as compared to } n \text{ or } \exp\left(\frac{E_v - E_f}{kT}\right)$$
$$\ll \exp\left(\frac{E_f - E_c}{kT}\right).$$

Therefore,

$$N_c \exp\left(\frac{E_f - E_c}{kT}\right) = \frac{N_D}{\{\exp[(E_f - E_D)/kT] + 1\}} \quad (4.95)$$

$$N_D = N_c \exp\left(\frac{2E_f - E_c - E_D}{kT}\right) + N_c \exp\left(\frac{E_f - E_c}{kT}\right)$$

or

$$e^{2E_f/kT}\left\{N_c \exp\left[-\frac{(E_c + E_D)}{kT}\right]\right\} + e^{E_f/kT}(N_c e^{-E_c/kT}) - N_D = 0. \quad (4.96)$$

Equation (4.96) is a quadratic equation in $e^{E_f/kT}$, and hence

$$e^{E_f/kT} = \frac{-N_c e^{-E_c/kT} \pm \{N_c^2 e^{-2E_c/kT} + 4N_c N_D \exp[-(E_D + E_c)/kT]\}^{1/2}}{2N_c \exp[-(E_D + E_c)/kT]}.$$

Simplifying,

$$e^{E_f/kT} = \frac{-1 \pm \{1 + (4N_D/N_c) \exp[(E_c - E_D)/kT]\}^{1/2}}{2e^{-E_D/kT}}.$$

Since the exponential cannot be negative, only the plus sign would be used for the root. Therefore,

$$e^{E_f/kT} = \frac{-1 + \{1 + (4N_D/N_c) \exp[(E_c - E_D)/kT]\}^{1/2}}{2e^{-E_D/kT}}. \quad (4.97)$$

E_f can be evaluated in two limits, still bearing in mind that the temperature is not too high.

1. Suppose

$$\frac{4N_D}{N_c} \exp\left(\frac{E_c - E_D}{kT}\right) \ll 1$$

corresponding to low N_D or high T, but not too high. Therefore,

$$e^{E_f/kT} = \frac{-1 + 1 + \tfrac{1}{2}(4N_D/N_c) \exp[(E_c - E_D)/kT]}{2e^{-E_D/kT}},$$

$$e^{E_f/kT} = \frac{N_D}{N_c} e^{E_c/kT},$$

or

$$E_f \cong E_c + kT \ln \frac{N_D}{N_c}, \quad (4.98)$$

and so from Eqs. (4.98) and (4.88),

$$n = N_c \exp\left(\frac{E_f - E_c}{kT}\right) = N_c e^{-E_c/kT} e^{E_f/kT}$$

or

$$n = N_c e^{-E_c/kT} \frac{N_D}{N_c} e^{E_c/kT} = N_D. \quad (4.99)$$

Equation (4.99) shows that the above assumption means that the total number of electrons in the conduction band equals the donor-atom concentration, or all the donor atoms are ionized and there is no contribution to electronic conduction, or there are no electrons excited from the valence band.

2. Suppose

$$\frac{4N_D}{N_c} \exp\left(\frac{E_c - E_D}{kT}\right) \gg 1$$

corresponding to large N_D or low temperature T. Hence

$$e^{E_f/kT} = \left(\frac{N_D}{N_c}\right)^{1/2} \frac{\exp\left[(E_c - E_D)/2kT\right]}{e^{-E_D/kT}}$$

or

$$e^{E_f/kT} = \left(\frac{N_D}{N_c}\right)^{1/2} \exp\left(\frac{E_c + E_D}{2kT}\right). \quad (4.100)$$

From Eqs. (4.100) and (4.88),

$$n = N_c \exp\left(\frac{E_f - E_c}{kT}\right) = N_c e^{-E_c/kT} \left(\frac{N_D}{N_c}\right)^{1/2} \exp\left(\frac{E_c + E_D}{2kT}\right)$$

or

$$n = (N_D N_c)^{1/2} \exp\left(\frac{E_D - E_c}{2kT}\right). \quad (4.101)$$

$E_D - E_c$ is the ionization energy of the donor. The concentration of electron varies as the square root of donor-atom concentration.

It might be mentioned once again that an exact value of E_f in an N-type material is determined by solving Eq. (4.94). If, however, the material is P-type, a similar procedure is followed with $N_D = 0$ in Eq. (4.90).

P-type material. Figure 4.22 shows the variation of Fermi level with temperature and concentration N_A for a P-type material. If $T = 0$,

$$E_f = \frac{E_v + E_A}{2},$$

and at $T \to$ large,

$$E_f = E_I.$$

The discussion of the Fermi level will be concluded by stating that it is

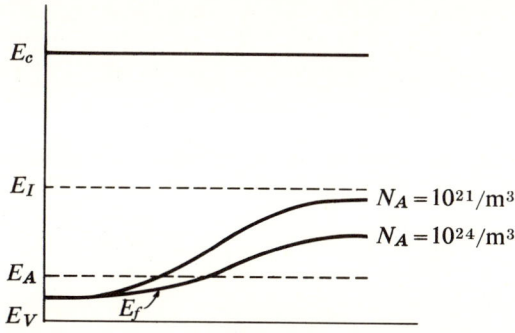

Fig. 4.22
Fermi level as a function of temperature for a P-type semiconductor

a function of both the concentration of impurities and the temperature of the semiconductor, and its exact determination results from Eq. (4.90).

Further discussion on extrinsic semiconductors. The concentration of electrons and holes for extrinsic semiconductors is given by Eqs. (4.88) and (4.89). The Fermi level in the two equations depends upon the concentration of impurities and the temperature. For an extrinsic semiconductor,

$$n = 2\left(\frac{2\pi m_e^* kT}{h^2}\right)^{3/2} \exp\left(\frac{E_f - E_c}{kT}\right) \tag{4.88}$$

or

$$n = N_c \exp\left(\frac{E_f - E_c}{kT}\right). \tag{4.102}$$

Similarly,

$$p = 2\left(\frac{2\pi m_h^* kT}{h^2}\right)^{3/2} \exp\left(\frac{E_v - E_f}{kT}\right) \tag{4.89}$$

or

$$p = N_v \exp\left(\frac{E_v - E_f}{kT}\right). \tag{4.103}$$

Multiplying Eqs. (4.102) and (4.103),

$$np = N_c N_v \exp\left(\frac{E_v - E_c}{kT}\right). \tag{4.104}$$

The right side of Eq. (4.104) is the same as Eq. (4.76), showing that for an extrinsic semiconductor the product of electron and hole concentrations is the same as the square of the charge carrier density if the semiconductor is an intrinsic semiconductor:

$$n_i^2 = np. \tag{4.105}$$

Equation (4.105) is an expression of the law of mass action, stating that if n is increased by a certain factor, p must be reduced by a reciprocal factor.

Carrier charge densities in terms of intrinsic carrier density and intrinsic Fermi level E_I. From Eq. (4.78) the intrinsic Fermi level is

$$E_I = \frac{E_c + E_v}{2} + kT \ln\left(\frac{N_v}{N_c}\right)^{1/2}. \tag{4.78}$$

Subtracting E_c from both sides of the above equations and transposing E_I on the right side,

$$-E_c = -E_I + \frac{E_v - E_c}{2} + \frac{kT}{2} \ln\left(\frac{N_v}{N_c}\right). \tag{4.106}$$

The number of electrons per unit volume in an extrinsic semiconductor is given by Eq. (4.102),

$$n = N_c \exp\left(\frac{E_f - E_c}{kT}\right). \tag{4.102}$$

Substituting the expression for $-E_c$ [Eq. (4.106)] in Eq. (4.102),

$$n = N_c \exp\left(\frac{E_f - E_I}{kT}\right) \exp\left(\frac{E_v - E_c}{2kT}\right)\left(\frac{N_v}{N_c}\right)^{1/2},$$

or, simplifying,

$$n = (N_c N_v)^{1/2} \exp\left(\frac{E_v - E_c}{2kT}\right) \exp\left(\frac{E_f - E_I}{kT}\right). \tag{4.107}$$

In Eq. (4.107),

$$(N_v N_c)^{1/2} \exp\left(\frac{E_v - E_c}{2kT}\right) = n_i, \tag{4.77}$$

and, therefore,

$$n = n_i \exp\left(\frac{E_f - E_I}{kT}\right). \tag{4.108}$$

Similarly,

$$p = n_i \exp\left(\frac{E_I - E_f}{kT}\right). \tag{4.109}$$

In Eqs. (4.108) and (4.109) the carrier concentrations depend upon the difference of the actual Fermi level and the intrinsic Fermi level and also the intrinsic carrier concentration n_i.

4.7 Photoconductivity, excitons, and luminescence

Until now we have considered the effect of temperature on the behavior of a semiconductor. An increase in temperature of a semiconductor makes the lattice vibrations more vigorous, and more charge carriers are thermally generated in both extrinsic and intrinsic semiconductors. In this section we shall be concerned with the effect of an incident radiation on a semiconductor. Let the radiation be light; therefore, we shall deal with optical phenomenon. Assuming the incident radiation to be monochromatic and of frequency ν and wavelength λ, the radiation is in the form of photons each of energy $h\nu$ or hc/λ, where c is the velocity of light. The photons are absorbed more by the parent atoms, because even in a heavily doped semiconductor, the ratio of the population of the parent atoms to impurity atoms is $10^3:1$. If the energy of the incident radiation photon hc/λ is greater than E_g, the forbidden energy of the semiconductor, a photon absorbed produces an electron-hole pair. The electrons and holes are free charge carriers with certain effective masses and are free to move in the crystal. Because the number of charge carriers increases, the conductivity of the semiconductor specimen increases.

Photoconductivity. Photoconductivity is a property of the semiconductor. It is the increase in electrical conductivity of a semiconductor as a result of radiation incident on the semiconductor. The photoconductivity may be the result of light, the wavelength of which may vary from ultraviolet to infrared, depending upon the semiconductor. In the case of silicon at room temperature, the forbidden energy gap at room temperature is 1.1 eV and hence if the wavelength of incident radiation is either $\lambda = hc/E_g$ or less, it would result in photoconductance. Therefore, wavelengths of incident radiation either 1.13×10^{-4} cm or lower than this value would produce photoconduction in silicon. The electron-hole pairs produced by the incident radiation would ultimately recombine. If the energy of the incident photon is less than the forbidden energy gap, the energy absorbed by a parent of semiconductor atom may be converted into thermal energy, making the lattice vibrations vigorous but not enough to generate an electron-hole pair. There is, however, a probability, however small, that photons of energy lower than the forbidden energy gap may be absorbed by an impurity atom and thus produce mobile electrons or holes, depending upon whether the impurity is pentavalent or trivalent. In this section we shall concern ourselves with bulk semiconductors; the effect of incident radiation on *P-N* junctions will be discussed in Sec. 5.5.

Photoconductivity in intrinsic semiconductors. Consider a uniform intrinsic semiconductor with light of wavelength $\lambda \leq hc/E_g$ falling on it. Let R be the number of photons absorbed per unit volume per unit time. Then R is the rate of generation of electrons n or holes p, both expressed per unit volume, assuming all the photons falling on the semiconductor are absorbed. In an intrinsic semiconductor

$$n = p = n_i.$$

The rate of recombination is proportional to the product of the number of electrons and holes, and hence the rate of recombination is $Anp = An_i^2$, where A is a constant. The net rate of generation of electrons or holes per unit volume is

$$\frac{dn}{dt} = \frac{dp}{dt} = \frac{dn_i}{dt} = \text{rate of generation} - \text{rate of recombination},$$

or

$$\frac{dn_i}{dt} = R - An_i^2. \tag{4.110}$$

In the steady state the net rate of generation of charge carriers is zero, or the rate of generation equals the rate of recombination. Therefore,

$$0 = R - An_i^2,$$

or

$$n_i = \left(\frac{R}{A}\right)^{1/2}.$$

Calling the steady-state concentration of charge carriers n_{i_0},

$$n_{i_0} = \left(\frac{R}{A}\right)^{1/2}. \quad (4.111)$$

The electrical conductivity due to light or photoconductivity σ_p is given by

$$\sigma_p = n_{i_0}(e\mu_e + e\mu_h) \quad \text{[from Eq. (4.52)]}.$$

Assuming $\mu_h < \mu_e$ (a fair assumption because $m_h^* > m_e^*$),

$$\sigma_p = n_{i_0} e \mu_e. \quad (4.112)$$

Substituting n_{i_0} from Eq. (4.111) in Eq. (4.112),

$$\sigma_p = \left(\frac{R}{A}\right)^{1/2} e\mu_e. \quad (4.113)$$

Equation (4.113) predicts that for a given electric field $\mathcal{E}$, the photocurrent

$$J_p = \sigma_p \mathcal{E} = \left(\frac{R}{A}\right)^{1/2} e\mu_e \mathcal{E}; \quad (4.114)$$

i.e., it varies as the square root of the intensity of the incident radiation. Let us now assume that light is suddenly switched off from a semiconductor in which the steady-state carrier concentration is n_{i_0} and the carrier decay. Therefore, from Eq. (4.110), because $R = 0$,

$$\frac{dn_i}{dt} = -An_i^2,$$

$$-\frac{dn_i}{n_i^2} = A\, dt.$$

Integrating,

$$1/n_i = At + B,$$

where B is the constant of integration.

If at $t = 0$, $n_i = n_{i_0}$,

$$\frac{1}{n_i} = \frac{1}{n_{i_0}} + At,$$

$$n_i = \frac{n_{i_0}}{1 + Atn_{i_0}}. \quad (4.115)$$

From Eq. (4.115) we find that the carrier concentration drops to one half of the steady-state value in time t_0 after the light has been switched off:

$$t_0 = \frac{1}{An_{i_0}} = \frac{n_{i_0}}{R}. \qquad (4.116)$$

Equation (4.116) suggests that for a given R or for a given illumination level, the response time is directly proportional to photoconductivity.

The ability of a crystal to detect the incident radiation is called its *photosensitivity*. It is defined as the ratio of the carrier photocurrent to the rate of generation of electron-hole charge carrier pairs by light:

$$G = \frac{J_p}{e} \frac{1}{Rd},$$

where J_p is the photocurrent density and Rd is the photon density in the crystal. The photocurrent flows as a result of the incident light and an applied electric field in the specimen:

$$J_p = n_{i_0} e \mu_e \mathcal{E}.$$

Referring to Fig. 4.23,

$$\mathcal{E} = V/d,$$

and according to Eq. (4.111),

$$n_{i_0} = \left(\frac{R}{A}\right)^{1/2}.$$

Therefore,

$$J_p = \left(\frac{R}{A}\right)^{1/2} e \mu_e \frac{V}{d}$$

$$\frac{J_p}{e} = \frac{V \mu_e Rd}{d^2 (RA)^{1/2}}.$$

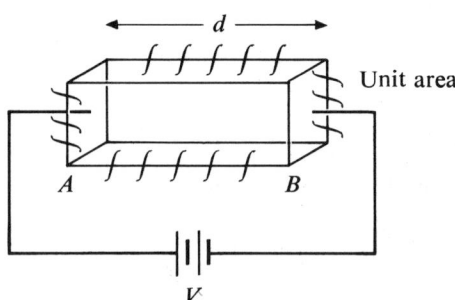

Fig. 4.23
Photoconductivity

Therefore,
$$G = \frac{V\mu_e}{d^2(RA)^{1/2}}. \tag{4.117}$$

If T_d is defined as the transit time of carriers between electrodes A and B,

$$T_d = \frac{d}{\text{drift velocity}} = \frac{d^2}{V\mu_e}.$$

The response time has already been defined as

$$t_0 = \frac{1}{An_{i_0}} = (AR)^{-1/2}. \tag{4.118}$$

Hence

$$\frac{t_0}{T_d} = \frac{(AR)^{-1/2}\mu_e V}{d^2} = \frac{\mu_e V}{d^2(RA)^{1/2}} = G. \tag{4.119}$$

The photosensitivity is the ratio of the response time to the transit time of a carrier between electrodes.

The above analysis for intrinsic (pure) semiconductors shows that the photocurrent varies as the square root of the incident radiation. Experimental observations show that the photocurrent is proportional to R^x, where x normally lies between 0.5 and 1.0 and is even greater than 1 in some crystals. Further, if t_0 is measured and T_d calculated, the gain computed from these values differs from the experimental values to a great extent. The discrepancy may be attributed to trapping.

Traps. Traps are defects such as missing or extra atoms or structural faults in the lattice. They are capable of capturing an electron or hole and the captured carrier may be reemitted at a later time. The energies associated with these defects are found experimentally to lie near the center of the forbidden gap.

In a crystal containing traps, the photosensitivity can be calculated by making the following assumptions [A. Rose, *RCA Rev.*, **12**, 361 (1951)]:

1. The holes are entirely trapped.
2. The fraction of electrons nominally in the conduction band is trapped at any time instant.
3. Thermal generation of electrons and holes is neglected.

Let N_T be the number of electron trap levels per unit volume. The generation rate of electrons $= R$. The recombination rate of electrons $= An(p + N_T)$. A is also assumed to be the constant of proportionality for the trapping rate of electrons. We have also assumed that the rate of the ionization of traps is negligible. Hence the net rate of generation of electrons is

$$\frac{dn}{dt} = R - An(p + N_T). \qquad (4.120)$$

In steady state $dn/dt = 0$, and if it is an intrinsic semiconductor,

$$n = p = n_i,$$

and in steady state $n_i = n_{i_0}$. Therefore,

$$n_{i_0}(n_{i_0} + N_T) = R/A.$$

If the intensity of illumination is small, n_{i_0} may be much smaller than the number of traps N_T. Hence

$$n_{i_0} N_T = \frac{R}{A}$$

or

$$n_{i_0} = \frac{R}{AN_T}. \qquad (4.121)$$

Equation (4.121) shows that for low-intensity illumination, the photoconductivity is proportional to the intensity of incident radiation. However, if the intensity of illumination is very large, the total number of traps N_T may be much smaller than n_{i_0}. Hence

$$n_{i_0} = \left(\frac{R}{A}\right)^{1/2}.$$

This is Eq. (4.111), which was derived for the case of a pure semiconductor without traps.

Response time in the presence of traps. If the incident radiation, after a steady state has been reached, is turned off, the rate of decay of charge carriers from Eq. (4.120) is

$$\frac{dn_i}{dt} = -An_i(n_i + N_T),$$

$$\frac{dn_i}{n_i(n_i + N_T)} = -A\,dt,$$

or

$$\frac{1}{N_T}\left(\frac{1}{n_i} - \frac{1}{n_i + N_T}\right) dn_i = -A\,dt.$$

Integrating,

$$\frac{1}{N_T}[\ln n_i - \ln(n_i + N_T)] = -At + B,$$

where B is the constant of integration.

At $t = 0$, $n_i = n_{i_0}$. Therefore,

$$B = \frac{1}{N_T} \ln \frac{n_{i_0}}{n_{i_0} + N_T},$$

$$\ln \frac{n_i}{n_i + N_T} - \ln \frac{n_{i_0}}{n_{i_0} + N_T} = -AtN_T,$$

or

$$\ln \frac{n_i}{n_{i_0}} \frac{n_{i_0} + N_T}{n_i + N_T} = -AtN_T. \qquad (4.122)$$

If the intensity of illumination is low,

$$n_{i_0} \ll N_T$$

and therefore

$$n_i \ll N_T.$$

Hence

$$n_i = n_{i_0} e^{-AtN_T}.$$

If the response time is defined as the time t_0 in which the carrier concentration and hence the photoconductivity drops to $1/e$ of its steady-state value,

$$t_0 = \frac{1}{AN_T}. \qquad (4.123)$$

Comparing Eq. (4.123) with Eq. (4.116) we find that the presence of traps affects the response time for low-intensity illumination, the response time being inversely proportional to the trap concentration.

The photosensitivity is proportional to the response time, and hence the photosensitivity at low-intensity levels is greatly affected by the presence of traps.

Excitons. In the process of photoconduction we have considered the absorption of photons of energy equal to or greater than the forbidden energy gap and hence discussed the generation of electron-hole pairs in which both the charge carriers were free charge carriers capable of moving through the crystal independently. However, if the photon energy is less than the forbidden energy gap, an exciton may result from the absorption of photon. An exciton is a coupled electron-hole pair. It is neutral and is an excited mobile state of a crystal. It can travel through only as a neutral entity and may give up its energy of formation. Because it is neutral it does not contribute to electrical conductivity. In the case of an exciton, the electron does not undergo a transition to a quantum state in the conduction band but remains in the vicinity of the hole and is therefore localized. If r is the separation between the electron and hole, the exciton is held together by a Coulomb's force of attraction, which is

$$F = \frac{e^2}{4\pi\epsilon_0\epsilon_r r^2},$$

where ϵ_r is the relative dielectric constant of the material (crystal). The Coulomb potential is

$$V = \int_\infty^r \frac{e^2}{4\pi\epsilon_0\epsilon_r r^2} \, dr = -\frac{e^2}{4\pi\epsilon_0\epsilon_r r}.$$

Following a procedure similar to the one used in the development of the Bohr theory and making the assumptions of the theory, we find that the energy of an electron bound to a hole to form an exciton is quantized and is given by

$$E_n = -\frac{m_0^* e^4}{8\epsilon_0^2 \epsilon_r^2 n^2 h^2},$$

where $n = 1, 2, 3, 4, \ldots, \infty$ and m_0^* is the mean effective mass, the harmonic mean between the effective masses of electron and hole:

$$\frac{1}{m_0^*} = \frac{1}{m_e^*} + \frac{1}{m_h^*}.$$

The exciton ground state corresponds to $n = 1$. The binding energy with $n = \infty$ is zero, meaning that the electron is no longer bound, is at the bottom of the conduction band, and is available as a free particle for conduction. Figure 4.24 shows the various energy levels of an electron in an exciton.

If $\epsilon_r = 6$ and $m_0^* = m$,

$$E_1 = 13.6/36 = 0.38 \text{ eV}.$$

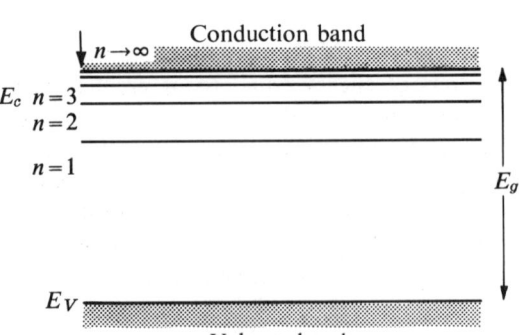

Fig. 4.24
Energy levels of an electron in an exciton

Absorption lines in the absorption spectra of many insulating crystals and semiconductors very close to the expected band may be caused by excitation of an electron to form an exciton. It is sometimes possible to observe a transition between the top of the valence band and an exciton band.

Luminescense. We have discussed the excitation of electrons in an atom from the ground state to a higher energy state either thermally or optically. If nothing else is done, the electron returns back to its ground state, thereby radiating the excess energy. The time taken by the electron to jump back to the ground state may vary from very small duration, such as microseconds, to relatively larger intervals, such as minutes. However, there exists the possibility that the electron may not lose its energy by radiation but may in turn dissipate it in interacting with the lattice vibrations. The process of emission of radiation by the excited electron is called *luminescence*. Luminescent materials contain an impurity referred to as an *activator*. The incorporation of an activator atom in crystalline solids gives rise to energy levels localized in the forbidden energy gap. The activators may be imperfections or foreign atoms.

Let A and B be two impurity states. In the ground state the state A has an electron and B is empty. In the excited state B is occupied and A is empty. Luminescence is exhibited during transition of the electron from state B to state A. Several methods may be employed from state A to B. If the excitation is achieved by photon absorption, it is called *photoluminescence*.

Fluorescence. If a crystal emits while it is being excited or has a very short lifetime, it is said to be *fluorescent* and the process is called *fluorescence*.

Phosphorescence. If a crystal emits after the excitation is switched off, it is said to be *phosphor* and the process is called *phosphorescence*.

Problem 1

The following relationship exists between E and K at the energy extrema:

$$E(K) = \left(\frac{h^2}{8\pi^2 m}\right) K^2 + \frac{1}{2}\left[\left(E_g^2 + 8\frac{K^2 P^2}{3}\right)^{1/2} - E_g\right],$$

where P is the momentum matrix element, m the free electron mass, and E_g the forbidden energy gap. Assuming P to be independent of K, show that $E_g \propto m^*$, if $m^* \ll m$. [See E. O. Kane, *J. Phys. Chem. Solids*, 1, 249 (1956).]

Problem 2

Show that when the effective mass tensor is characterized by principal values m_1^*, m_2^*, and m_3^* along three orthogonal axes, the density of state effective mass is given by $(m_1^*, m_2^*, m_3^*)^{1/3}$.

Problem 3

Using the definition of effective mass m^*, show that $m^* = m$ for a truly free electron.

Problem 4

The ratio of arsenic atoms to parent atoms in a specimen of germanium is 1 to 10^4. Assuming all donors are ionized, compute the resistivity of the specimen at room temperature and also determine the voltage required for a current of 10 A to flow through a specimen of area 1 cm^2 and length 1 cm. Assume $E_D = 0.01$; $m_e^* = 0.2m$; $\mu_e = 3600$ cm^2/V sec.   *Ans.* $\rho = 0.666 \times 10^{-3}$ Ω cm; $V = 0.666 \times 10^{-2}$ V

Problem 5

Silicon has a forbidden energy band gap of width 1.1 eV. Compute the position of the Fermi level in the intrinsic material at $T = 8$ and 300°K. Assume $m_h^* = 6m_e^*$. Also compute the density of free electrons at $T = 300$ and 600°K. Assume $m_e^* = 0.8m$, where m is the mass of the free electron.   *Ans.* 0.55 eV; 0.516 eV; 2.24 $\times 10^{10}$ cm^{-3}; 3.17 $\times 10^{15}$ cm^{-3}

Problem 6

a. The resistivity of intrinsic germanium at room temperature is 47 Ω cm. Assume electron and hole mobilities to be 0.39 and 0.19 m^2/V sec, respectively, and compute the intrinsic carrier density at room temperature.

b. The resistivity of intrinsic silicon at room temperature is 2.3 $\times 10^5$ Ω cm. Assume electron and hole mobilities to be 0.135 m^2/V sec and 0.048 m^2/V sec, respectively, and compute the intrinsic carrier density at room temperature.
     Ans.   **a.** 2.4 $\times 10^{13}$ cm^{-3} **b.** 1.5 $\times 10^{10}$ cm^{-3}

Problem 7

The intrinsic conductivities of Ge and Si are found to vary as $e^{-4300/T}$ and $e^{-6400/T}$, respectively. Compute the forbidden energy gap of the two materials at room temperature if the effective masses of the electrons and holes are assumed constant.   *Ans.* 0.74 eV; 1.1 eV

Problem 8

Estimate the mobility for a typical metal with 10^{23} free electrons per cm^3 having a resistivity of 10^{-5} Ω cm.   *Ans.* 6.67 cm^2/V sec

Problem 9

An insulator contains a small amount of impurity which contributes a small number of states below the conduction band to the energy-level structure. Show that at $T = 0$ the Firmi level lies between the impurity levels and the conduction band, if the impurity levels all contain electrons at $T = 0$.

5

Continuity Equations and P-N Junctions

5.1 Electrochemical potential

The *electrochemical potential* can be defined by considering a few concepts of thermostatics and thermodynamics. The internal energy of a body can be changed by the flow of heat to it or by doing work on it.

If dQ = amount of heat which enters a body, dW = amount of work done by a body. The change in internal energy according to the law of conservation of energy is

$$dE = dQ - dW. \tag{5.1}$$

This is the first law of thermodynamics.

It can be easily conceived that the energy E of a body can be a function of entropy, volume, concentration, and charge.

$$E = E(S, V, N, q), \tag{5.2}$$

where S is the entropy, V the volume, N the concentration, and q the charge. Taking the differential of Eq. (5.2),

$$dE = \left.\frac{\partial E}{\partial S}\right)_{V,N,q} dS + \left.\frac{\partial E}{\partial V}\right)_{S,N,q} dV + \left.\frac{\partial E}{\partial N}\right)_{S,V,q} dN + \left.\frac{\partial E}{\partial q}\right)_{S,V,N} dq. \tag{5.3}$$

The partial derivatives are given certain names:

$$\left(\frac{\partial E}{\partial S}\right)_{V,N,q} = T = \text{absolute temperature,} \tag{5.4}$$

$$-\left(\frac{\partial E}{\partial V}\right)_{S,N,q} = P = \text{pressure,} \tag{5.5}$$

$$\left(\frac{\partial E}{\partial N}\right)_{S,V,q} = K = \text{chemical potential,} \tag{5.6}$$

$$\left(\frac{\partial E}{\partial q}\right)_{S,V,N} = \theta = \text{electrostatic potential.} \tag{5.7}$$

Therefore,
$$dE = T\,dS - P\,dV + K\,dN + \theta\,dq. \tag{5.8}$$

If there is no change in concentration and the particles are not charged,
$$dN = dq = 0,$$
$$dE = T\,ds - P\,dV, \tag{5.9}$$

where
$$dW = P\,dV, \tag{5.10}$$
$$dQ = T\,dS. \tag{5.11}$$

The amount of heat dQ which enters a body is the product of its temperature and change in entropy of the body.

In case the particles are electrons of charge $-e$,
$$dq = -e\,dN,$$
and hence Eq. (5.8) may be rewritten
$$dE = T\,dS - P\,dV + (K - e\theta)\,dN, \tag{5.12}$$

where $(K - e\theta)$ is the electrochemical potential $\bar{\mu}$.

The electrochemical potential can also be described in terms of the Helmholtz function. It is known from basic thermodynamics that
$$A = E - TS, \tag{5.13}$$

where T is the temperature, A the Helmholtz function, and S the entropy. From Eq. (5.13), differentiating,
$$dA = dE - T\,dS - S\,dT. \tag{5.14}$$

Substituting the value of dE from Eq. (5.12),
$$dA = -P\,dV - S\,dT + \bar{\mu}\,dN. \tag{5.15}$$

If the concentration of the constituents in the system does not change, $dN = 0$, and if also the temperature is constant, $dT = 0$, then

$$dA = -P\,dV$$

and dA is the amount of work done on the system if there is no change in concentration and temperature.

From Eq. (5.15),

$$\bar{\mu} = \left(\frac{\partial A}{\partial N}\right)_{T,V}.$$

The electrochemical potential is defined as the change in Helmholtz function with concentration if the temperature and volume are constant. From thermodynamic principles it follows that for charge carriers obeying Fermi-Dirac statistics,

$$\bar{\mu} = \left(\frac{\partial A}{\partial N}\right)_{T,V} = E_f,$$

the Fermi energy level. A physical interpretation of $E_f = \bar{\mu}$ in a solid will be discussed in the following sections.

5.2 Diffusion; Modified Ohm's law

In semiconductors a situation might exist in which the concentration of charge carriers, electrons and holes, may not be the same throughout the material. The presence of a concentration gradient results in the flow of charge carriers and hence an electric current. Figure 5.1 shows an arbitrary one-dimensional concentration gradient in the x direction. According to diffusion theory, charge carriers would flow in the direction of the arrow or from higher concentration to lower concentration. Considering a unit area YZ plane, the net flow of charge carriers through the YZ plane is propor-

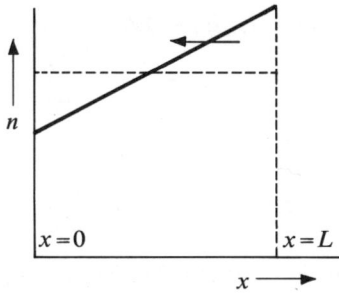

Fig. 5.1
One-dimensional concentration gradient

tional to $(-dn/dx)$. It is assumed that n is the number of electrons per unit volume. Calling the current density $J_{n(\text{diffusion})}$, it follows that

$$J_{n(\text{diffusion})} \propto \left(-\frac{dn}{dx}\right)(-e),$$

where $-e$ is the charge of an electron, or

$$J_{n(\text{diffusion})} = +D_n e \frac{dn}{dx}, \tag{5.16}$$

where D_n is called the *diffusion constant for electrons*.

Equation (5.16) shows that an electric current might flow in a semiconductor even if there is no applied electric field. The current flow evidently in such a case is due to the concentration gradient and is called the *diffusion current*.

However, if both an electric field $\mathcal{E}_x$ is applied and a concentration gradient dn/dx exists, the two currents may be superposed to give

$$J_n = ne\mu_e \mathcal{E}_x + D_n e \frac{dn}{dx}. \tag{5.17}$$

The first term on the right side of Eq. (5.17) is the electric current density due to electrons as free charge carriers in the presence of an applied electric field and is called the *drift current density* due to electrons:

$$J_{n(\text{drift})} = ne\mu_e \mathcal{E}_x$$

or

$$J_n = J_{n(\text{drift})} + J_{n(\text{diffusion})}.$$

An expression similar to Eq. (5.17) may be easily written for the current density due to holes:

$$J_p = J_{p(\text{drift})} + J_{p(\text{diffusion})},$$

where

$$J_{p(\text{drift})} = pe\mu_h \mathcal{E}_x,$$

$$J_{p(\text{diffusion})} = -eD_h \frac{dp}{dx}.$$

D_h is the diffusion constant for holes. Hence

$$J_p = pe\mu_h \mathcal{E}_x - eD_h \frac{dp}{dx}. \tag{5.18}$$

In the case of the diffusion current density, the sign is negative because

$$J_{p(\text{diffusion})} \propto -dp/dx,$$

and the charge of a hole is $+e$.

Having discussed Eqs. (5.17) and (5.18) for electron and hole current densities, respectively, it now remains to discuss the diffusion constants D_e and D_h.

Diffusion constant D_h. Referring to Eq. (5.18), let us assume that there exists a situation in which both an electric field and a concentration gradient are present in a material and the resultant hole current density is zero. Therefore,

$$0 = pe\mu_h \mathcal{E}_x - eD_h \frac{dp}{dx}$$

or

$$pe\mu_h \mathcal{E}_x = eD_h \frac{dp}{dx}. \tag{5.19}$$

The presence of the electric field produces a potential-energy variation along the x direction. This causes the hole density to vary with position:

$$p(x) = Ae^{-e\theta(x)/kT}, \tag{5.19a}$$

where A is a constant and

$$\mathcal{E}_x = -\frac{d\theta(x)}{dx}.$$

Equation (5.19a) holds good for nondegenerate materials. Differentiating Eq. (5.19a),

$$\frac{dp}{dx} = -\frac{Ae}{kT}\frac{d\theta}{dx}(e^{-e\theta/kT})$$

or

$$\frac{dp}{dx} = -\frac{pe}{kT}\frac{d\theta}{dx} = \frac{pe}{kT}\mathcal{E}_x. \tag{5.20}$$

Substituting the value of dp/dx from Eq. (5.20) in Eq. (5.19),

$$pe\mu_h\mathcal{E}_x = \frac{e^2 D_h p}{kT}\mathcal{E}_x$$

or

$$D_h = \left(\frac{kT}{e}\right)\mu_h. \tag{5.21}$$

Equation (5.21) is a relation between the mobility and diffusion constant for holes known as the *Einstein relation*.

Diffusion constant D_e. Following a treatment similar to the derivation of Eq. (5.21), it can be easily shown that

$$D_e = \left(\frac{kT}{e}\right)\mu_e, \tag{5.22}$$

where μ_e is the mobility of electrons.

Modified Ohm's law and physical interpretation of the electrochemical potential. We have seen that besides an applied electric field, a charge-carrier concentration gradient can produce a current density. Ohm's law, $J = \sigma \mathcal{E}$, assumed no charge-carrier concentration gradient. Equations (5.17) and (5.18), which express the current density in terms of applied electric field and the concentration gradient, are often referred to as a *modified Ohm's law*. Another way of expressing the modified Ohm's law is to rewrite Eqs. (5.17) and (5.18) in terms of the chemical-potential gradient. In Sec. 4.6 we developed expressions for the charge-carrier densities in an extrinsic semiconductor in terms of intrinsic carrier density,

$$n = n_i \exp\left(\frac{E_f - E_I}{kT}\right), \tag{4.108}$$

$$p = n_i \exp\left(\frac{E_I - E_f}{kT}\right), \tag{4.109}$$

where n_i is the intrinsic carrier density, E_f is the Fermi level of the material (extrinsic), and E_I is the Fermi level if it were intrinsic. It was discussed in Sec. 5.1 that

$$\bar{\mu} = \begin{cases} E_f, \\ K - e\theta, \end{cases}$$

or

$$\bar{\mu} + e\theta = K.$$

The introduction of impurities results in the Fermi level being different from the case of intrinsic materials. If K, the chemical potential, is a measure of the difference of the Fermi level caused by the introduction of impurities,

$$-e\theta = E_I \quad \text{and} \quad \bar{\mu} = E_f.$$

Equations (4.108) and (4.109) may now be rewritten in terms of the chemical potential,

$$K = E_f - E_I,$$

$$n = n_i e^{K/kT}, \tag{5.23}$$

$$p = n_i e^{-K/kT}. \tag{5.24}$$

With the help of Eqs. (5.23) and (5.24), the modified Ohm's law of Eqs. (5.17) and (5.18) may now be expressed in terms of the chemical-potential gradient.

Differentiating Eq. (5.23) with respect to x,

$$\frac{dn}{dx} = \frac{n_i}{kT} e^{K/kT} \frac{dK}{dx}$$

or

$$\frac{dn}{dx} = \frac{n}{kT}\frac{dK}{dx}. \tag{5.25}$$

Similarly,
$$\frac{dp}{dx} = -\frac{p}{kT}\frac{dK}{dx}. \qquad (5.26)$$

Substituting the values of dn/dx and dp/dx in terms of the chemical-potential gradient from Eqs. (5.25) and (5.26) in Eqs. (5.17) and (5.18), respectively,

$$J_n = ne\mu_e \mathcal{E}_x + D_n \frac{en}{kT}\frac{dK}{dx}, \qquad (5.27)$$

$$J_p = pe\mu_h \mathcal{E}_x + D_h \frac{ep}{kT}\frac{dK}{dx}. \qquad (5.28)$$

Simplifying further, from Einstein's relations,

$$ne\mu_e = \sigma_e,$$

$$pe\mu_h = \sigma_h,$$

$$D_e \frac{en}{kT} = n\mu_e,$$

$$D_h \frac{ep}{kT} = p\mu_h,$$

$$J_n = \sigma_e \mathcal{E}_x + \frac{\sigma_e}{e}\frac{dK}{dx}, \qquad (5.29)$$

$$J_p = \sigma_h \mathcal{E}_x + \frac{\sigma_e}{e}\frac{dK}{dx}. \qquad (5.30)$$

Equations (5.29) and (5.30) represent the flow of electron and hole currents, respectively, when both electrical and chemical-potential gradients are present or when both electrical and chemical forces are acting. Equations (5.29) and (5.30) can be expressed in terms of the electrochemical potential $\bar{\mu}$:

$$\bar{\mu} = \begin{cases} K - e\theta & \text{for electrons,} \\ K + e\theta & \text{for holes.} \end{cases}$$

Therefore,
$$J_n = \frac{\sigma_e}{e}\frac{d\bar{\mu}}{dx}, \qquad (5.31)$$

$$J_h = -\frac{\sigma_h}{e}\frac{d\bar{\mu}}{dx}. \qquad (5.32)$$

Equations (5.31) and (5.32) are often referred to as a modified Ohm's law for electron and hole currents, respectively.

5.3 Continuity equations

At any given temperature in a semiconductor both electrons and holes are present. In an intrinsic semiconductor the number of holes equals the number of electrons:

$$n = p = n_i.$$

In an *N*-type semiconductor, electrons are in the majority and holes in the minority. The concentration of *electrons* in an *N*-type material is called the *majority carrier concentration* and the concentration of holes the *minority carrier concentration*:

$$n \gg p.$$

According to the law of mass action given by Eq. (4.106),

$$np = n_i^2. \qquad (4.106)$$

Similarly, in a *P*-type semiconductor, holes are in the majority and electrons in the minority. The concentration of *holes* in a *P*-type material is called the *majority carrier concentration* and the concentration of electrons the *minority carrier concentration*:

$$p \gg n.$$

Again, according to the law of mass action,

$$pn = n_i^2.$$

If maintained free of any external effects and influences, charge carriers are being generated in the semiconductor at a certain rate. In addition, charge carriers are being recombined or trapped. The net rate of change of charge-carrier concentration is given by the following equation:

rate of change of carrier concentration = rate of generation
− rate of recombination.

Because in most cases of interest external injection or creation of charge carriers causes a greater percentage change in the number of minority carriers than in majority carriers, we shall discuss only the rate of change of minority carrier concentration.

Considering a *P*-type semiconductor,

$$\frac{dn}{dt} = g - R, \qquad (5.33)$$

where n is the minority carrier concentration, g the rate of generation of minority carrier, and R the recombination rate of the minority carrier. In the absence of any external effects or connections, charge carriers are produced only by thermal excitation, which means g is a function of temperature and hence $g(T)$.

Recombination. Before discussing recombination equations, it will be worthwhile to discuss the process of recombination. Electrons and holes which are either generated thermally or through some external process may be reunited or may recombine. It has been found that in germanium and silicon, recombination of electrons and holes may occur by the following processes:

1. *Direct recombinations:* The electron may drop down to a hole directly, thereby recombining and emitting a photon to conserve energy. The probability of such recombinations is very small.

2. *Indirect recombinations:* Indirect recombinations may occur in the structure at imperfections of the lattice. The impurities and imperfections represent discontinuities in the periodic structure of the lattice. We have already called these impurities or imperfections *traps*. Electrons are trapped by them and hence the charge carriers cannot contribute to conduction any more or cannot be considered as free charge carriers any more. The trap, which is normally neutral, becomes ionized upon capturing a charge, carrier and releases its excess energy to the lattice. After staying awhile at the traps, the electrons may recombine with holes, thereby completing the process of recombination.

Lifetime of the charge carriers. The average time that electrons and holes as free charge carriers before either recombining directly or being captured by a trap is called the *lifetime* of the charge carriers. If the recombination occurs directly, both electrons and holes are removed simultaneously, and therefore the lifetime of the electron equals the lifetime of the holes. However, if the electrons are first trapped, the lifetimes of electrons are different from holes.

If p_P and n_P are the normal equilibrium concentrations of holes and electrons in a P-type material, the probability of their recombination is proportional to the product of n_P and p_P.

The time rate of electron-hole pair recombination $R = rn_P p_P$, where r is called the *recombination coefficient* and is the constant of proportionality. In equilibrium $dn/dt = 0$, and therefore

$$0 = g(T) - rn_P p_P$$

$$g(T) = rn_P p_P \tag{5.34}$$

Similarly, in an N-type material in equilibrium without any external effects,

$$g(T) = rp_N n_N, \tag{5.35}$$

where p_N is the equilibrium concentration of holes in the N-type material and n_N is the equilibrium concentration of electrons in the N-type material. If Δn is the number of additional electrons injected into a P-type material, the equilibrium concentration would change:

$$n = n_P + \Delta n,$$

where n is the concentration of electrons (equilibrium plus injection). A semiconductor of either type P or N type must remain electrically neutral at all times, so holes must also enter the semiconductor. The concentration of entering holes Δp must equal the concentration of injected electrons,

$$\Delta p = \Delta n.$$

This is a result of the space-change principle. Therefore, the concentration of holes has increased and the new concentration of holes is

$$p = p_P + \Delta p.$$

The rate of change of minority carrier concentration is

$$\frac{dn}{dt} = g(T) - R. \tag{5.33}$$

If the temperature remains the same, $g(T)$ remains unchanged, and according to Eq. (5.35),

$$g(T) = r p_P n_P.$$

The recombination rate R must be proportional to the product of electron and hole concentration after injection,

$$R = r(p_P + \Delta p)(n_P + \Delta n).$$

Therefore,

$$\frac{dn}{dt} = r p_P n_P - r(p_P + \Delta p)(n_P + \Delta n).$$

Because

$$\Delta p = \Delta n,$$

$$\frac{dn}{dt} = r p_P n_P - r(p_P + \Delta n)(n_P + \Delta n),$$

$$\frac{dn}{dt} = r p_P n_P - r p_P n_P - r p_P \Delta n - r n_P \Delta n - \Delta n^2,$$

or

$$\frac{dn}{dt} = -r[(p_P + n_P)\Delta n + \Delta n^2].$$

If the carrier density of injected electrons is small as compared to equilibrium concentration of electrons and holes, Δn^2 may be neglected as compared to $\Delta n(p_P + n_P)$. Therefore,

$$\frac{dn}{dt} = -r\,\Delta n(p_P + n_P). \tag{5.33a}$$

Now, $n = n_P + \Delta n$. and hence

$$\frac{d}{dt}(n_P + \Delta n) = -r\,\Delta n(p_P + n_P).$$

n_P is constant and, therefore,

$$\frac{d}{dt}(\Delta n) = -r\,\Delta n(p_P + n_P). \tag{5.36}$$

Equation (5.36) shows the variation of electrons Δn injected.

Solving Eq. (5.36),

$$\Delta n(t) = (\Delta n)_0 e^{-t/\tau_n}. \tag{5.37}$$

$(\Delta n)_0$ is the initially injected number of electrons at $t = 0$ and $\Delta n(t)$ is the number of injected electrons left at time t. τ_n is the lifetime of electrons in seconds:

$$\tau_n = \frac{1}{r(p_P + n_P)}. \tag{5.38}$$

The injected minority carrier density decays exponentially with the lifetime of minority carriers.

We have discussed the rate of change of minority carrier concentration in bulk P- and N-type semiconductors with discontinuous minority carrier injection. Consider now an infinite medium of a P-type material in which the electrons are being injected continuously at $x = 0$. To start with, the injection of electrons at $x = 0$ (Fig. 5.2) increases the minority carrier

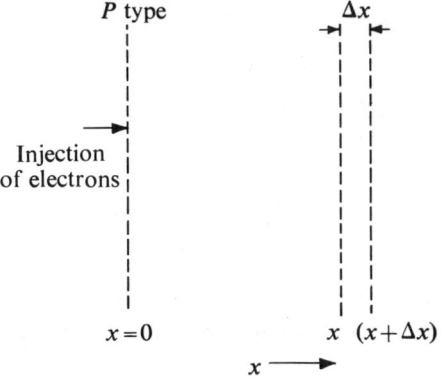

Fig. 5.2
Injection of electrons

concentration, the rest of the material having an equilibrium minority carrier concentration n_P. There would, therefore, be a concentration gradient and minority carriers would diffuse from $x = 0$ to increasing values of x. Taking a thickness of the material Δx at a distance x and considering a unit area of cross section of the plane, the net rate of change of minority carrier concentration in the volume Δx is decided by the following processes:

1. Thermal rate of generation of carriers.
2. Recombination rate of minority carriers.
3. Rate of change of charge carriers due to flow of carriers as a result of concentration gradient if there is no externally applied field. However, if there exists an externally applied field, the rate of change of charge carriers in this volume would depend upon both drift and diffusion current.

The first two processes have already been discussed. Now, considering the third process, let $J_n(x)$ be the electronic current density at x and $J_n(x + \Delta x)$ be the electronic current density at $x + \Delta x$. The number of electrons entering the volume at x is $[-1/eJ_n]$, and the number of electrons leaving the volume at $x + \Delta x$ is $[-1/eJ_n(x + \Delta x)]$. The net rate of increase of electrons in the volume Δx is $(-1/e)[J_n(x) - J_n(x + \Delta x)]$:

$$\left.\frac{\partial n}{\partial t}\right)_{\text{flow of carriers}} \Delta x = \left(-\frac{1}{e}\right)[J_n(x) - J_n(x + \Delta x)],$$

$$\left.\frac{\partial n}{\partial t}\right)_{\text{flow of carriers}} = \left(-\frac{1}{e}\right)\frac{J_n(x) - J_n(x + \Delta x)}{\Delta x}.$$

Taking $\Delta x \to 0$,

$$\lim_{\Delta x \to 0} \frac{J_n(x) - J_n(x + \Delta x)}{\Delta x} = -\frac{dJ_n}{dx}.$$

Hence

$$\left.\frac{\partial n}{\partial t}\right)_{\text{flow of carriers}} = \frac{1}{e}\frac{dJ_n}{dx}. \tag{5.39}$$

It can easily be shown in a similar manner that for an N-type material, the rate of change of concentration of charge carriers due to flow of minority carriers of holes is

$$\left.\frac{\partial p}{\partial t}\right)_{\text{flow of carriers}} = -\frac{1}{e}\frac{dJ_p}{dx}. \tag{5.40}$$

The net rate of change of minority carrier concentration for a *P*-type material may now be written by making use of Eqs. (5.33) and (5.39):

$$\left(\frac{dn}{dt}\right) = g(T) - R + \frac{1}{e}\frac{dJ_n}{dx}.$$

From Eq. (5.33a),

$$g(T) - R = -r\,\Delta n(p_P + n_P),$$

and from Eq. (5.38),

$$\tau_n = \frac{1}{r(p_P + n_P)}.$$

Therefore,

$$\left(\frac{dn}{dt}\right) = -\frac{\Delta n}{\tau_n} + \frac{1}{e}\frac{dJ_n}{dx}.$$

Δn is the change in concentration,

$$\Delta n = n - n_P.$$

Hence

$$\frac{dn}{dt} = \frac{n_P - n}{\tau_n} + \frac{1}{e}\frac{dJ_n}{dx}. \tag{5.41}$$

Equation (5.41) is called the *continuity equation* of electrons in a *P*-type material in one dimension. The equation for a three-dimensional case may now be written

$$\frac{dn}{dt} = \frac{n_P - n}{\tau_n} + \frac{1}{e}\nabla\cdot\mathbf{J}_n. \tag{5.42}$$

In a similar way, the continuity equation of holes in an *N*-type material can be written

$$\frac{dp}{dt} = \frac{p_N - p}{\tau_p} - \frac{1}{e}\nabla\cdot\mathbf{J}_p. \tag{5.43}$$

This equation can be written for the one-dimensional case as

$$\frac{dp}{dt} = \frac{p_N - p}{\tau_p} - \frac{1}{e}\frac{dJ_p}{dx}. \tag{5.44}$$

The electron current density J_n and the hole current density J_p used in Eqs. (5.41) and (5.44) are given for a one-dimensional case by Eqs. (5.17) and (5.18), respectively:

$$J_n = ne\mu_e\mathcal{E}_x + D_n e\frac{dn}{dx}, \tag{5.17}$$

$$J_p = pe\mu_h\mathcal{E}_x - D_h e\frac{dp}{dx}. \tag{5.18}$$

CONTINUITY EQUATIONS AND P-N JUNCTIONS

In three-dimensional diffusion, the equations of the electron and hole current densities may be expressed

$$\mathbf{J}_n = ne\mu_e \vec{\mathcal{E}} + D_n e \nabla n, \quad (5.45)$$

$$\mathbf{J}_p = pe\mu_h \vec{\mathcal{E}} - D_h e \nabla p. \quad (5.46)$$

Equations (5.17), (5.18), (5.41), and (5.42) help us understand various phenomena in semiconductors.

Diffusion length. In the one-dimensional example of Sec. 5.2, the distance the carriers penetrate before their density drops to $1/e$ of its initial value due to recombination is defined as the diffusion length L. If the charge carriers are electrons, the diffusion length is denoted L_n and for holes as charge carries it is denoted L_h.

The diffusion length can be understood if we consider the penetration of injected minority carriers. Referring back to Fig. 5.2, let $(\Delta n)_0$ be the concentration of injected carriers. We assume that there is no externally applied field, $\mathcal{E}_x = 0$.

For a P-type material, the rate of change of carrier concentration at any time t at a position x is

$$\frac{dn}{dt} = \frac{n_P - n}{\tau_n} + \frac{1}{e}\frac{dJ_n}{dx}. \quad (5.41)$$

In the absence of an externally applied electric field,

$$J_n = D_n e \frac{dn}{dx} \quad \text{[from Eq. (5.17)]}.$$

Substituting the value of J_n from Eq. (5.17) in Eq. (5.41),

$$\frac{dn}{dt} = \frac{n_P - n}{\tau_n} + D_n \frac{d^2 n}{dx^2}. \quad (5.47)$$

In an equilibrium situation, the rate of change of carrier concentration at any time t at any position x is zero. Therefore,

$$\frac{dn}{dt} = 0 = \frac{n_P - n}{\tau_n} + D_n \frac{d^2 n}{dx^2}. \quad (5.48)$$

Simplifying,

$$D_n \tau_n \frac{d^2 n}{dx^2} - n + n_P = 0. \quad (5.49)$$

Denoting $D_n \tau_n = L_n^2$,

$$L_n^2 \frac{d^2 n}{dx^2} - n + n_P = 0, \quad (5.50)$$

where n_P is the equilibrium minority carrier concentration and hence a constant. The general solution of Eq. (5.50) is of the form

$$n = Ae^{-x/L_n} + Be^{x/L_n} + n_P, \quad (5.51)$$

where A and B are constants of integration and can be determined by boundary conditions. As $x \to \infty$, $n \to n_P$, because at large distances from $x = 0$ the presence of the injected carriers is hardly felt, because of recombination. Therefore $B = 0$; otherwise n would increase with increasing x. Hence

$$n = Ae^{-x/L_n} + n_P. \quad (5.52)$$

At $x = 0$, $n - n_P = (\Delta n)_0$ (assumed). Therefore,

$$A = (\Delta n)_0$$
$$\Delta n = n - n_P = (\Delta n)_0 e^{-x/L_n}. \quad (5.53)$$

Equation (5.53) shows that if the carriers are injected at $x = 0$, their concentration decays exponentially with increasing x and drops to $1/e$ of its value at $x = 0$ at a distance L_n defined as the diffusion length of electrons in a P-type material.

Similarly, in an N-type material,

$$\Delta p = p - p_N = (\Delta p)_0 e^{-x/L_p}, \quad (5.54)$$

where L_p is the diffusion length of holes in the material:

$$L_n = (D_n \tau_n)^{1/2}, \quad (5.55)$$
$$L_p = (D_p \tau_p)^{1/2}. \quad (5.56)$$

5.4 P-N junction theory

A P-N junction is a junction of P-type and N-type semiconductors. A junction cannot be formed by just bringing the two types of materials in contact, because a surface is a discontinuity in the regular structure of the crystal and prevents the flow of charge carriers from one material to another. There are several ways to fabricate P-N junctions. We shall now discuss briefly the alloying and diffusion techniques.

Alloying. A P-type dopant such as a solder dot of indium is placed on a disc or wafer of N-type germanium. Both the dopant and N-type germanium in this assembly are heated together to a temperature of about 500 to 600°C, preferably in an inert atmosphere, for several minutes. Indium and some of the germanium in contact with the indium melts and forms a liquid indium-germanium mixture. In the molten matter the process of diffusion distributes a high concentration of indium atoms throughout the indium-germanium liquid. Upon cooling, germanium recrystallizes in a single crystal, but because it is heavily doped with indium atoms which are acceptors, the upper portion of the basically *N*-type material is *P* type. The

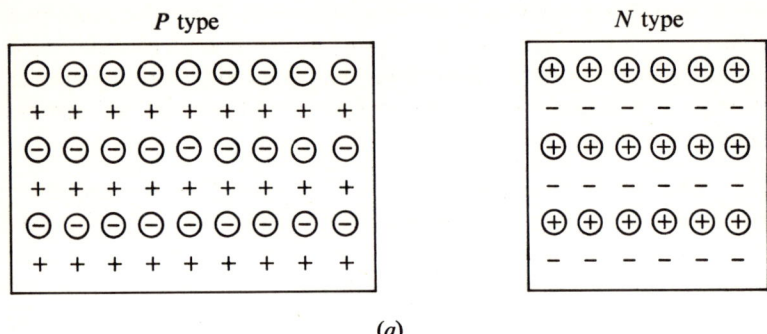

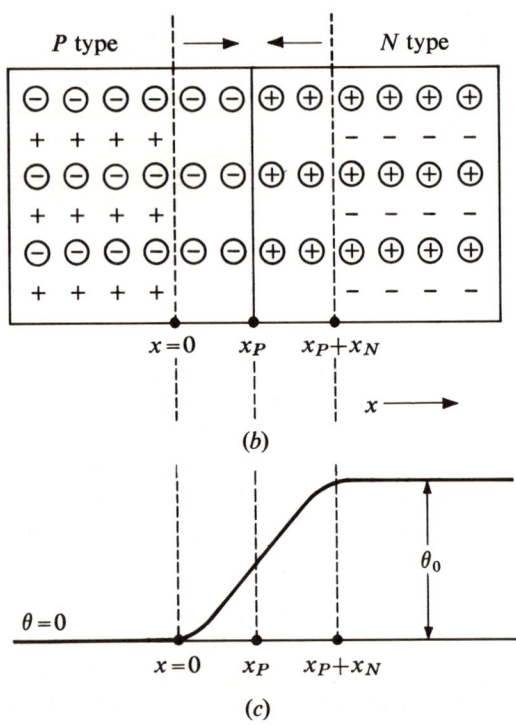

Fig. 5.3
(*a*) Carrier concentration in *P*- and *N*-type materials before the junction is formed. (*b*) Space-charge layer in a *P-N* junction. (*c*) Potential profile of a *P-N* junction (for holes)

boundary between the two regions, the regrown P type and the original N type, is the P-N junction.

Diffusion. If a material (semiconducting) is heated in the presence of a donor or acceptor gas, the gas atoms diffuse into the material and thus alter the conductivity properties of the material at the surface, forming a diffused junction. The depth of penetration of impurity gas atoms is determined by the time and temperature of the diffusion process and increases with both time and temperature.

Basic theory of the junction. Let us consider the situation in P- and N-type materials before the junction has been fabricated or formed. Figure 5.3(a) shows the two types of materials separately. The P-type material has an excess of holes, and electrons are the minority carriers. The number of holes is much larger than the number of electrons, so only acceptors having negative charges and holes are shown in the figure. This fact can be substantiated as follows.

In a P-type germanium of resistivity $\frac{1}{100}$ Ω m at room temperature,

$$p = 3.68 \times 10^{21} \text{ m}^{-3},$$

$$n = 1.70 \times 10^{17} \text{ m}^{-3}.$$

In other words, if an electron is to be shown in the P-type material of Fig. 5.3(a) we would have to represent approximately 10,000 holes, because for each 10^4 holes we have an electron. Similarly, in the N-type material shown in Fig. 5.3(a), only the presence of electrons and ionized donors is indicated. Let us now consider that a junction has been fabricated and that the transition from P-type to N-type material is abrupt. Alloyed junctions are normally abrupt. As soon as the junction is fabricated, charge carriers can flow from one material to another. Electrons that are in the majority in N-type material will diffuse into P-type material, and holes that are in the majority in P-type material will diffuse into N-type material. Electrons crossing over from N type to P type meet holes after crossing the boundary, recombine with holes, and leave behind a sheet of ionized acceptors in the vicinity of the boundary.

Similarly, holes crossing over to N-type material recombine with electrons, leaving behind ionized donors. Equilibrium is reached when only a few majority charge carriers can cross in either direction because of the space charge of opposite sign opposing the crossover. Besides the majority charge carriers, those minority carriers that can reach the junction find favorable conditions because of space charge at the junction and hence can cross over easily, but minority carriers are few. In equilibrium, therefore, there is a flow of majority and minority carriers from both materials, but the sum of the currents due to charge carriers is zero.

If I_1 is the hole current from P to N type, I_2 the electron current from N to P type, I_3 the electron current from P to N type, and I_4 the hole current from N to P type, the condition in equilibrium in the P-N junction not connected externally is

$$I_1 + I_2 + I_3 + I_4 = 0.$$

Figure 5.3(b) shows the space-charge layer in a P-N junction. The space-charge layer is also known as the *depletion region.* Let us consider the space-charge layer and let x_P be the thickness of the space-charge layer in P-type material and x_N be the thickness of the space-charge layer in N-type material. The space-charge layers give rise to a potential barrier which can be determined by solving Poisson's equations in one dimension for the two regions. If N_A is the number of acceptor atoms per unit volume in P-type material, the charge density in the reg on from $x = 0$ to $x = x_P$ is $(-eN_A)$, because the acceptors are all negatively ionized in this region. Similarly, if N_D is the number of donor atoms per unit volume in N-type material, the charge density in the region

$$x = x_P \text{ to } x_P + x_N \text{ is } (eN_D).$$

If θ_1 is the potential in the region from $x = 0$ to $x = x_P$ and θ_2 is the potential in the region from $x = x_P$ to $x = x_P + x_N$, Poisson's equations for the two regions are: From $x = 0$ to $x = x_P$,

$$\frac{d^2\theta_1}{dx^2} = -\frac{\rho}{\epsilon} = \frac{eN_A}{\epsilon}, \tag{5.57}$$

where ρ is the volume charge density and ϵ is the dielectric constant of the parent material. From $x = x_P$ to $x_P + x_N$,

$$\frac{d^2\theta_2}{dx^2} = -\frac{eN_D}{\epsilon}, \tag{5.58}$$

Integrating Eq. (5.57) once,

$$\frac{d\theta_1}{dx} = \frac{eN_A}{\epsilon}x + A,$$

where A is the integrating constant.

Assuming the potential of the P-type material to be zero and considering the fact that the entire P-type material on the left of $x = 0$ is neutral, $d\theta_1/dx = 0$ at $x = 0$. Hence $A = 0$, and, therefore,

$$\frac{d\theta_1}{dx} = \frac{eN_A x}{\epsilon}. \tag{5.59}$$

Integrating Eq. (5.59) once more,

$$\theta_1 = \frac{eN_Ax^2}{2\epsilon} + B,$$

where B is the integration constant. $\theta_1 = 0$ at $x = 0$ (assumed) leads us to the fact that $B = 0$. Hence

$$\theta_1 = \frac{eN_Ax^2}{2\epsilon}. \tag{5.60}$$

Equation (5.60) shows that in the region from $x = 0$ to $x = x_P$ the potential rises quadratically with distance. This has been plotted in Fig. 5.3(c). The potential at $x = x_N$ is

$$\theta_1(x = x_P) = \frac{eN_Ax_P^2}{2\epsilon}. \tag{5.61}$$

The potential in the region $x = x_P$ to $x = x_N + x_P$ can be determined by integrating Eq. (5.58) twice with proper boundary conditions.

Integrating Eq. (5.58) once gives

$$\frac{d\theta_2}{dx} = -\frac{eN_D}{\epsilon}x + C, \tag{5.62}$$

where C is the constant of integration. The electric field must be continuous in the space-charge region and hence

$$\frac{d\theta_1}{dx}(x = x_P) = \frac{d\theta_2}{dx}(x = x_P). \tag{5.63}$$

The left-hand side of Eq. (5.63) is given by Eq. (5.59) if $x = x_P$,

$$\text{l.h.s.} = \frac{eN_Ax_P}{\epsilon}.$$

The right-hand side of Eq. (5.63) is given by Eq. (5.62) if $x = x_P$,

$$\text{r.h.s.} = -\frac{eN_Dx_P}{\epsilon} + C.$$

Hence

$$C = \frac{ex_P}{\epsilon}(N_A + N_D),$$

and therefore substituting the value of C in Eq. (5.62),

$$\frac{d\theta_2}{dx} = -\frac{eN_Dx}{\epsilon} + \frac{ex_P}{\epsilon}(N_A + N_D). \tag{5.64}$$

Integrating Eq. (5.64) with respect to x,

$$\theta_2(x) = -\frac{eN_Dx^2}{2\epsilon} + \frac{ex_P}{\epsilon}(N_A + N_D)x + D, \tag{5.65}$$

where D is the constant of integration.

The potential θ_2 at $x = x_P$ must be the same as θ_1 at $x = x_P$. Therefore,

$$\theta_2(x = x_P) = -\frac{eN_D x_P^2}{2\epsilon} + \frac{ex_P^2}{\epsilon}(N_A + N_D) + D, \tag{5.66}$$

and the right sides of Eqs. (5.61) and (5.66) must be equal:

$$\theta_1(x = x_P) = \frac{eN_A x_P^2}{2\epsilon} = \theta_2(x = x_P) = -\frac{eN_D x_P^2}{\epsilon} + -\frac{ex_P^2}{\epsilon}(N_A + N_D) + D.$$

Simplifying,

$$D = -\frac{ex_P^2}{2\epsilon}(N_A + N_D).$$

The potential θ_2 at any point in the region $x = x_P$ to $x = x_P + x_N$ may now be obtained by substituting for the constant D in Eq. (5.65):

$$\theta_2(x) = -\frac{eN_D x^2}{2\epsilon} + \frac{ex_P}{\epsilon}(N_A + N_D)x - \frac{ex_P^2}{2\epsilon}(N_A + N_D). \tag{5.67}$$

The potential at the point $x = x_N + x_P$ has been shown to be θ_0 in Fig. 5.3(c). This potential, with its reference zero potential at $x = 0$, is called the *contact potential*. As we have seen, the contact potential has developed as a result of the formation of the junction:

$$\theta_0 = \theta_2(x = x_N + x_P) = -\frac{eN_D(x_N + x_P)^2}{2\epsilon} + \frac{ex_P}{\epsilon}(N_A + N_D)(x_N + x_P)$$

$$- \frac{ex_P^2}{2\epsilon}(N_A + N_D). \tag{5.68}$$

In Eq. (5.68) x_P can be expressed in terms of the total thickness of the depletion region by noting that at $x \geq x_N + x_P$, the N-type material is neutral and therefore the associated electric field is zero.

Therefore, from Eq. (5.64),

$$\frac{d\theta_2}{dx}(x = x_P + x_N) = 0 = -\frac{eN_D(x_P + x_N)}{\epsilon} + \frac{ex_P(N_A + N_D)}{\epsilon},$$

and hence

$$x_P = \frac{N_D}{N_D + N_A}(x_P + x_N). \tag{5.69}$$

Equation (5.69) can be further simplified to show that

$$x_P N_A = x_N N_D. \tag{5.70}$$

Equation (5.70) shows the equality of charge transferred from donors to acceptors. Substituting x_P in terms of $x_P + x_N$ from Eq. (5.69) in Eq. (5.68) leads to

$$\theta_0 = \frac{e}{2\epsilon}\left(\frac{N_D N_A}{N_D + N_A}\right)(x_P + x_N)^2. \tag{5.71}$$

The contact potential for given N_D and N_A is given by

$$x_P + x_N = \left(\frac{2\epsilon\theta_0}{e}\right)^{1/2}\left(\frac{1}{N_A} + \frac{1}{N_D}\right)^{1/2}. \tag{5.72}$$

If $N_A \gg N_D$,

$$x_P + x_N = \left(\frac{2\epsilon\theta_0}{e}\frac{1}{N_D}\right)^{1/2}. \tag{5.73}$$

Equation (5.72) shows that the junction gets narrower as the impurity concentration is increased. It can also be seen that if the P region is heavily doped, the potential is confined mainly to the N region.

Having established the fact that there exists a potential barrier as a result of the formation of the P-N junction, it is now desired to estimate the operation of the junction when an external voltage source is connected across the junction.

In Fig. 5.3(c) the potential diagram is for holes. Holes crossing from P-type material have to cross over the hill before entering N-type material. A similar potential hill would exist for electrons crossing from N-type material to P-type material.

Consider once more the condition when no external voltage is applied to the junction. We have already discussed above that the net flow of current across the junction I in this case is zero, and hence

$$I = 0 = I_1 + I_2 + I_3 + I_4.$$

I_1 is defined as the hole current from the P side to the N side. If p_P is the equilibrium hole density in P-type material, only a small fraction of holes, with energy equal to or greater than $e\theta_0$, will be in a position to cross over, and hence

$$I_1 = C_1 p_P e^{-e\theta_0/kT},$$

where C_1 is a constant.

I_2 is defined as the electron current from the N side to the P side. If n_N is the equilibrium electron density in N-type material, only those electrons with energy equal to or greater than $e\theta_0$ will be in a position to cross over, and hence

$$I_2 = C_2 n_N e^{-e\theta_0/kT}.$$

I_3 and I_4 are the current contributions due to minority carriers crossing from N to P type and vice versa. The minority carriers can cross over the junction as they approach because the space-charge region is a potential well for them and hence their current contribution is proportional to the density of minority carriers, so

$$I_3 = C_3 n_P,$$

$$I_4 = C_4 p_N,$$

where C_3 and C_4 are constants. n_P and p_N are the minority carrier concentrations in the P-type and the N-type materials, respectively.

Considering the flow of holes from P type to N type as positive current,

$$I = C_1 p_P e^{-e\theta_0/kT} + C_2 n_N e^{-e\theta_0/kT} - C_3 n_P - C_4 p_N = 0$$

or

$$C_3 n_P + C_4 p_N = C_1 p_P e^{-e\theta_0/kT} + C_2 n_N e^{-e\theta_0/kT}. \tag{5.74}$$

If a positive potential θ is applied to the P side, the potential hill for majority carriers crossing the junction gets lower and becomes $(-\theta + \theta_0)$; the number of majority carriers crossing the junction increases but the number of minority carriers crossing the junction remains the same, and hence

$$I = -(C_3 n_P + C_4 p_N) + (C_1 p_P + C_2 n_N) \exp\left[-\frac{e(\theta_0 - \theta)}{kT}\right]. \tag{5.75}$$

If a negative potential $-\theta$ is applied to the P side, the potential hill for majority carriers crossing the junction gets higher and becomes $\theta + \theta_0$, and the net current flowing across the junction will be

$$I = -(C_3 n_P + C_4 p_N) + (C_1 p_P + C_2 n_N) \exp\left[-\frac{e(\theta_0 + \theta)}{kT}\right]. \tag{5.76}$$

If the negative potential connected to the P side is of sufficiently large magnitude, the second term on the right side of Eq. (5.76) is negligible and

$$I = -(C_3 n_P + C_4 p_N) = -I_0, \tag{5.77}$$

where I_0 is the maximum reverse current or the reverse saturation current. It consists only of the current flow due to minority carriers, the majority carrier current contribution being negligible because of the high potential hill the majority carriers have to climb in the presence of a negative potential of sufficiently large magnitude. Using Eqs. (5.74) and (5.77), Eq. (5.75) may be rewritten

$$I = I_0(e^{e\theta/kT} - 1). \tag{5.78}$$

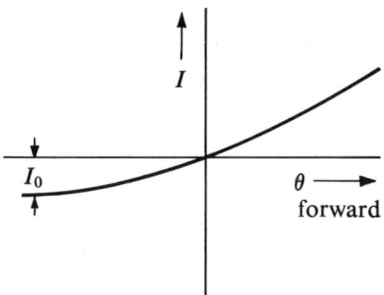

Fig. 5.4
Current-voltage characteristics of *P-N*-junction diode

Equation (5.78) is called the *junction diode equation*. Figure 5.4 shows a plot of I versus V.

Capacitance of the P-N Junction. The space-charge double layer in the *P-N* junction has properties similar to a charged condenser. If C is the capacitance per unit area, one finds that

$$C = \frac{\epsilon}{x_P + x_N}, \tag{5.79}$$

a formula which is the same as the capacitance of a parallel-plate capacitor. Substituting the value of $x_P + x_N$ from Eq. (5.72) in Eq. (5.79),

$$C = \epsilon \left(\frac{e}{2\epsilon\theta_0}\right)^{1/2} \left(\frac{N_A N_D}{N_A + N_D}\right)^{1/2} = \left[\frac{e\epsilon N_A N_D}{2\theta_0(N_A + N_D)}\right]^{1/2}. \tag{5.80}$$

Equation (5.72) shows that the thickness of the junction is proportional to $\theta_0^{1/2}$. A forward bias of θ reduces the height of the effective potential barrier between *P*- and *N*-type material to $\theta_0 - \theta$; hence the thickness of the depletion region decreases with increasing forward bias and is proportional to $(\theta_0 - \theta)^{1/2}$. A reverse bias of θ volts similarly increases the height of the potential hill. The thickness of the junction increases with increasing reverse bias.

Equation (5.80) may therefore be modified to define the capacitance of a biased *P-N* junction,

$$C = \left[\frac{\epsilon N_A N_D e}{2(\theta_0 \pm \theta)(N_A + N_D)}\right]^{1/2}, \tag{5.81}$$

where the minus sign in the denominator of Eq. (5.81) holds for forward bias and the plus sign for reverse bias.

For a heavily doped *P*-type material in the depletion region, $N_A \gg N_D$,

$$C = \left[\frac{\epsilon e N_A}{2(\theta_0 \pm \theta)}\right]^{1/2}. \tag{5.82}$$

A junction capacitor varies in capacitance with applied signals. The above equations of the capacitance of a *P-N* junction hold good only for an ideal abrupt junction. When the junction is not abrupt, the capacitance of the junction may be expressed

$$C \propto (\theta_0 \pm \theta)^{-x}, \tag{5.83}$$

where x depends upon the transition from N to P type. As already discussed, $x = \frac{1}{2}$ for abrupt junctions. It may be shown by a similar treatment that for a *P-N* junction made with a linear transition in the depletion region from P to N type, $x = \frac{1}{3}$.

Theory of the junction from continuity equations. To obtain the boundary conditions required to solve the continuity equations for the junction, it is necessary to consider the junction from the point of view of energy bands. In Fig. 5.5 the energy-band diagrams of the separate *P*- and *N*-type materials are shown. At any temperature $T > 0$, the Fermi level in the *N*-type material lies somewhere between $(E_c + E_D)/2$ and E_{IN}, depending upon the temperature and concentration of impurities. E_{IN} is the intrinsic Fermi level in the *N*-type material. Similarly, the Fermi level in *P*-type material at any temperature $T > 0$ lies somewhere between $(E_v + E_A)/2$ and E_{IP}, depending upon the temperature and concentration of impurities. E_{IP} is the intrinsic Fermi level of the *P*-type material. The actual Fermi levels of the two materials have been shown as E_{fN} and E_{fP} in the *N*- and *P*-type materials, respectively, in Fig. 5.5. When these two dissimilar materials are brought together to form a junction in the manner discussed previously, a *P-N* junction is formed. Charge carriers from the *N*-type material having a higher Fermi level will spill over into the *P*-type material of lower Fermi level to equalize the Fermi levels.

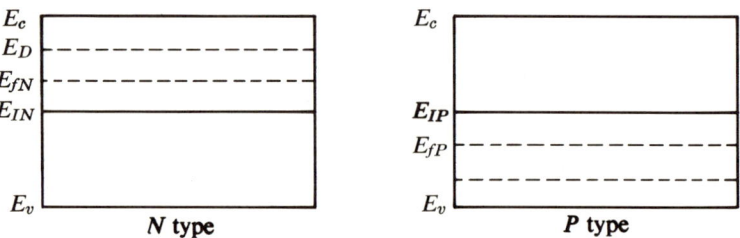

Fig. 5.5
Energy diagram for *P*- and *N*-type materials before the junction is formed

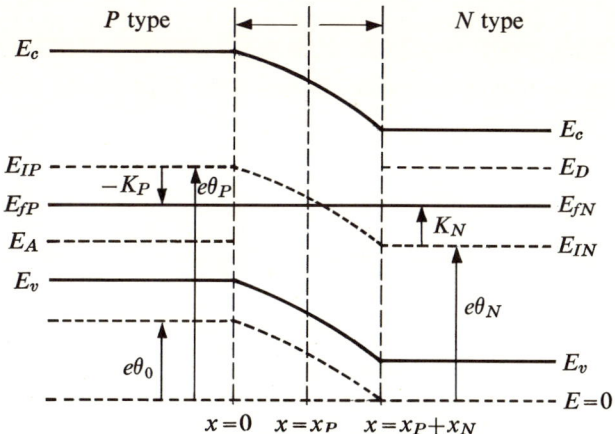

Fig. 5.6
Energy bands for a *P-N* junction

Now, considering the junction to be in equilibrium, E_f is uniform over the junction, E_{fN} is near the top of the forbidden energy gap in the *N*-type material, and E_{fP} is near the bottom of the forbidden energy gap. The only way for these conditions to be satisfied simultaneously is for the energy bands to adjust themselves in equilibrium as shown in Fig. 5.6 The *N*-type material is taken as the reference. Considering holes as charge carriers, $E_{fP} = E_{fN}$. Substituting the definition of the Fermi level in terms of the electrochemical potential,

$$E_{fP} = e\theta_P - K_P,$$

$$E_{fN} = e\theta_N + K_N,$$

where K_P is the chemical potential of *P*-type material and K_N is the chemical potential of *N*-type material. Therefore,

$$e\theta_P - K_P = e\theta_N + K_N$$

or

$$K_P + K_N = e(\theta_P - \theta_N), \tag{5.84}$$

where θ_P and θ_N are the electrostatic potentials in *P*- and *N*-type materials:

$$K_P = kT \ln \frac{p_P}{n_i}, \tag{5.85}$$

$$K_N = kT \ln \frac{n_i}{p_N}. \tag{5.86}$$

Equations (5.85) and (5.86) are derived from Eqs. (5.23) and (5.24). p_p is the carrier concentration of holes in P-type material and p_N is the hole concentration in N-type material. Substituting Eqs. (5.85) and (5.86) in Eq. (5.84),

$$kT \ln \frac{p_P}{n_i} + kT \ln \frac{n_i}{p_N} = e(\theta_P - \theta_N).$$

Simplifying,

$$kT \ln \frac{p_P}{p_N} = e(\theta_P - \theta_N),$$

and hence

$$p_P = p_N \exp\left[\frac{e(\theta_P - \theta_N)}{kT}\right]. \qquad (5.87a)$$

$\theta_P - \theta_N$ is called the *contact potential* θ_0, and hence

$$p_P = p_N e^{e\theta_0/kT}. \qquad (5.87b)$$

Referring to Fig. 5.6,

$$p(x = x_P + x_N) = p_N = p_P e^{-e\theta_0/kT}, \qquad (5.88)$$

$$p(x = 0) = p_P = p_N e^{e\theta_0/kT}. \qquad (5.89)$$

If the junction shown again in Fig. 5.7 is biased in the forward direction by connecting the P side to the positive potential, the potential barrier is reduced to $\theta_0 - \theta$, and hence

$$p(x = x_P + x_N) = p_P \exp\left[-\frac{e(\theta_0 - \theta)}{kT}\right].$$

Using Eq. (5.88),

$$p(x = x_P + x_N) = p_N e^{e\theta/kT}. \qquad (5.90)$$

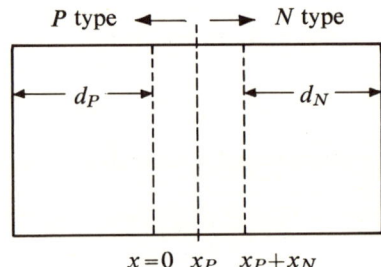

Fig. 5.7
P-N junction

P-N JUNCTION THEORY

Equation (5.90) suggests that the concentration of holes or minority carriers increases from its equilibrium value p_N if a forward-bias potential θ is applied to the junction. In a similar way it can be shown that the concentration of electrons increases from its equilibrium value n_P at $x = 0$ if the junction is forward-biased:

$$n(x = 0) = n_P e^{e\theta/kT}. \tag{5.91}$$

The minority carriers are injected if a forward bias is applied. These excess injected minority carriers diffuse away from the junction. The flow of minority carriers is explained by the continuity equation and results in the diode equation (5.78).

The continuity equation for minority carriers in a P-type material in equilibrium is given by Eq. (5.48),

$$0 = \frac{n_P - n}{\tau_n} + D_n \frac{d^2 n}{dx^2}. \tag{5.48}$$

It may be mentioned that this equation was derived with no externally applied field. We have concluded above that an applied field changes the minority concentration at $x = 0$, and we shall use this result as a boundary condition for the solution of the above equation.

If we connect highly conducting metal to both P- and N-type materials at $x = d_N + x_N + x_P$ and $x = -d_P$,

$$\frac{dn}{dx} = 0 \quad \text{at } x = -d_P,$$

$$\frac{dp}{dx} = 0 \quad \text{at } x = d_N + x_N + x_P.$$

Equation (5.48) can be solved with the help of the boundary conditions given for the minority carrier concentration $n(x)$.

The electron current density is due to those electrons that cross the junction:

$$J_n = eD_n \left(\frac{dn}{dx}\right)_{x=0}.$$

Solving Eq. (5.48), we obtain

$$n = Ae^{-x/L_n} + Be^{x/L_n} + n_P, \tag{5.92}$$

where A and B are constants of integration. At $x = 0$,

$$n = n_P e^{e\theta/kT}. \tag{5.91}$$

Therefore,

$$n_P e^{e\theta/kT} = (A + B) + n_P$$

or

$$A + B = n_P(e^{e\theta/kT} - 1). \tag{5.93}$$

From Eq. (5.92), differentiating with respect to x,

$$\frac{dn}{dx} = -\frac{A}{L_n}e^{-x/L_n} + \frac{B}{L_n}e^{x/L_n}. \tag{5.94}$$

At $x = -d_P$, $dn/dx = 0$:

$$0 = -\frac{A}{L_n}e^{d_P/L_n} + \frac{B}{L_n}e^{-d_P/L_n}. \tag{5.95}$$

From Eqs. (5.93) and (5.95) the constants A and B can be evaluated:

$$A = -B + n_P(e^{e\theta/kT} - 1),$$

$$0 = +\frac{B}{L_n}e^{d_P/L_n} - \frac{n_P}{L_n}e^{d_P/L_n}(e^{e\theta/kT} - 1) + \frac{B}{L_n}e^{-d_P/L_n}.$$

Hence

$$B = \frac{(n_P/L_n)e^{d_P/L_n}(e^{e\theta/kT} - 1)}{(1/L_n)(e^{-d_P/L_n} + e^{d_P/L_n})}. \tag{5.96}$$

Simplifying,

$$B = \frac{n_P}{1 + e^{-2d_P/L_n}}(e^{e\theta/kT} - 1),$$

$$A = \frac{-n_P}{1 + e^{-2d_P/L_n}}(e^{e\theta/kT} - 1) + n_P(e^{e\theta/kT} - 1).$$

Simplifying,

$$A = n_P(e^{e\theta/kT} - 1)\left(\frac{e^{-2d_P/L_n}}{1 + e^{-2d_P/L_n}}\right). \tag{5.97}$$

Substituting A and B from Eqs. (5.96) and (5.97) in Eq. (5.94),

$$\frac{dn}{dx} = -\frac{n_P}{L_n}(e^{e\theta/kT} - 1)\left(\frac{e^{-2d_P/L_n}}{1 + e^{-2d_P/L_n}}\right)e^{-x/L_n} + \frac{n_P}{L_n}(e^{e\theta/kT} - 1)\frac{1}{1 + e^{-2d_P/L_n}}. \tag{5.98}$$

From Eq. (5.98),

$$\left.\frac{dn}{dx}\right)_{x=0} = -\frac{n_P}{L_n}(e^{e\theta/kT} - 1)\frac{e^{-2d_P/L_n}}{1 + e^{-2d_P/L_n}} + \frac{n_P}{L_n}(e^{e\theta/kT} - 1)\frac{1}{1 + e^{-2d_P/L_n}}.$$

Simplifying further,

$$\left.\frac{dn}{dx}\right)_{x=0} = \frac{n_P}{L_n}(e^{e\theta/kT} - 1)\frac{1 - e^{-2d_P/L_n}}{1 + e^{-2d_P/L_n}}$$

or

$$\left.\frac{dn}{dx}\right)_{x=0} = \frac{n_P}{L_n}(e^{e\theta/kT} - 1)\tanh\frac{d_P}{L_n}. \tag{5.99}$$

The current contribution due to electrons crossing from P-type to N-type material is given by

$$J_n = eD_n\left.\frac{dn}{dx}\right)_{x=0}.$$

Substituting $dn/dx)_{x=0}$ from Eq. (5.99) in the above equation,

$$J_n = +\frac{eD_n n_P}{L_n}\left(\tanh \frac{d_P}{L_n}\right)(e^{e\theta/kT} - 1). \tag{5.100}$$

In a similar manner the contribution of holes crossing the junction can be obtained by solving the continuity equation for holes:

$$0 = \frac{p_N - p}{\tau_p} + D_p \frac{d^2p}{dx^2}. \tag{5.101}$$

For the boundary conditions,

$$p(x = x_N + x_P) = p_N e^{e\theta/kT},$$

$$\frac{dp}{dx} = 0 \quad \text{at } x = d_N + x_N + x_P.$$

The hole current density results from those holes crossing the junction,

$$J_p = -eD_p \frac{dp}{dx}\bigg)_{x = x_N + x_P}.$$

Assuming that d_N is much larger than the depletion region,

$$d_N \gg x_N + x_P,$$

and the depletion region d_N is much smaller than the diffusion length of holes,

$$L_p \gg x_N + x_P.$$

It can be easily shown that

$$J_p = \frac{eD_p p_N}{L_p} \tanh \frac{d_N}{L_p}(e^{e\theta/kT} - 1). \tag{5.102}$$

The total current density for a forward bias θ is obtained by summing up Eqs. (5.100) and (5.102),

$$J = J_p + J_n$$

or

$$J = \left(\frac{eD_n n_P}{L_n} \tanh \frac{d_P}{L_n} + \frac{eD_p p_N}{L_p} \tanh \frac{d_N}{L_p}\right)(e^{e\theta/kT} - 1). \tag{5.103}$$

Putting

$$J_0 = \frac{eD_n n_P}{L_n} \tanh \frac{d_P}{L_n} + \frac{eD_p p_N}{L_p} \tanh \frac{d_N}{L_p}, \tag{5.104}$$

Eq. (5.103) may be rewritten

$$J = J_0(e^{e\theta/kT} - 1). \tag{5.105}$$

The only difference between Eqs. (5.78) and (5.105) is that the former is expressed in terms of current and the latter in terms of current density. If θ is negative and large,

$$J = -J_0,$$

where $-J_0$ is the reverse saturation current density under back-biased conditions. According to Eq. (5.104),

$$J_0 = \left(\frac{eD_n n_P}{L_n} \tanh \frac{d_P}{L_n} + \frac{eD_p p_N}{L_p} \tanh \frac{d_N}{L_p}\right). \qquad (5.104)$$

If we restrict our discussion to diodes in which the thickness of the P and N layers is much greater than the diffusion length of minority carriers,

$$d_P \gg L_n,$$
$$d_N \gg L_p.$$

Therefore $\tanh(d_P/L_n) \simeq 1$ and $\tanh(d_N/L_p) \simeq 1$. Hence

$$J_0 = \frac{eD_n n_P}{L_n} + \frac{eD_p p_N}{L_p}. \qquad (5.106)$$

Equation (5.106) is commonly encountered in the literature as the equation describing the reverse saturation current density. It must always be considered that the expression describes the reverse saturation current density only if $d_P \gg L_n$ and $d_N \gg L_p$.

According to measurements, the lifetime of minority carriers is of the order of 100 μsec for some germanium samples. So if

$$\tau_n = \tau_p = 100\mu \text{ sec},$$
$$\mu_n = 0.39 \text{ m}^2/\text{V sec},$$
$$\mu_p = 0.19 \text{ m}^2/\text{V sec},$$
$$L_p = \left(\frac{kT}{e}\right)^{1/2} \mu_p \tau_p \cong 0.10 \text{ cm},$$
$$L_n = \left(\frac{kT}{e}\right)^{1/2} \mu_n \tau_n \cong 0.07 \text{ cm},$$

$d_N \gg 0.10$ cm and $d_P \gg 0.07$ cm.

Equation (5.106) may be obtained directly from the solution of the continuity equation if the assumption stated above is incorporated in the boundary conditions of the solution of the continuity equations:

$$d_N \gg L_p \quad \text{and} \quad d_P \gg L_n,$$

which mean that for $x \to \infty$,

$$p = p_N,$$

and for $x \to -\infty$,

$$n = n_P.$$

Equations (5.78) and (5.105) are called the *Shockley diode equations* for ideal junctions. Impurities and imperfections in the materials lead to a junction whose $V - I$ characteristics differ from those of an ideal junction.

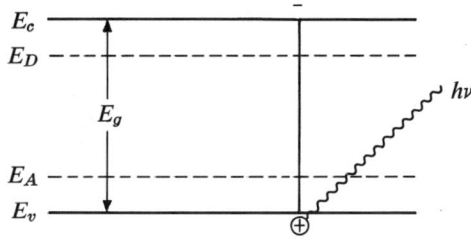

Fig. 5.8
Energy diagram of a semiconductor with donor and acceptor levels

5.5 Optical effects in *P-N* junctions

In this section we shall be concerned with the effect of incident electro-radiation on a *P-N* junction. In Fig. 5.8 the energy diagram of a semimagnetic conductor with donor and acceptor levels is shown. In equilibrium there is a certain carrier concentration of charge carriers which could be calculated by the procedure given in Sec. 4.6. If a photon of energy $h\nu$ falls on the semiconductor material and is absorbed by the material, the effect of the absorbed photon in the semiconductor depends upon the energy of the incident photon. Even a heavily doped material contains 10^3 to 10^4 parent atoms for every impurity atom, so the probability of a photon being absorbed by a parent atom is much larger than the probability of a photon being absorbed by an impurity atom. For example, 10^{19} impurity atoms per cm^3 is a very heavy doping. The number of atoms of germanium in a crystal is approximately 10^{22} cm^{-3}, showing that the number of Ge atoms would be much greater than the number of impurity atoms, even in a highly doped specimen of germanium of either *P* or *N* type.

The result of the absorption of a photon by a semiconductor depends upon the energy of the photon. If $h\nu < E_g$, the photon may be absorbed by a phonon, thus heating the semiconductor. If $h\nu \geq E_g$, the absorption of a photon usually results in the formation of electron-hole pairs. The valence electron gains sufficient energy to jump into the conduction band, leaving behind a hole in the valence band. An exact amount of energy E_g is required for the electron to jump from the valence band into the conduction band, so the difference in energy is $h\nu - E_g$, if $h\nu > E_g$ would be used by the electron as additional thermal energy.

The generation of electron-hole pairs by incident radiation increases the carrier concentration. The percentage increase in minority carriers is much more appreciable than in majority carriers. In a semiconductor bulk material the increase in carrier concentration results in photoconduc-

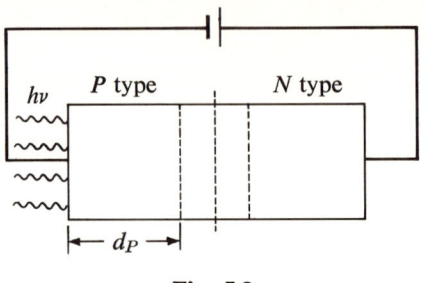

Fig. 5.9
Photo cell

tivity. The increase of carrier concentration in a P-N junction due to incident radiant energy helps us to use the junction in photo cells and in radiant-energy-conversion devices.

Photo cells. Photo cells are photoelectric detectors of incident radiation. The exposed layer in Fig. 5.9 is made very thin. In fact $d_P < L_n$, the diffusion length of minority carriers in P-type material, so electron-hole pairs formed in the surface may diffuse to the junction before recombining. Once the minority carriers reach the junction because of the potential well at the junction, they can easily cross over the junction. The resistance of an unexposed P-N junction is a function of the applied voltage. This can be shown as follows:

$$I = I_0(e^{e\theta/kT} - 1). \tag{5.78}$$

Differentiating Eq. (5.78) with respect to I,

$$I = I_0 e^{\theta e/kT} \frac{e}{kT} \frac{d\theta}{dI}$$

or

$$R = \frac{d\theta}{dI} = \left(\frac{kT}{eI_0}\right) e^{-e\theta/kT}. \tag{5.107}$$

The resistance of the P-N junction is very large in the reverse-biased case (Fig. 5.10). In photo cells the junction is back-biased and exposed to radiation. Incident radiation creates hole-electron pairs, and the minority carriers cross over the junction and increase the reverse saturation current. The resistance of the junction is very large, so the change in conductivity due to absorption of photons is very large and the incident radiation is easily detected.

Photovoltaic Cells. Photovoltaic cells are P-N junctions and serve to convert incident radiant energy into useful electrical energy. No external voltage source is needed. The junction itself turns into a voltage source or a

OPTICAL EFFECTS IN P-N JUNCTIONS

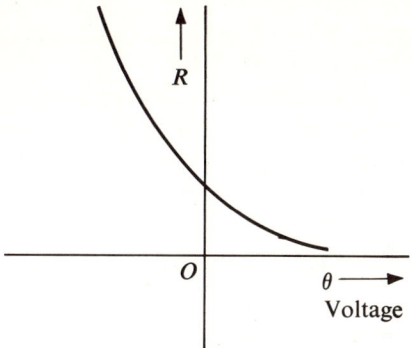

Fig. 5.10
Variation of the resistance of an unexposed P-N junction as a function of applied voltage

generator of electrical power if it is exposed to radiation of frequency ν such that $h\nu \geq E_g$. The minority carriers generated in the P-type material by absorption of incident radiation cross over the junction and can flow through an external resistance R_L (Fig. 5.11), thus converting the incident radiant energy into electrical energy. The current flowing through the resistance R_L causes a potential drop across this resistor, which opposes the flow of the light-generated current. Taking the direction of the current generated by minority carriers crossing the junction to be positive (this is just the opposite of a P-N junction biased externally), the net current density J of the current through the resistance R_L is given by

$$J = J_R - J_0(e^{e\theta_L/kT} - 1), \tag{5.108}$$

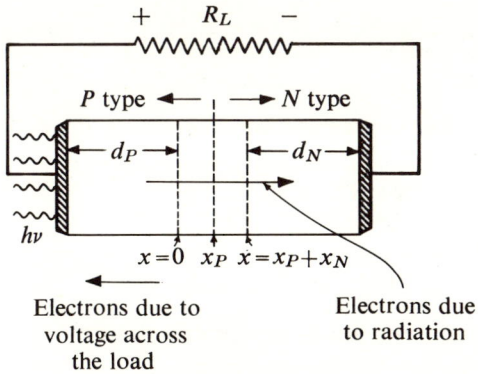

Fig. 5.11
Photovoltaic cell

where J_R is the current density generated by absorption of incident radiation and θ_L is the voltage across the load.

Equation (5.108) may be derived by writing down the continuity equation and solving it for the minority carrier concentration. The continuity equation for the minority carriers in a P-type interval in the presence of incident electromagnetic radiation may be written

$$\frac{dn}{dt} = g(T) - R + g(x) + D_n \frac{d^2n}{dx^2}, \qquad (5.109)$$

where it was shown in Sec. 5.3 that $g(T)$ is the thermal rate of generation of electrons, R is the recombination rate of holes, and $g(T) - R = (n_P - n)/\tau_n$. In equilibrium $dn/dt = 0$, and Eq. (5.109) may be written

$$\frac{n_P - n}{\tau_n} + g(x) + D_n \frac{d^2n}{dx^2} = 0, \qquad (5.110)$$

where $g(x)$ is the generation rate of electrons due to incident radiation at a distance x. The absorption coefficient of the P-type material is different for different wavelengths, so it would be more exact to solve the equation for monochromatic incident radiation.

If Q_λ is the number of photons of wavelength λ falling per unit area on P-type material and $(L_\lambda)^{-1}$ is the absorption coefficient for wavelength λ, the generation rate due to light is, by definition of the absorption coefficient,

$$g(x) = \frac{Q_\lambda}{L_\lambda} \exp\left(-\frac{x + d_P}{L_\lambda}\right). \qquad (5.111)$$

Substituting Eq. (5.111) in Eq. (5.110),

$$\frac{n_P - n}{\tau_n} + \frac{Q_\lambda}{L_\lambda} \exp\left(-\frac{x + d_P}{L_\lambda}\right) + D_n \frac{d^2n}{dx^2} = 0. \qquad (5.112)$$

Equation (5.112) may be solved for $n(x)$ using the following boundary conditions:

$$\frac{dn}{dx} = 0 \qquad \text{at } x = -d_P,$$

$$n = n_P e^{\,e\theta_L/kT} \qquad \text{at } x = 0$$

The electron current density is the contribution of those electrons that cross the junction,

$$J_n = eD_n \left(\frac{dn}{dx}\right)_{x=0}.$$

The N-type material is not exposed to radiation, so Eq. (5.101) holds even in this case. Solving Eq. (5.101) we obtain the hole current density,

$$J_p = -eD_p \left.\frac{dp}{dx}\right)_{x=x_N+x_P}.$$

The solution of Eq. (5.112) is rather involved, so we shall assume that in equilibrium the number of electron-hole pairs generated by incident radiation in the entire P-type material is constant and equals g_R (g_R is the number of electrons generated by radiation per unit volume per second in P-type material and is constant). Equation (5.112) may, therefore, be rewritten

$$\frac{n_P - n}{\tau_n} + g_R + D_n \frac{d^2n}{dx^2} = 0. \tag{5.113}$$

Solving Eq. (5.113) we obtain

$$n = Ae^{-x/L_n} + B e^{x/L_n} + n_P + g_R \tau_n, \tag{5.114}$$

where A and B are constants of integration and

$$D_n \tau_n = L_n^2.$$

At $x = 0$, $n = n_P e^{e\theta_L/kT}$.

Therefore,

$$A + B = n_P(e^{e\theta_L/kT} - 1) - g_R \tau_n. \tag{5.115}$$

Differentiating Eq. (5.114) with respect to x,

$$\frac{dn}{dx} = -\frac{A}{L_n} e^{-x/L_n} + \frac{B}{L_n} e^{x/L_n}. \tag{5.116}$$

At $x = -d_P$, $dn/dx = 0$. Therefore,

$$0 = -\frac{A}{L_n} e^{d_P/L_n} + \frac{B}{L_n} e^{-d_P/L_n}. \tag{5.117}$$

Constants A and B may be evaluated from Eqs. (5.115) and (5.117),

$$A = -B + n_P(e^{e\theta_L/kT} - 1) - g_R \tau_n,$$

$$0 = \frac{B}{L_n} e^{d_P/L_n} - \frac{n_P}{L_n} e^{d_P/L_n}(e^{e\theta_L/kT} - 1) + \frac{g_R \tau_n}{L_n} e^{d_P/L_n} + \frac{B}{L_n} e^{-d_P/L_n},$$

and hence

$$B = \frac{(n_P/L_n)e^{d_P/L_n}(e^{e\theta_L/kT} - 1) - (g_R \tau_n/L_n)e^{d_P/L_n}}{(1/L_n)(e^{-d_P/L_n} + e^{d_P/L_n})}.$$

Simplifying,

$$B = \frac{n_P(e^{e\theta_L/kT} - 1) - g_R\tau_n}{1 + e^{2d_P/L_n}}, \quad (5.118)$$

$$A = \frac{-n_P(e^{e\theta_L/kT} - 1) + g_R\tau_n}{1 + e^{2d_P/L_n}} + n_P(e^{e\theta_L/kT} - 1) - g_R\tau_n.$$

Simplifying further,

$$A = \frac{n_P(e^{e\theta_L/kT} - 1) - g_R\tau_n}{1 + e^{2d_P/L_n}} e^{2d_P/L_n}. \quad (5.119)$$

Substituting A and B from Eqs. (5.118) and (5.119) in Eq. (5.116),

$$\frac{dn}{dx} = \frac{g_R\tau_n - n_P(e^{e\theta_L/kT} - 1)}{e^{2d_P/L_n} + 1} e^{2d_P/L_n} \frac{1}{L_n} e^{-x/L_n} + \frac{[n_P(e^{e\theta_L/kT} - 1) - g_R\tau_n]}{L_n(1 + e^{2d_P/L_n})} e^{x/L_n}. \quad (5.120a)$$

From Eq. (5.120a),

$$\left.\frac{dn}{dx}\right)_{x=0} = \frac{g_R\tau_n}{L_n}\left(\frac{e^{2d_P/L_n} - 1}{e^{2d_P/L_n} + 1}\right) + \frac{n_P}{L_n}(e^{e\theta_L/kT} - 1)\frac{1 - e^{2d_P/L_n}}{1 + e^{2d_P/L_n}}. \quad (5.120b)$$

Simplifying Eq. (5.120b),

$$\left.\frac{dn}{dx}\right)_{x=0} = \frac{g_R\tau_n}{L_n} \tanh\frac{d_P}{L_n} - \frac{n_P}{L_n}\left(\tanh\frac{d_P}{L_n}\right)(e^{e\theta_L/kT} - 1). \quad (5.121)$$

$$J_n = eD_n \left.\frac{dn}{dx}\right)_{x=0}.$$

Substituting the value of $dn/dx)_{x=0}$ from Eq. (5.121) in the above equation,

$$J_n = \frac{g_R\tau_n D_n e}{L_n}\left(\tanh\frac{d_P}{L_n}\right) - \frac{eD_n n_P}{L_n}\left(\tanh\frac{d_P}{L_n}\right)(e^{e\theta_L/kT} - 1). \quad (5.122)$$

Similarly, the current density due to holes from an N-type unexposed layer can be determined:

$$J_p = -\frac{eD_p p_N}{L_p}\left(\tanh\frac{d_N}{L_p}\right)(e^{e\theta_L/kT} - 1). \quad (5.123)$$

In deriving Eq. (5.123), it has been assumed that the penetration of radiation in the unexposed layer is negligible. The total current density is

$$J = J_n + J_p. \quad (5.124)$$

Substituting Eqs. (5.122) and (5.123) in Eq. (5.124),

$$J = \frac{g_R \tau_n D_n e}{L_n} \left(\tanh \frac{d_P}{L_n} \right) - \left(\frac{e D_n n_P}{L_n} \tanh \frac{d_P}{L_n} + \frac{e D_p p_N}{L_p} \tanh \frac{d_N}{L_p} \right)(e^{e\theta_L/kT} - 1). \quad (5.125)$$

Calling

$$J_R = \frac{e g_R \tau_n D_n}{L_n} \left(\tanh \frac{d_P}{L_n} \right) \quad (5.126)$$

and making use of the definition of J_0,

$$J_0 = \frac{e D_n n_P}{L_n} \tanh \frac{d_P}{L_n} + \frac{e D_p p_N}{L_p} \tanh \frac{d_N}{L_p}.$$

Equation (5.125) may be rewritten

$$J = J_R - J_0(e^{e\theta_L/kT} - 1),$$

which is Eq. (5.108), already discussed.

In the expression for J_R in Eq. (5.126), it can be noted that J_R is directly proportional to g_R, the number of electrons generated by radiation per unit volume in P-type material per second. If the probability of a photon producing an electron-hole pair is assumed to be unity, g_R would then be just the number of photons per second per unit volume absorbed by the P-type semiconductor. In case the thickness of the P-type material is much less than the diffusion length of electrons,

$$d_P < L_n,$$

$$\tanh \frac{d_P}{L_n} \simeq \frac{d_P}{L_n},$$

$$J_R \simeq e g_R \frac{\tau_n D_n}{L_n^2} d_P = e g_R d_P. \quad (5.127)$$

The photovoltaic cell may therefore be termed a *constant current generator*.

The generator current density J_R is proportional to the number of photons absorbed per second per unit volume. With this concept in mind (Fig. 5.12),

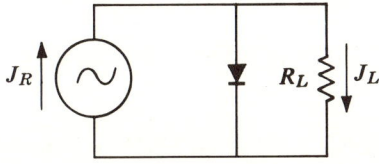

Fig. 5.12
Equivalent circuit of a photovoltaic cell

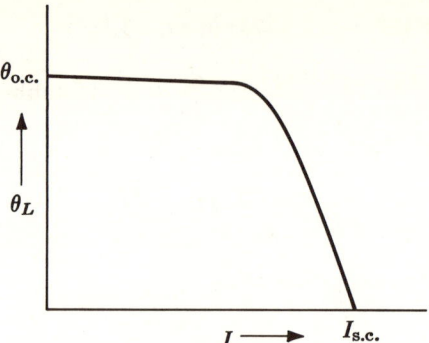

Fig. 5.13
V-I characteristics of a photovoltaic cell

the operating performance of the cell can be calculated. Equation (5.108) may be rewritten in terms of actual currents by multiplying both sides of the equation by the area of cross section of the cell:

$$I = I_R - I_0(e^{e\theta_L/kT} - 1). \tag{5.128}$$

If the cell is short-circuited $\theta_L = 0$, and the short-circuit current $I_{s.c.}$ will be given by

$$I_{s.c.} = I_R. \tag{5.129}$$

If the cell is in open circuit $I = 0$, and the open-circuit voltage $\theta_{o.c.}$ is given by

$$\theta_{o.c.} = \theta_L = \frac{kT}{e} \ln\left(\frac{I_R}{I_0} + 1\right). \tag{5.130}$$

The load characteristic, called the θ-I *characteristic*, is given by the curve shown in Fig. 5.13. The power obtained from the cell is

$$P = I_L^2 R_L = [I_R - I_0(e^{e\theta_L/kT} - 1)]^2 R_L, \tag{5.131}$$

where R_L is the load resistance. Maximum power would be obtained if R_L is appropriately adjusted. The condition for maximum power is

$$dP/dR_L = 0.$$

Differentiating Eq. (5.131) and equating it to zero, remembering that θ_L is a function of R_L,

$$\frac{dP}{dR_L} = 0 = [I_R - I_0(e^{e\theta_L/kT} - 1)]^2 - 2R_L[I_R - I_0(e^{e\theta_L/kT} - 1)]I_0 e^{e\theta_L/kT}$$

$$\times \frac{e}{kT}\frac{d\theta_L}{dR_L}.$$

Because $[I_R - I_0(e^{e\theta_L/kT} - 1)] \neq 0$,

$$I_R - I_0(e^{e\theta_L/kT} - 1) = 2R_L I_0 e^{e\theta_L/kT} \frac{e}{kT} \frac{d\theta_L}{dR_L}. \quad (5.132)$$

Now $\theta_L = [I_R - I_0 (e^{e\theta_L/kT} - 1)] R_L$, and hence

$$\frac{d\theta_L}{dR_L} = [I_R - I_0(e^{e\theta_L/kT} - 1)] - R_L I_0 e^{e\theta_L/kT} \frac{e}{kT} \frac{d\theta_L}{dR_L}.$$

Therefore,

$$\frac{d\theta_L}{dR_L} = \frac{I_R - I_0(e^{e\theta_L/kT} - 1)}{1 + I_0 R_L e^{e\theta_L/kT}(e/kT)}. \quad (5.133)$$

Substituting Eq. (5.133) in Eq. (5.132),

$$I_R - I_0(e^{e\theta_L/kT} - 1) = \frac{2R_L(I_0 e/kT)e^{e\theta_L/kT}[I_R - I_0(e^{e\theta_L/kT} - 1)]}{1 + I_0 R_L(e/kT)e^{e\theta_L/kT}}. \quad (5.134)$$

Simplifying Eq. (5.134),

$$R_L = \frac{kT}{eI_0} e^{-e\theta_{L(m.p.)}/kT}, \quad (5.135)$$

where $\theta_{L(m.p.)}$ is the voltage at the load at maximum power:

$$\theta_{L(m.p.)} = \frac{kT}{e} \ln \frac{kT}{eI_0 R_L}. \quad (5.136)$$

The load current for maximum power output $I_{L(m.p.)}$ is given by

$$I_{L(m.p.)} = \frac{kT}{eR_L} \ln \frac{kT}{eI_0 R_L}. \quad (5.137)$$

The maximum power can be obtained by the product of Eqs. (5.136) and (5.137). From Eq. (5.136),

$$\frac{e\theta_{L(m.p.)}}{kT} = \ln \frac{kT}{eI_0 R_L} \quad (5.138)$$

Further, from Eq. (5.130),

$$\frac{e\theta_{o.c.}}{kT} = \ln\left(\frac{I_R}{I_0} + 1\right),$$

$$\frac{e}{kT} \theta_{L(m.p.)} = \frac{e}{kT}[I_R - I_0(e^{e\theta_{L(m.p.)}/kT} - 1)]R_L$$

$$= \frac{eI_R R_L}{kT} + \frac{eI_0 R_L}{kT} - \frac{eI_0 R_L}{kT} e^{e\theta_{L(m.p.)}/kT},$$

or

$$\frac{e}{kT}\theta_{L(m.p.)} + 1 = \frac{eI_R R_L}{kT} + \frac{eI_0 R_L}{kT} - \frac{eI_0 R_L}{kT} e^{e\theta_{L(m.p.)}/kT} + 1. \quad (5.139)$$

In Eq. (5.139),
$$\frac{eI_0R_L}{kT}e^{e\theta_{L(m.p.)}/kT} = 1 \quad \text{[according to Eq. (5.135)]}.$$
Therefore,
$$\frac{e}{kT}\theta_{L(m.p.)} + 1 = \frac{eI_RR_L}{kT} + \frac{eI_0R_L}{kT}. \tag{5.140}$$

Multiplying the right side by kT/eI_0R_L and the left side by $e^{e\theta_{L(m.p.)}/kT}$, the two being equal according to Eq. (5.135),

$$e^{e\theta_{L(m.p.)}/kT}\left[\frac{e\theta_{L(m.p.)}}{kT} + 1\right] = \frac{I_R}{I_0} + 1. \tag{5.141}$$

The right side of Eq. (5.141) is equal to $e^{e\theta_{o.c.}/kT}$, and hence

$$e^{e\theta_{L(m.p.)}/kT}\left[\frac{e\theta_{L(m.p.)}}{kT} + 1\right] = e^{e\theta_{o.c.}/kT}. \tag{5.142}$$

The maximum power-load voltage is related to the open-circuit voltage by Eq. (5.130), and hence the maximum power depends upon the open-circuit voltage. The open-circuit voltage according to Eq. (5.130) depends upon I_R, which equals the short-circuit current.

Silicon cells are commonly used for conversion of radiant energy of the sun into electrical energy and have been fabricated with about 15 to 16 percent efficiency with total incident energy as a measure of input power. Germanium cells show great potential for conversion of radiant energy from incandescent sources to electrical energy.

Problem 1

Show that the junction capacitance of a *P-N* junction is proportional to $(\theta_0 \pm \theta)^{-1/3}$ if the concentration of impurity atoms in the depletion region varies linearly with position. θ_0 is the contact potential and $\pm\theta$ is the applied potential (positive sign for reverse bias and negative sign for forward bias).

Problem 2

The resistances of the *P* and *N* layers of a silicon abrupt *P-N* junction are 10^{-2} and $10\,\Omega$, respectively. The capacitance of the junction at zero applied bias voltage was 200 $\mu\mu$F. If the thicknesses of the *P* and *N* layers are 1 mm each and the junction has a rectangular cross section 1mm²:
a. Find the concentration of impurity atoms on each side of the junction.
b. Determine the width of the depletion region at zero applied bias.
c. Determine the contact potential.
Assume $\mu_e = 1.45 \times 10^3$ cm²/V sec and $\mu_h = 0.5 \times 10^3$ cm²/V sec.
Ans. a. $N_A = 1.25 \times 10^{19}$ cm⁻³; $N_D = 4.8 \times 10^{15}$ cm⁻³   b. 0.53×10^{-4} cm
c. 0.19 V

Problem 3

An intrinsic semiconductor is doped with acceptor atom. The concentration of acceptor atoms varies with position exponentially:

$$N_A(x) = N_{A_0} e^{-bx},$$

where N_{A_0} is the concentration of acceptors at $x = 0$. Assuming charge neutrality at each point, show that the electrostatic field is constant and is given by

$$\varepsilon = -b\frac{kT}{q}.$$

Problem 4

a. Show that the reverse saturation current density J_0 in a P-N junction is proportional to n_i^2, the square of the intrinsic carrier concentration, if $d_N \ll L_P$ and $d_P \ll L_n$.

b. Show that J_0 is a function of temperature.

Problem 5

Consider the germanium diode shown in Fig. B. The data for the diode are:
$n_P = 1.9 \times 10^8$ cm^{-3},   area of cross section of the diode $= 10^{-2}$ cm^2,
$p_N = 3.9 \times 10^{11}$ cm^{-3}, x increasing P type,
$L_p = 2.2 \times 10^{-2}$ cm, x' decreasing N type,
$L_n = 3.2 \times 10^{-2}$ cm,
$\tau_n = 10$ μsec,
$\tau_p = 10$ μsec,

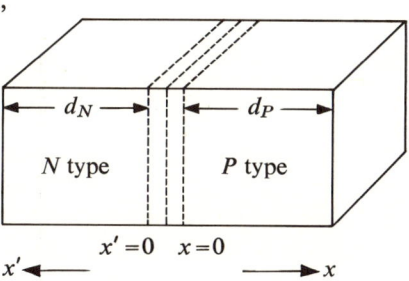

For a forward bias of 150 mV and at room temperature:
a. Evaluate the excess hole density at $x' = 0$ and the excess electron density at $x = 0$.
b. Determine an expression for the excess hole density as a function of x'.
c. Determine an expression for the excess electron density as a function of x.
d. Find the electronic current in P-type material at $x = 0$.
e. Find the hole current in N-type material at $x' = 0$.
f. Determine the total current flowing through the diode.

  Ans. **a.** 1.25×10^{14} cm^{-3}, 6.1×10^{10} cm^{-3} **d.** 0.05 μA   **e.** 4.4×10^{-4} A
  f. 0.44 mA

Problem 6

Compute the reverse saturation current for the *P-N* junction of Problem 5 if $d_P \ll L_n$ and $d_N \gg L_p$.   *Ans.* 1.37×10^{-6} A

Problem 7

Using the data of Problems 5 and 6 determine at room temperature the forward-bias voltage for a forward current of 5 mA.   *Ans.*   0.213 V

Problem 8

a. If the junction described in Problem 5 is exposed to monochromatic electromagnetic radiation of frequency v such that $hv > E_g$ and the radiation current is 1 A, determine the open-circuit voltage of the junction as a photovoltaic energy converter.

b. Using the junction as an energy converter, determine the load voltage under maximum-power-output conditions.

Ans.   a. 0.35 V   b. 0.285 V

6

Effect of Electric Fields on Insulators

Capacitors are very widely used by electrical engineers. It is the purpose of the following discussion to look into the properties of materials which form the dielectric media in capacitors. For simplicity let us consider a parallel-plate capacitor. The capacitance C of a parallel-plate capacitor is given by

$$C = \frac{\epsilon_0 \epsilon_r A}{d} \quad \text{F,} \tag{6.1}$$

where ϵ_0 is the dielectric constant or permittivity of vacuum, $\epsilon_0 = 8.854 \times 10^{-12}$ F/m, A is the area of the plate in square meters, d is the separation of the parallel plates, and ϵ_r is the relative dielectric constant or relative permittivity.

It is evident from Eq. (6.1) that the capacitance depends upon the dimensions of the parallel plates and their separation and can be changed by changing A and d. ϵ_0 is a fundamental constant. Equation (6.1) indicates that for given dimensions and separation of the plates, the capacitance of a parallel-plate capacitor may be changed by ϵ_r, which is a property of the material. Different materials have different values of ϵ_r, its values being unity for vacuum: air (S.T.P.) 1.0006; silicon, 12; germanium, 16; and distilled water, 80. The capacitance is directly proportional to ϵ_r, so any variations in ϵ_r as a result of temperature, pressure, or frequency of the applied field would affect the operation of the capacitor as a circuit element. The relative dielectric constant ϵ_r can be best understood in terms of atomic or microscopic quantities.

6.1 Results of field theory

An electric charge is considered to give rise to electric flux streaming away from the charge. The electric flux density D is the number of flux lines crossing a surface normal to the lines, divided by the surface area. It is a vector quantity and its direction at a point is in the direction of flux lines at that point. Assuming the point charge to be of magnitude Q, the electric flux density at a distance r from the point charge is

$$\mathbf{D} = \frac{Q}{4\pi r^2} \mathbf{a}_r, \tag{6.2}$$

where $4\pi r^2$ is the surface area of a sphere of radius r and $\mathbf{a}_r$ is a unit vector directed radially outward.

The electric field $\mathcal{E}$ is defined as the ratio of the force exerted on a positive test charge divided by the magnitude of the fixed charge as the magnitude of the test charge approaches zero:

$$\mathcal{E} = \lim_{Q_1 \to 0} \frac{\mathbf{F}}{Q_1}, \tag{6.3}$$

where $\mathbf{F}$ is the force exerted on the test positive charge Q_1.

The force exerted by charge Q on charge Q_1 separated by a distance r is given by *Coulomb's law*, which states that the force between two small charges in a uniform homogeneous medium separated by a distance r which is large compared to the size of the charges is proportional to the product of the two charges and inversely proportional to the square of the distance between the charges.

$$\mathbf{F} = \frac{QQ_1}{4\pi\epsilon_0\epsilon_r r^2} \mathbf{a}_r, \tag{6.4}$$

where $\mathbf{a}_r$ is a unit vector in the direction from charge Q to Q_1. Hence in Fig. 6.1 if Q_1 is placed at a distance r from Q, $\mathbf{a}_r$ would be a unit vector directed radially outward. Therefore, using Eq. (6.3),

$$\mathcal{E} = \frac{Q}{4\pi\epsilon_0\epsilon_r r^2} \mathbf{a}_r. \tag{6.5}$$

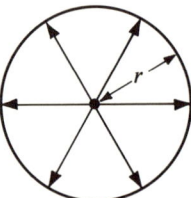

Fig. 6.1
A point charge

From Eqs. (6.2) and (6.5) it can be concluded that

$$\mathbf{D} = \epsilon_0 \epsilon_r \mathbf{\mathcal{E}}. \tag{6.6}$$

The above treatment holds good for isotropic materials, which are materials in which ϵ_r, the relative dielectric constant of the material in which the charge is placed, is independent of the direction. If ϵ_r is dependent upon the direction of measurement, $\mathbf{D}$ and $\mathbf{\mathcal{E}}$ may not be parallel and ϵ_r is a tensor. We shall be dealing in this chapter with isotropic materials.

Gauss's theorem. The electric flux passing through any closed surface is equal to the total charge enclosed by the surface. This law is discussed in detail in books on field theory. We shall state its mathematical form.

$$\oint_s \mathbf{D} \cdot d\mathbf{s} = \int_{vol} \rho \, dv. \tag{6.7}$$

The left side of Eq. (6.7) is the closed surface integral. $d\mathbf{s}$ is the vector surface element directly outward at the point where D is measured, ρ the volume charge density, and dv the volume element.

Dipole moment and polarization. Dipole moment and polarization are important concepts for correlating ϵ_r with the atomic or microscopic properties of the material. These quantities are defined by field theory and we shall review them below.

Dipole moment. The electric dipole moment of a neutral system of i point charges is defined by a vector quantity $\mathbf{\mu}$, where

$$\mathbf{\mu} = \sum_i Q_i \mathbf{r}_i, \tag{6.8}$$

where $\mathbf{r}_i$ is a vector drawn from the origin to charge Q_i. Referring to Fig. 6.2, a system of two charges $-Q$ and Q separated by a distance d is

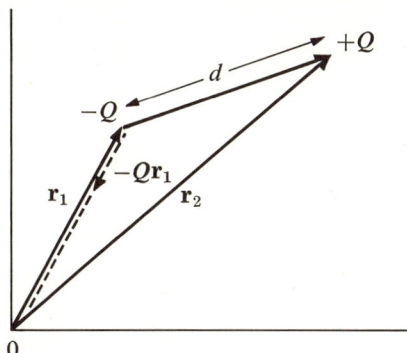

Fig. 6.2
System of two charges $-Q$ and Q separated by a distance d

considered. A coordinate system is chosen such that r_1 is the vector joining the origin with $(-Q)$ and r_2 is the vector joining the origin with (Q). By definition, the dipole moment of the system of these two charges is

$$\mathbf{\mu} = -Q\mathbf{r}_1 + Q\mathbf{r}_2 = Q(\mathbf{r}_2 - \mathbf{r}_1). \tag{6.9}$$

In Eq. (6.9), $-Q\mathbf{r}_1$ is the product of the charge $-Q$ and $\mathbf{r}_1$ and is represented by the dashed line in Fig. 6.2.

It can be easily seen that in Fig. 6.2 the magnitude of this simple dipole is

$$\mu = Qd. \tag{6.10}$$

It can be proved easily that the definition of the dipole moment given by Eq. (6.8) is independent of the selection of the coordinate system. Commonly an electric dipole or simply a dipole is a system of two point charges of opposite sign separated by a distance that is small compared to the distance to the observer.

Field due to a dipole. Consider a point dipole made up of two charges $-Q$ and Q separated by a distance d as shown in Fig. 6.3. The field at a point A due to this dipole can be determined by finding the potential at A due to charges $-Q$ and Q using the definition that

$$\mathcal{E} = -\nabla\theta,$$

where θ is the potential at A due to the dipole. In a two-dimensional polar coordinate system

$$\nabla\theta = \frac{\partial\theta}{\partial r}\mathbf{a}_r + \frac{1}{r}\frac{\partial\theta}{\partial\phi}\mathbf{a}_\phi,$$

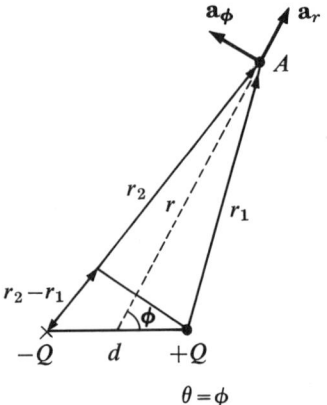

Fig. 6.3
Calculation of the dipole field

where $\mathbf{a}_r$ is the unit vector in the direction of $\mathbf{r}$ outward and $\mathbf{a}_\phi$ is the unit vector in the direction of ϕ. Hence

$$\mathcal{E} = -\left(\frac{\partial \theta}{\partial r}\mathbf{a}_r + \frac{1}{r}\frac{\partial \theta}{\partial \phi}\mathbf{a}_\phi\right) = \mathcal{E}_r + \mathcal{E}_\phi, \tag{6.11}$$

where

$$\mathcal{E}_r = -\frac{\partial \theta}{\partial r}\mathbf{a}_r, \tag{6.12}$$

$$\mathcal{E}_\phi = \frac{1}{r}\frac{\partial \theta}{\partial \phi}\mathbf{a}_\phi. \tag{6.13}$$

The potential $\theta(r, \phi)$ can be calculated as follows:

$\theta(r, \phi)$ = potential due to charge $(-Q)$ + potential due to charge Q.

Potential due to charge $-Q$.

$$\theta_-(r, \phi) = -\int_\infty^{r_2} \mathcal{E} \cdot d\mathbf{r} = \int_\infty^{r_2} + \frac{Q}{4\pi\epsilon_0\epsilon_r r^2}\, dr = -\frac{Q}{4\pi\epsilon_0\epsilon_r r_2}.$$

Potential due to charge Q is

$$\theta_+(r, \phi) = -\int_\infty^{r_1} \mathcal{E} \cdot d\mathbf{r} = -\int_\infty^{r_1} \frac{Q}{4\pi\epsilon_0\epsilon_r r^2}\, dr = \frac{Q}{4\pi\epsilon_0\epsilon_r r_1},$$

where the potential due to both these charges is assumed to be zero as $r \to \infty$. Hence

$$\theta(r, \phi) = -\frac{Q}{4\pi\epsilon_0\epsilon_r r_2} + \frac{Q}{4\pi\epsilon_0\epsilon_r r_1} = \frac{Q}{4\pi\epsilon_0\epsilon_r}\frac{r_2 - r_1}{r_1 r_2}.$$

If the dipole is a point dipole, r_1 and r_2 are approximately parallel and $r_2 - r_1 \simeq d \cos \phi$. Further, $r_1 - r_2$, and hence $r_1 r_2 \simeq r^2$. Therefore,

$$\theta(r, \phi) = \frac{Qd \cos \phi}{4\pi\epsilon_0\epsilon_r r^2}. \tag{6.14}$$

In Eq. (6.14) $Qd = \mu$, and hence

$$\theta(r, \phi) = \frac{\mu \cos \phi}{4\pi\epsilon_0\epsilon_r r^2}. \tag{6.15}$$

From Eq. (6.15) the components of the resultant electric field due to the dipole can be determined by making use of Eqs. (6.12) and (6.13),

$$\mathcal{E}_r = \frac{1}{4\pi\epsilon_0\epsilon_r}\frac{2\mu \cos \phi}{r^3}\mathbf{a}_r, \tag{6.16}$$

$$\mathcal{E}_\phi = \frac{1}{4\pi\epsilon_0\epsilon_r}\frac{\mu \sin \phi}{r^3}\mathbf{a}_\phi, \tag{6.17}$$

$$\mathcal{E} = \frac{1}{4\pi\epsilon_0\epsilon_r}\frac{2\mu \cos \phi}{r^3}\mathbf{a}_r + \frac{1}{4\pi\epsilon_0\epsilon_r}\frac{\mu \sin \phi}{r^3}\mathbf{a}_\phi. \tag{6.18}$$

Equation (6.18) gives the resultant field at any position (r,ϕ) due to a dipole and will be seen to be very helpful in calculating the interaction of dipoles.

Polarization. The dipole moment per unit volume is called polarization P:

$$\mathbf{P} = \frac{\mathbf{\mu}}{\text{volume}} = \frac{\sum_i Q_i \mathbf{r}_i}{\text{volume}}. \tag{6.19}$$

To express the polarization in terms of the applied electric field and the relative dielectric constant ϵ_r of a material, we consider the case of two parallel plates of area A separated by a distance d. Let us consider the two plates to have a vacuum between them and consider that the voltage applied results in an electric field $\mathcal{E}$. The electric flux between the plates is given by

$$\mathbf{D} = \epsilon_0 \mathcal{E}, \tag{6.20}$$

where $\mathcal{E} = V/d$, because $\epsilon_r = 1$ for vacuum.

Let us now introduce between the parallel plates a block of dielectric completely filling the space between plates (see Fig. 6.4). Let ϵ_r be the relative dielectric constant of the material. The voltage is held constant, so

$$\mathcal{E} = \mathcal{E}_l,$$

where $\mathcal{E}_i$ is the electric field inside the material of relative dielectric constant ϵ_r. But

$$\mathbf{D}_i = \epsilon_0 \epsilon_r \mathcal{E}_i, \tag{6.21}$$

and hence from Eqs. (6.20) and (6.21),

$$\mathbf{D} = \mathbf{D}_i / \epsilon_r$$

because $\mathcal{E} = \mathcal{E}_l$.

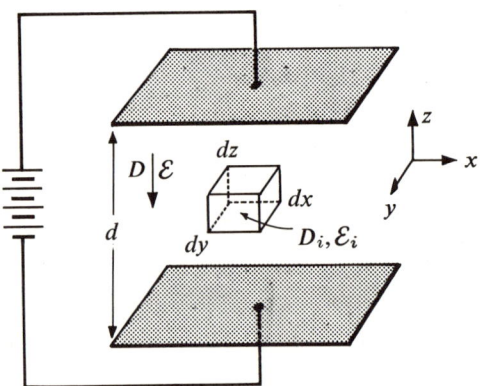

Fig. 6.4
Model for calculation of the polarization in a dielectric material

ELECTRONIC, IONIC, AND ORIENTATIONAL POLARIZATION

The difference in **D** in the two cases must be a result of polarization induced in the dielectric material, and it can be shown that

$$\mathbf{P} = \frac{\mu}{\text{volume}} = (\mathbf{D}_i - \mathbf{D}) = (\epsilon_r - 1)\mathbf{D},$$

but $\mathbf{D} = \epsilon_0 \mathcal{E}$ [according to Eq. (6.20)]. Therefore,

$$\mathbf{P} = \epsilon_0(\epsilon_r - 1)\mathcal{E}. \tag{6.22}$$

Equation (6.22) tells us that if $\epsilon_r \neq 1$, an application of electric field $\mathcal{E}$ would produce a polarization in the dielectric. This result of the field theory is used as a connecting link between ϵ_r and the atomic properties.

From Eq. (6.22),

$$\mathbf{P} = \epsilon_0 \epsilon_r \mathcal{E} - \epsilon_0 \mathcal{E},$$

$$\mathbf{P} = \mathbf{D} - \epsilon_0 \mathcal{E},$$

or

$$\mathbf{D} = \mathbf{P} + \epsilon_0 \mathcal{E}. \tag{6.23}$$

Equation (6.23) is a fundamental relationship between **D**, **P**, and $\mathcal{E}$ and shows that if any dielectric material is introduced in an $\mathcal{E}$ field, the dielectric material would have a polarization

$$\mathbf{P} = \mathbf{D} - \epsilon_0 \mathcal{E},$$

indicating that polarization is the difference in flux densities with and without the dielectric material. Polarization is expressed in C/m^2.

6.2 Electronic, ionic, and orientational polarization in static fields

Basic field theory suggested that there would be a polarization in the direction of the applied field if a dielectric material is subjected to an electric field. According to Eq. (6.22), if the field were removed, the polarization would be zero. Polarization by definition is the total dipole moment per unit volume and hence one can conclude that the electric field must be inducing or creating dipoles. There are, however, materials in which there are permanent dipole moments present. In the absence of an applied electric field, the dipoles already present are randomly distributed and the effect of the applied electric field in these materials is to align some or all of them in the direction of the applied electric field. In this section we shall assume that the interaction of one dipole with another is negligible. We shall see in Sec. 6.3 that if this interaction of dipoles becomes large we would have spontaneous polarization.

Polarization may therefore be referred to as a temporary electrical strain in the material resulting from the applied electric field. The atoms of all materials are strained by the application of an electric field. This strain is called *electronic polarization*. In ionic crystals the position of positive ions with respect to negative ions is affected by the application of the electric field; this strain is called *ionic polarization*. Materials with permanent dipole moments have a third type of strain, called *orientational polarization*. The electric field exerts forces on the permanent dipole moments and tries to align them in its direction, the direction being that of minimum energy. The total polarization is the sum of these three-component polarizations.

Electronic polarization. Electronic polarization has been defined as strain in the atom. To study strain in the atom due to applied electric field, a model of the atom has to be chosen. We shall first take up the classical model and then see the modification of the results that would take place by replacing the classical model with the quantum-mechanical model. We are assuming no interaction of adjacent dipoles, so the treatment is only valid for gases in which the interatomic distances are very large and interatomic interactions may be neglected.

Classical model. Consider a single atom subjected to an electric field $\mathcal{E}$. According to the Bohr theory the nucleus of the atom would have a net positive charge Ze, where Z is the atomic number of the atom. To keep the neutrality of the atom, Z electrons are assumed to revolve around the nucleus. In the quantum-mechanical treatment it was shown that there is no fixed position of the electrons and that there is a certain probability of finding an electron at a certain distance from the nucleus. We would choose an intermediate classical model for the atom. Let us assume that the total electronic charge $-Ze$ outside the nucleus is uniformly distributed in the volume of a sphere of radius R (Fig. 6.5), where R is the radius of the atom. In spite of the crudeness of this model we shall see that the results obtained are in close agreement with experimental fact.

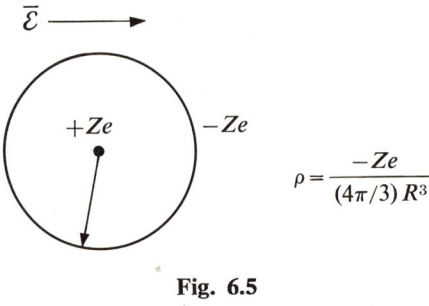

Fig. 6.5
An atom

The volume charge density of the electronic charges ρ is given by

$$\rho = -\frac{Ze}{4(\pi/3)R^3}.$$

When an atom with this atomic model is placed in an electric field, the atom, being that of an insulator, has no free electrons and hence there will be no flow of current as a result of electronic motion. The nucleus and the electron clouds are of opposite charge and hence experience Lorentz forces in opposite directions. The nucleus experiences the force in the direction of the electric field and the electron cloud in a direction opposite to the field. The electron cloud moves and the nucleus and electron cloud are pulled apart. As the two are pulled apart, a Coulomb force develops between them, which tends to bring the electron cloud back to its original position, where the nucleus is at the center of the electron cloud. In equilibrium the Lorentz force equals the Coulomb force and the nucleus is shifted slightly relative to the electron cloud by a distance x (Fig. 6.6).

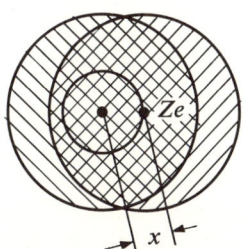

Fig. 6.6
Schematic representation of the displacement of the electron orbit relative to the nucleus due to externally applied field

The Lorentz force on the electron cloud is $-Ze\mathcal{E}$. The Coulomb force is only exerted on that portion of the electron cloud which does not surround the positive nucleus Ze. This is a consequence of the Gauss theorem. The portion of the electron cloud taking part in the Coulomb force is contained in the sphere of radius x. The charge contained in a sphere of this radius is

$$= \rho(4\pi/3)x^3 = -\frac{Ze(4\pi/3)x^3}{(4\pi/3)R^3} = -\frac{Zex^3}{R^3}.$$

Therefore the Coulomb force is

$$\frac{[Ze(x^3/R^3)](Ze)}{4\pi\epsilon_0 x^2}.$$

For equilibrium

$$Ze\mathcal{E} = \frac{[Ze(x^3/R^3)](Ze)}{4\pi\epsilon_0 x^2}. \tag{6.24}$$

Simplifying Eq. (6.24),

$$x = \frac{4\pi\epsilon_0 R^3}{Ze}\mathcal{E}. \tag{6.25}$$

According to Eq. (6.25) we find that the displacement of the electron cloud is proportional to the applied electric field. The displacement x is a strain and is proportional to the applied field $\mathcal{E}$ or the stress. As long as we do not have a very high electric field, the stress is proportional to the strain and Hooke's law of elastic forces is applicable.

The displacement of the electron cloud is a separation of the negative charge from the positively charged nucleus. This results in a dipole moment:

$$\mathbf{\mu} = Zex = 4\pi\epsilon_0 R^3 \mathcal{E}. \tag{6.26}$$

The quantity $4\pi\epsilon_0 R^3$ is called the *electronic polarizability* α_e. Hence

$$\mathbf{\mu} = \alpha_e \mathcal{E}. \tag{6.27}$$

The dipole moment μ is the induced dipole moment and is proportional to the applied field. There are N such atoms per unit volume and, because the atoms are far apart, the dipole moments in all atoms are induced in the same direction. Hence

$$\mathbf{P}_e = N\mathbf{\mu} = N\alpha_e \mathcal{E}, \tag{6.28}$$

or

$$\mathbf{P}_e = 4\pi\epsilon_0 N R^3 \mathcal{E}. \tag{6.29}$$

He, A, and Ne are good examples of gases that have only electronic polarization when subjected to an electric field. They are inert gases and do not take part in chemical actions.

Consider the case of Ar and determine the displacement x for a field of 10^6 V/m. For argon $Z = 18$, $R \simeq 10^{-10}$ m. Therefore,

$$x = \frac{4\pi\epsilon_0 R^3 \mathcal{E}}{Ze} = \frac{4\pi \times 8.854 \times 10^{-12} \times 10^{-30} \times 10^6}{18 \times 1.6 \times 10^{-19}},$$

$$x \simeq 4 \times 10^{-17} \text{ m}.$$

The above computations indicate that the perturbing influence of even a large applied electric field of magnitude 10^6 V/m is very small, and the center of the electron cloud is close to the nucleus.

Semi-quantum-mechanical model. The classical model assumed was a model intermediate between the purely classical Bohr model and

ELECTRONIC, IONIC, AND ORIENTATIONAL POLARIZATION

the quantum-mechanical model. One of the very important results of this intermediate model is the fact that the perturbing influence of the applied field is confined to a region close to the nucleus. For simplicity we consider the quantum-mechanical treatment of the hydrogen atom in the ground state discussed in Chapter 1. According to Eq. (1.66) the volume charge density of an electron at a distance r from the nucleus in this case is

$$\rho(r) = -\frac{e}{\pi r_1^3} e^{-2r/r_1}, \tag{1.66}$$

where r_1 is the first Bohr radius.

Assuming that the application of an electric field does not change the form of the density variation, and because charges are confined to very small values of r, we can approximate Eq. (1.66) for the case $r \ll r_1$:

$$\rho(r) = -\frac{e}{\pi r_1^3}. \tag{6.30}$$

Equation (6.30) shows that we can take the charge density to be constant in the close vicinity of the nucleus. If the electron is shifted by a distance x, the charge contained in the region $r = 0$ to $r = x$ is given by

$$\int_0^x \rho(r) 4\pi r^2 \, dr = -\frac{e}{\pi r_1^3} \frac{4\pi}{3} x^3 = -\frac{e}{r_1^3} \frac{4}{3} x^3.$$

The Coulomb force of attraction is

$$\frac{[(e/r_1^3)\tfrac{4}{3}x^3]e}{4\pi\epsilon_0 x^2}$$

for the hydrogen atom, and hence in equilibrium

$$e\mathcal{E} = \frac{4}{3}\frac{e}{r_1^3}\frac{x^3 e}{4\pi\epsilon_0 x^2} \tag{6.31}$$

Simplifying Eq. (6.31),

$$x = \frac{3\pi r_1^3 \epsilon_0}{e} \mathcal{E}. \tag{6.32}$$

The induced dipole moment is

$$\mu = ex = 3\pi r_1^3 \epsilon_0 \mathcal{E}. \tag{6.33}$$

Comparing Eqs. (6.33) and (6.26) the only difference in the induced dipole moment as a result of a more exact model of the atom has been use of the factor 3 in place of 4, but this does not affect the order of magnitude of μ. Hence for the small contribution of the electronic polarization for $\mathcal{E} = 10^6$ V/m, $N = 10^{25}$ m^{-3},

$$P_e(\text{He, Ar}) \simeq 10^{-9} \text{ C/m}^2,$$

not much difference would be created by the semiclassical model. The electronic polarization is determined only by the atomic structure and is therefore independent of temperature.

Ionic polarization. Ionic polarization is defined as the electrical strain in chemical compounds having ionic formation produced by an applied electric field. Assume that the ionic compound is liquid, the interatomic distance is so large that the interaction of one dipole with another is negligible, and the field seen by the atoms is the same as the applied field. Let us consider a typical ionic compound NaCl. Sodium has 7 electrons around the nucleus and chlorine has 17 electrons around the nucleus (Fig. 6.7). The outermost sodium electron spends most of the time in the

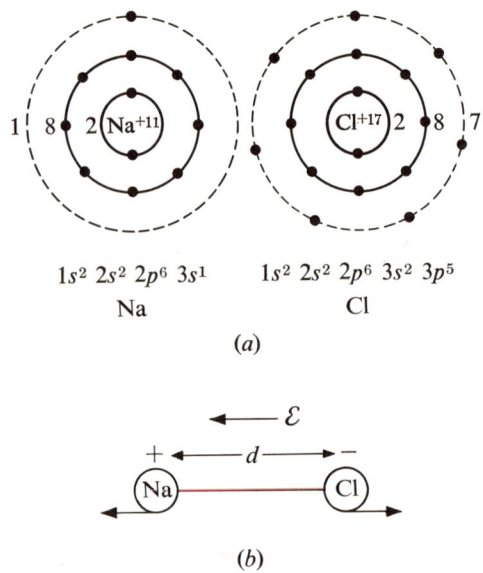

Fig. 6.7
(a) Electronic configuration of NaCl. (b) Ionic polarization

vicinity of the chlorine nucleus, thereby causing sodium to be electropositive and chlorine electronegative in the compound sodium chloride. The bond between Na^+ and Cl^- is ionic in character. Even in the absence of an applied field the Na^+ and Cl^- atoms separated by a distance d form a dipole moment, called a *permanent dipole moment*. A molecule consisting of more than two atoms has a permanent dipole moment if the vector sum of the moments associated with the bonds in the molecule has a finite magnitude and direction. We shall discuss the effect of electric fields on permanent dipole moments in the next subsection.

Referring to the molecule Na^+Cl^-, the applied field $\mathcal{E}$ moves the positive ion in the direction of the applied field and the negative ion in a direction opposite to the applied field. A change in separation d [Fig. 6.7(b)] between the two ions causes a dipole moment to be induced. The induced dipole moment in ionic compounds may be expressed by an equation similar to (6.27),

$$\mathbf{\mu} = \alpha_i \mathcal{E}, \tag{6.34}$$

where α_i is known as *ionic polarizability*. It follows from reasoning similar to that employed with electronic polarizability that the ionic polarizability α_i is independent of temperature:

$$\mathbf{P}_i = N\alpha_i \mathcal{E}, \tag{6.35}$$

where N is the number of molecules per m³.

Orientational polarization $\mathbf{P}_0$. We shall start once more with the basic assumption that the interatomic distance is large, so that there is no interaction of one dipole moment with another. Hence we are concerning ourselves in this section with liquids and gases and assume that the field seen by an atom or molecule is the applied field $\mathcal{E}$. Orientational polarization $\mathbf{P}_0$ comes into play in those materials which have permanent dipole moments. Let μ_P be the permanent dipole moment of a molecule.

Absence of an electric field. Consider a system of N dipole moments per unit volume each of magnitude μ_P at a temperature T. In the absence of an electric field there is no preferred direction and the vector sum of all the μ_P's in any direction is zero. Hence the polarization in the absence of an electric field is zero. We shall try to prove this statement and thereby develop a procedure to calculate the polarization in the presence of an electric field. Consider a sphere of unit radius in a system of N dipole moments per unit volume. There is no preferred direction, so the probability of finding a dipole in any direction is the same. The number of dipole moments in a direction between ϕ and $\phi + d\phi$ is proportional to the surface area lying between ϕ and $\phi + d\phi$, with x as the reference axis (see Fig. 6.8). If $f(\phi)\,d\phi$ is the number of dipole moments between ϕ and $\phi + d\phi$ in a unit volume,

$$f(\phi)\,d\phi \propto 2\pi \sin\phi\,d\phi,$$

$$f(\phi)\,d\phi = 2\pi A \sin\phi\,d\phi, \tag{6.36}$$

where A is a constant of proportionality. The constant A can be deter-

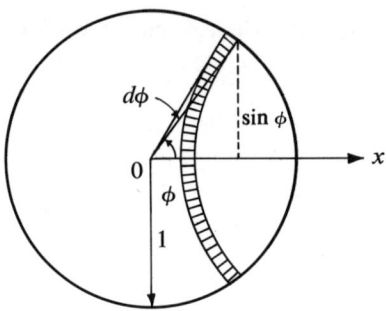

Fig. 6.8
Schematic diagram for calculating $f(\phi) \, d\phi$

mined by the normalization condition stating that the total number of dipole moments is obtained by integrating $f(\phi) \, d\phi$ from $\phi = 0$ to $\phi = \pi$. Therefore,

$$N = \int_0^\pi f(\phi) \, d\phi = \int_0^\pi 2\pi A \sin \phi \, d\phi.$$

Solving, $A = N/4\pi$.

The total polarization is the contribution of all the dipole moments in the direction of x. A dipole moment μ_P lying between ϕ and $\phi + d\phi$ would have a component $\mu_P \cos \phi$ in the direction of x. Hence dP, the polarization contribution of dipole moments lying between ϕ and $\phi + d\phi$ in the direction of x, is the product of the number of the dipole moments in that direction and the component of each:

$$dP = f(\phi) \, d\phi \, \mu_P \cos \phi = \frac{N}{4\pi} 2\pi \sin \phi \cos \phi \, \mu_P \, d\phi$$

or

$$dP = \frac{N\mu_P}{2} \sin \phi \cos \phi \, d\phi = \frac{N\mu_P}{4} \sin 2\phi \, d\phi. \tag{6.37}$$

The total polarization is obtained by integrating Eq. (6.37) from $\phi = 0$ to $\phi = \pi$:

$$P = \int dP = \int_0^\pi \frac{N\mu_P}{4} \sin 2\phi \, d\phi. \tag{6.38}$$

Solving the integral in Eq. (6.38) we find that $P = 0$, which shows that the polarization in the absence of an electric field is zero.

Presence of an electric field $\mathcal{E}$. Consider that in this system of N dipole moments per unit volume, each of magnitude μ_P at a temperature T,

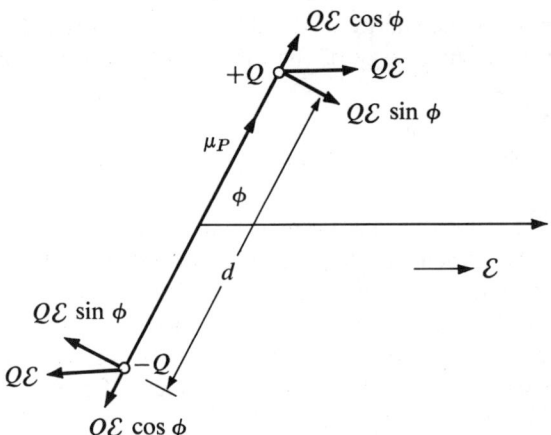

Fig. 6.9
Torque exerted on a dipole by an external field

an electric field $\mathcal{E}$ is applied in the direction of x. Before considering the effect of the applied field on the entire system, it would be necessary to determine and discuss the effect of an applied electric field on a dipole moment μ_P in a direction ϕ with reference to the direction of the applied field. Let the dipole moment μ_P be made up of $-Q$ and Q charges separated by a distance d:

$$\mu_P = Qd.$$

The positive charge Q experiences a force in the presence of the applied field $\mathcal{E}$ in the direction of the field. The magnitude of this Lorentz force is $Q\mathcal{E}$. The negative charge also experiences a force $-Q\mathcal{E}$ but in a direction opposing the field $\mathcal{E}$. The forces on the charges can be resolved in components parallel and perpendicular to the dipole moment. The components of the forces $Q\mathcal{E} \cos \phi$ on positive and negative charges are in opposite directions and induce a dipole in the basic dipole. This is similar to the dipole moment in ionic compounds. The component of forces perpendicular to μ_P are $Q\mathcal{E} \sin \phi$ and $-Q\mathcal{E} \sin \phi$ on positive and negative charges, respectively (Fig. 6.9). These forces could be compared to a couple of forces on a lever arm and tend to orient the dipole in the direction of the field.

The torque experienced by the dipole is

$$Q\mathcal{E} \sin \phi \, d = \mu_P \mathcal{E} \sin \phi = T(\phi). \tag{6.39}$$

and
$$T(\phi) = \boldsymbol{\mu}_P \times \boldsymbol{\mathcal{E}}$$

If we arbitrarily set the energy of the dipole in the presence of the field at $\phi = 90$ to be zero, the energy of the dipole in the presence of the field at an angle ϕ is

$$\int_{\phi=90}^{\phi} T(\phi)\, d\phi = \int_{\phi=90}^{\phi} \mu_P \mathcal{E} \sin\phi\, d\phi = -\mu_P \mathcal{E} \cos\phi = -\pmb{\mu}_P \cdot \pmb{\mathcal{E}}.$$

The torque experienced by the dipole is creating an order in the dipoles by trying to align the dipoles in the direction of the field. This is opposed by the energy associated with the temperature T trying to create disorder by distributing the dipoles randomly. The probability of finding a dipole at an angle ϕ in the presence of the field is proportional to $\exp(\mu_P \mathcal{E} \cos\phi / kT)$, in accordance with Maxwell-Boltzmann statistics. If $p(\phi)$ is the probability of finding a dipole at an angle ϕ,

$$p(\phi) \propto \exp\left(\frac{\mu_P \mathcal{E} \cos\phi}{kT}\right).$$

Hence $n(\phi)\, d\phi$, the number of dipole moments in a direction between ϕ and $\phi + d\phi$, is given by

$$n(\phi)\, d\phi = B f(\phi) p(\phi)\, d\phi, \tag{6.40}$$

where B is the constant of proportionality.

The constant of proportionality B is determined by the normalization condition, stating that the total number of dipole moments per unit volume is still N:

$$N = \int n(\phi)\, d\phi = \int_0^\pi B f(\phi) p(\phi)\, d\phi = B \int_0^\pi 2\pi \sin\phi \exp\left(\frac{\mu_P \mathcal{E} \cos\phi}{kT}\right) d\phi,$$

and hence

$$B = \frac{N}{\int_0^\pi 2\pi \sin\phi \exp(\mu_P \mathcal{E} \cos\phi / kT)\, d\phi}. \tag{6.41}$$

Hence dP_0, the orientational polarization contribution of dipole moments lying between ϕ and $\phi + d\phi$ in the direction x, is the product of the component of dipole moments in that direction and the number of dipoles:

$$dP_0 = \mu_P \cos\phi\, n(\phi)\, d\phi. \tag{6.42}$$

Substituting,

$$n(\phi)\, d\phi = B 2\pi \sin\phi \exp\left(\frac{\mu_P \mathcal{E} \cos\phi}{kT}\right) d\phi \quad \text{[in Eq. (6.42)]},$$

$$dP_0 = B 2\pi \sin\phi\, \mu_P \cos\phi \exp\left(\frac{\mu_P \mathcal{E} \cos\phi}{kT}\right) d\phi. \tag{6.43}$$

ELECTRONIC, IONIC, AND ORIENTATIONAL POLARIZATION

Substituting B from Eq. (6.41) in Eq. (6.43),

$$dP_0 = \frac{N\, 2\pi \sin\phi\, \mu_P \cos\phi \exp(\mu_P \mathcal{E} \cos\phi/kT)\, d\phi}{\int_0^\pi 2\pi \sin\phi \exp(\mu_P \mathcal{E} \cos\phi/kT)\, d\phi}. \tag{6.44}$$

The total orientational polarization P_0 is obtained by integrating Eq. (6.44) from $\phi = 0$ to $\phi = \pi$:

$$P_0 = \frac{N\int_0^\pi \mu_P \cos\phi\, 2\pi \sin\phi \exp(\mu_P \mathcal{E} \cos\phi/kT)\, d\phi}{\int_0^\pi 2\pi \sin\phi \exp(\mu_P \mathcal{E} \cos\phi/kT)\, d\phi}. \tag{6.45}$$

Simplifying Eq. (6.45),

$$P_0 = N\mu_P \frac{\int_0^\pi \cos\phi \sin\phi \exp(\mathcal{E}\mu_P \cos\phi/kT)\, d\phi}{\int_0^\pi \sin\phi \exp(\mu_P \mathcal{E} \cos\phi/kT)\, d\phi}. \tag{6.46}$$

Introducing a new variable,

$$z = \alpha \cos\phi, \tag{6.47}$$

where α is a constant,

$$\alpha = \frac{\mu_P \mathcal{E}}{kT}. \tag{6.48}$$

Therefore,

$$z = \frac{\mu_P \mathcal{E}}{kT} \cos\phi.$$

Differentiating,

$$dz = -\frac{\mu_P \mathcal{E}}{kT} \sin\phi\, d\phi.$$

Hence

$$P_0 = \frac{NkT}{\mathcal{E}} \frac{\int_\alpha^{-\alpha} z e^z\, dz}{\int_\alpha^{-\alpha} e^z\, dz}. \tag{6.49}$$

Solving Eq. (6.49),

$$P_0 = \frac{NkT}{\mathcal{E}} \frac{[ze^z - e^z|_{-\alpha}^{\alpha}]}{[e^z|_{-\alpha}^{\alpha}]} = \frac{NkT}{\mathcal{E}}\left(\alpha \frac{e^\alpha + e^{-\alpha}}{e^\alpha - e^{-\alpha}} - 1\right),$$

and therefore

$$P_0 = N\mu_P\left(\coth\alpha - \frac{1}{\alpha}\right). \tag{6.50}$$

The function $[\coth \alpha - (1/\alpha)]$ is called the *Langevin function* $L(\alpha)$, and Eq. (6.50) may be rewritten

$$P_0 = N\mu_P L(\alpha). \qquad (6.51)$$

The Langevin function is plotted graphically in Fig. 6.10. For larger values of α, $L(\alpha)$ approaches unity and

$$P_0 = N\mu_P. \qquad (6.52)$$

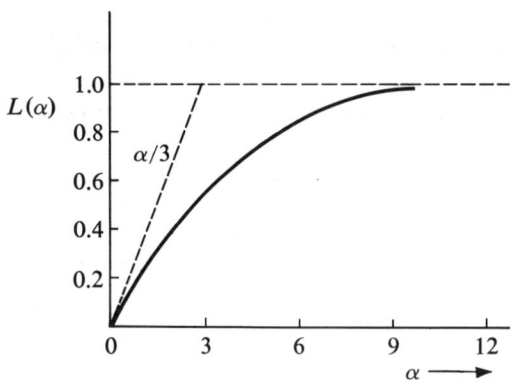

Fig. 6.10
Langevin function $L(\alpha)$; for $\alpha \ll 1$, the slope is $1/3$

The physical significance of α approaching a very large value is that in the expression for $\alpha = \mu_P \mathcal{E}/kT$, either $\mathcal{E}$ is very, very large or the temperature T is very, very low. The effect of the applied field $\mathcal{E}$ is to create an order by aligning the dipoles in the direction of the field, and the effect of the temperature is to create a disorder. With α very large, the disturbance is small and the order-creating effects large, and we may expect all the dipoles to be aligned. This situation in the limit would lead to a saturation with all the N dipole moments per unit volume aligned in the direction of the field. In practice the situation of saturation is never encountered in gases for all practical fields and temperatures. In practice

$$\alpha = \frac{\mu_P \mathcal{E}}{kT} \ll 1.$$

In such a case

$$L(\alpha) = \coth \alpha - \frac{1}{\alpha} = \frac{1 + (\alpha^2/2!) + (\alpha^4/4!)}{\alpha + (\alpha^3/3!) + (\alpha^5/5!)} - \frac{1}{\alpha},$$

and hence

$$L(\alpha) = \frac{\alpha + (\alpha^3/2!) + (\alpha^5/4!) + \cdots - \alpha - (\alpha^3/3!) - (\alpha^5/5!) \cdots}{\alpha^2 + (\alpha^4/3!) + (\alpha^6/5!) + \cdots}.$$

Taking terms through α^3 and neglecting higher-order terms,
$$L(\alpha) \simeq \alpha/3.$$
Therefore, for all practical purposes
$$P_0 = N\mu_P \frac{\alpha}{3}$$
or
$$P_0 = \frac{N\mu_P^2 \mathcal{E}}{3kT}. \tag{6.53}$$

The orientational polarization is directly proportional to the applied field $\mathcal{E}$ and inversely proportional to the temperature. The orientational polarization is also proportional to the square of the permanent dipole moment.

From Eq. (6.53) the orientational polarizability α_0 is given by
$$\alpha_0 = \frac{\mu_P^2}{3kT}. \tag{6.54}$$

μ_P is expressed in Debye units, where $1\,D = 3.33 \times 10^{-30}$ Coulomb meter. Measurements show that for HCl, $\mu_P = 1.04$ Debye unit. Therefore,
$$\alpha_0(\text{HCl}) = \frac{(1.04 \times 3.33 \times 10^{-30})^2}{3 \times 1.38 \times 10^{-23} \times 300} = 0.98 \times 10^{-39}\ \text{Farads meters}^2$$

at room temperature. If $N = 10^{27}\ \text{m}^{-3}$,
$$\mathcal{E} = 10^6\ \text{V/m},$$
$$P_0 = N\alpha_0 \mathcal{E} = 10^{27} \times 0.98 \times 10^{-39} \times 10^6 \simeq 10^{-6}\ \text{C/m}^2.$$

The total polarization is made up of three parts: electronic, ionic, and orientational polarizations:
$$\mathbf{P} = \mathbf{P}_e + \mathbf{P}_i + \mathbf{P}_0.$$

Substituting the expressions for $\mathbf{P}_e$, $\mathbf{P}_i$, and $\mathbf{P}_0$ in the above equation,
$$\mathbf{P} = (N\alpha_e + N\alpha_i + N\alpha_0)\mathcal{E}$$
or
$$\mathbf{P} = N(\alpha_e + \alpha_i + \alpha_0)\mathcal{E}. \tag{6.55}$$

The total polarizability α may be written
$$\alpha = \alpha_e + \alpha_i + \alpha_0$$
or
$$\alpha = \alpha_e + \alpha_i + \frac{\mu_P^2}{3kT}. \tag{6.56}$$

Equation (6.56) is known as the *Langevin-Debye equation* and is very helpful in the interpretation of molecular structures. This will be discussed below.

Equation (6.55) may be rewritten

$$\mathbf{P} = N\left(\alpha_e + \alpha_i + \frac{\mu_P^2}{3kT}\right)\boldsymbol{\varepsilon}. \tag{6.57}$$

In the above expression α_e and α_i are independent of temperature. From field theory it is known that **P** is related to $\boldsymbol{\varepsilon}$ in terms of ϵ_r by Eq. (6.22):

$$\mathbf{P} = \epsilon_0(\epsilon_r - 1)\boldsymbol{\varepsilon}. \tag{6.22}$$

Comparing the right sides of Eqs. (6.22) and (6.57),

$$\epsilon_0(\epsilon_r - 1) = N\left(\alpha_e + \alpha_i + \frac{\mu_P^2}{3kT}\right). \tag{6.58}$$

It may be mentioned that Eq. (6.57) and therefore Eq. (6.58) hold if and only if the separation between atoms is so large that there is no interaction between the dipoles and hence the field seen by a dipole is just the applied field. Equation (6.58) is therefore valid only for liquids and gases. Equation (6.58) suggests that the variation of $(\epsilon_r - 1)$ with $1/T$ is a straight line.

The presence of permanent dipole moments can be seen by measuring the relative dielectric constant of the gas or liquid at two or three temperatures. $\epsilon_0(\epsilon_r - 1)$ is then plotted against $1/T$, the slope of the line giving the quantity $N\mu_P^2/3k$. N can be easily determined and k is constant, so a knowledge of the slope of the straight line helps us to determine μ_P or the presence of permanent dipole moments (Fig. 6.11). The intercept at $1/T = 0$ is a measure of $\alpha_e + \alpha_i$. The presence of a permanent dipole

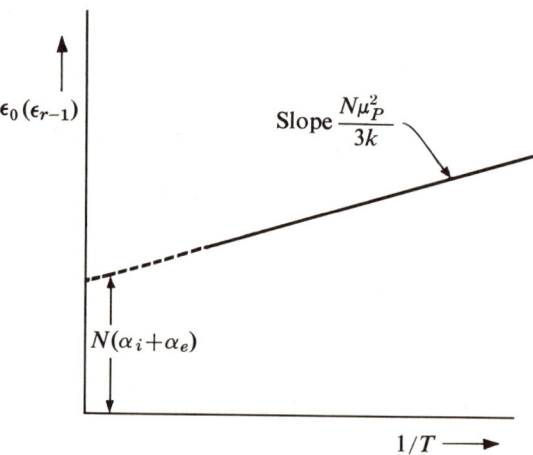

Fig. 6.11
Variation of $\epsilon_0(\epsilon_r - 1)$ with reciprocal temperature

moment helps us arrive at certain conclusions regarding the structure of molecules. The value of $\mu_P = 1.04$ D for HCl was determined in this way. Measurements for CS_2 indicate that $\mu_P = 0$ and hence the molecular structure for CS_2 is S = C = S, the sulfur bonds making a 180° angle with carbon.

Internal field in solids and liquids. So far in our discussion of polarization, the distance between the atoms has been considered to be large, and, as such, the discussion primarily holds for gases and some liquids. In solids the distance between atoms is of the order of a few angstroms and an atom or molecule not only sees the externally applied field but also the fields produced by the dipoles of other atoms or molecules. If $\mathcal{E}_i \neq \mathcal{E}$, where $\mathcal{E}_i$ is the field seen by an atom or molecule, called the *internal field*, and $\mathcal{E}$ is the externally applied field, Eq. (6.57) has to be modified:

$$\mathbf{P} = N\left(\alpha_e + \alpha_i + \frac{\mu_P^2}{3kT}\right)\mathcal{E}_i. \tag{6.59}$$

According to field theory

$$\mathbf{P} = \epsilon_0(\epsilon_r - 1)\mathcal{E}. \tag{6.22}$$

Therefore,

$$\epsilon_0(\epsilon_r - 1)\mathcal{E} = N\left(\alpha_e + \alpha_i + \frac{\mu_P^2}{3kT}\right)\mathcal{E}_i. \tag{6.60}$$

From Eq. (6.60) we see that ϵ_r can be expressed in terms of microscopic properties α_e, α_i, and α_0 if a relationship can be found between $\mathcal{E}_i$ and $\mathcal{E}$. Both in dielectric and magnetic theory it is very important to calculate the local field at an atom or ion as affected by the polarization of the specimen as a whole.

In 1936 Onsager showed that in liquids which have permanent dipole moments, the internal field which tends to orient the dipoles along the direction of the external field may generally be different from the internal field causing electronic and ionic polarizations. Equation (6.60) may therefore be correctly rewritten

$$\epsilon_0(\epsilon_r - 1)\mathcal{E} = N(\alpha_e + \alpha_i)\mathcal{E}_{i_1} + \frac{N\mu_P^2}{3kT}\mathcal{E}_{i_2}, \tag{6.61}$$

where $\mathcal{E}_{i_1}$ is the internal field causing ionic and electronic polarizations and $\mathcal{E}_{i_2}$ is the internal field tending to orient the dipoles. In the following treatment we shall assume $\mathcal{E}_{i_1} = \mathcal{E}_{i_2}$. The problem of the calculation of the internal field is best understood by taking up a one-dimensional model of the crystal. This model was suggested by Epstein and is very useful in

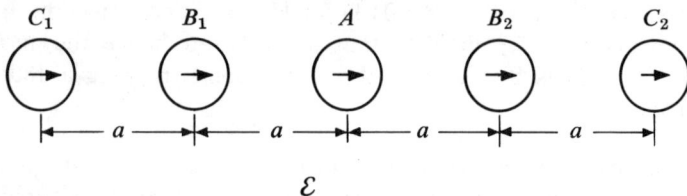

Fig. 6.12
Schematic representation of the effect of internal field on the resultant electric field

the understanding of internal fields. Figure 6.12 shows an infinite array of atoms separated by a distance a from each other. Consider an applied electric field $\mathcal{E}$ in the direction parallel to the infinite array and the problem is to determine the field seen by an atom $\mathcal{E}_i$.

If μ_{ind} is the induced dipole moment in each atom due to the applied field $\mathcal{E}$, the internal field seen by an atom A (Fig. 6.12) is the sum of the applied field and the fields produced by μ_{ind} of all other atoms $B_1, C_1, \ldots$ lying on the left of A and $B_2, C_2, \ldots$ lying on the right of A.

Referring to Eqs. (6.16) and (6.17), the components of the field produced by a dipole moment in the direction of unit vectors $\mathbf{a}_r$ and $\mathbf{a}_\phi$ are

$$\mathcal{E}_r = \frac{1}{4\pi\epsilon_0\epsilon_r} \frac{2\mu \cos\phi}{r^3} \mathbf{a}_r, \qquad (6.16)$$

$$\mathcal{E}_\phi = \frac{1}{4\pi\epsilon_0\epsilon_r} \frac{\mu \sin\phi}{r^3} \mathbf{a}_\phi. \qquad (6.17)$$

The field produced by μ_{ind} at B_1 at A is obtained by the above equations if $r = a$ and $\phi = 0$. Therefore,

$$\mathcal{E}_{AB_1} = \frac{1}{4\pi\epsilon_0\epsilon_r} \frac{2\mu_{ind}}{a^3}.$$

Similarly,

$$\mathcal{E}_{AB_2} = \frac{1}{4\pi\epsilon_0\epsilon_r} \frac{2(-\mu_{ind})}{(-a)^3} = \frac{1}{4\pi\epsilon_0\epsilon_r} \frac{2\mu_{ind}}{a^3} \qquad (\phi = 180° \text{ and } r = -a).$$

The contribution of the field due to atoms B_1 and B_2 at A is given by adding $\mathcal{E}_{AB_1}$ and $\mathcal{E}_{AB_2}$:

$$\mathcal{E}_{AB_1} + \mathcal{E}_{AB_2} = \frac{\mu_{ind}}{\pi\epsilon_0\epsilon_r a^3}.$$

In a similar way,

$$\mathcal{E}_{AC_1} + \mathcal{E}_{AC_2} = \frac{\mu_{ind}}{\pi\epsilon_0\epsilon_r (2a)^3}.$$

ELECTRONIC, IONIC, AND ORIENTATIONAL POLARIZATION

The internal field at atom A is

$$\mathcal{E}_i(A) = \mathcal{E} + \frac{\mu_{ind}}{\pi\epsilon_0\epsilon_r a^3} + \frac{\mu_{ind}}{\pi\epsilon_0\epsilon_r(2a)^3} + \frac{\mu_{ind}}{\pi\epsilon_0\epsilon_r(3a)^3} + \cdots,$$

$$\mathcal{E}_i(A) = \mathcal{E} + \frac{\mu_{ind}}{\pi\epsilon_0\epsilon_r a^3}\left(1 + \frac{1}{2^3} + \frac{1}{3^3} + \frac{1}{4^3} + \cdots\right),$$

or

$$\mathcal{E}_i(A) = \mathcal{E} + \frac{\mu_{ind}}{\pi\epsilon_0\epsilon_r a^3}\sum_{n=1}^{\infty}\frac{1}{n^3}, \quad (6.62)$$

where in Eq. (6.62) n is an integer 1, 2, 3, ...

In an infinite string of atoms, the same internal field as given by Eq. (6.62) would also be at any other atom. Besides there is vacuum between atoms and hence $\epsilon_r = 1$. Therefore,

$$\mathcal{E}_i = \mathcal{E} + \frac{\mu_{ind}}{\pi\epsilon_0 a^3}\sum_{n=1}^{\infty}\frac{1}{n^3}. \quad (6.63)$$

The internal field seen by an atom in an infinite array of atoms is larger than the applied field $\mathcal{E}$. As

$$\sum_{n=1}^{\infty}\frac{1}{n^3} \simeq 1.2,$$

Eq. (6.63) may be rewritten

$$\mathcal{E}_i = \mathcal{E} + \frac{1.2}{\pi}\frac{1}{\epsilon_0}\frac{\mu_{ind}}{a^3}. \quad (6.64)$$

In a three-dimensional case the calculation of the internal field would be very complicated and would depend upon the crystal structure. By analogy with Eq. (6.64) one expects that the internal field in a crystal would involve similar terms. In a three-dimensional case $1/3$ may be replaced by N, the number of atoms per unit volume, and $1.2/\pi$ by a constant γ, which depends upon the type of the structure. Hence

$$\mathcal{E}_i = \mathcal{E} + \frac{\gamma}{\epsilon_0}N\mu_{ind}. \quad (6.65)$$

As $N\mu_{ind} = \mathbf{P}$,

$$\mathcal{E}_i = \mathcal{E} + \frac{\gamma}{\epsilon_0}\mathbf{P}. \quad (6.66)$$

γ is called the *internal field constant*.

Lorentz in 1909 suggested a method to determine the internal field in a solid subjected to an electric field and found that for a cubic crystal structure $\gamma = \frac{1}{3}$. We shall discuss below the procedure adopted by Lorentz. He suggested that the simplest way to determine the value of the internal field was to introduce an imaginary cavity spherical around the atom, the radius of the cavity being large compared with the size of the atom (Fig. 6.13). The internal field at A is the sum of four fields:

$$\mathcal{E}_i = \mathcal{E}_1 + \mathcal{E}_2 + \mathcal{E}_3 + \mathcal{E}_4, \tag{6.67}$$

$\mathcal{E}_1$ is the field intensity due to charge density on the plates of the capacitor shown in Fig. 6.13:

$$\mathcal{E}_1 = \mathbf{D}/\epsilon_0,$$

but according to Eq. (6.23),

$$\mathbf{D} = \mathbf{P} + \epsilon_0 \mathcal{E},$$

and, therefore,

$$\mathcal{E}_1 = \mathcal{E} + \frac{\mathbf{P}}{\epsilon_0}. \tag{6.68}$$

$\mathcal{E}_2$ is the field intensity at the atom due to charge density induced on the two sides of the cubic dielectric opposite to the plates:

$$\mathcal{E}_2 = -\frac{\mathbf{D}}{\epsilon_0} = -\frac{\mathbf{P}}{\epsilon_0} \quad [\text{taking } \mathcal{E} = 0 \text{ in Eq. (6.23)}].$$

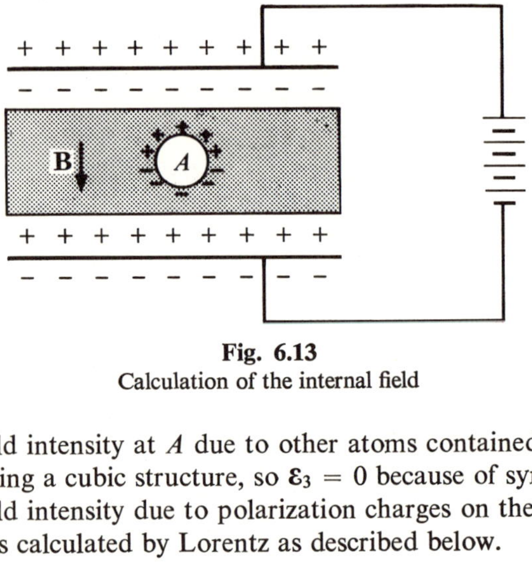

Fig. 6.13
Calculation of the internal field

$\mathcal{E}_3$ is the field intensity at A due to other atoms contained in the cavity. We are assuming a cubic structure, so $\mathcal{E}_3 = 0$ because of symmetry.

$\mathcal{E}_4$ is the field intensity due to polarization charges on the surface of the cavity and was calculated by Lorentz as described below.

Computation of $\mathcal{E}_4$. Figure 6.14 shows an enlarged view of the cavity. If dA is the surface area of the sphere of radius r lying between ϕ and $\phi + d\phi$, where ϕ is the direction with reference to the direction of the applied field,

$$dA = 2\pi r \sin \phi \, r \, d\phi$$

or

$$dA = 2\pi r^2 \sin \phi \, d\phi.$$

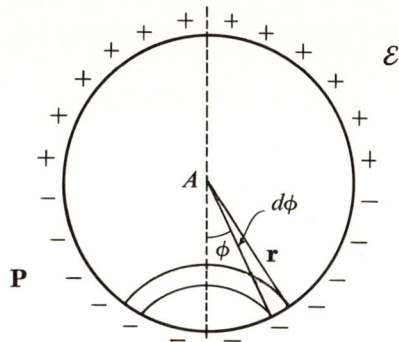

Fig. 6.14
Enlarged view of the cavity shown in Fig. 6.13

The charge dq on the surface dA is equal to the normal component of the polarization multiplied by the surface area. Therefore,

$$dq = P \cos \phi \, dA = 2\pi P r^2 \sin \phi \cos \phi \, d\phi. \tag{6.69}$$

The field due to charge dq at the point A denoted by $d\mathcal{E}_4$ in the direction of $\phi = 0$ is given by

$$d\mathcal{E}_4 = \left(\frac{dq}{4\pi \epsilon_0 r^2}\right)(\cos \phi). \tag{6.70}$$

Substituting dq from Eq. (6.69) in Eq. (6.70),

$$d\mathcal{E}_4 = \frac{P \cos \phi}{4\pi \epsilon_0 r^2} 2\pi r^2 \sin \phi \cos \phi \, d\phi.$$

Simplifying,

$$d\mathcal{E}_4 = \frac{P}{2\epsilon_0} \cos^2 \phi \sin \phi \, d\phi. \tag{6.71}$$

The total field $\mathcal{E}_4$ due to charges on the surface of the entire cavity is obtained by integrating Eq. (6.71) from $\phi = 0$ to $\phi = \pi$:

$$\mathcal{E}_4 = \int_0^\pi \frac{P}{2\epsilon_0} \cos^2 \phi \sin \phi \, d\phi$$

or

$$\mathcal{E}_4 = -\frac{P}{6\epsilon_0}\left(\cos^3 \phi \Big|_0^\pi\right) = \frac{P}{3\epsilon_0}. \tag{6.72}$$

The Lorentz field $\mathcal{E}_4$ is given by Eq. (6.72). Hence in a cubic crystal

$$\mathcal{E}_i = \mathcal{E}_1 + \mathcal{E}_2 + \mathcal{E}_3 + \mathcal{E}_4,$$

and substituting the expressions for $\mathcal{E}_1$, $\mathcal{E}_2$, $\mathcal{E}_3$, and $\mathcal{E}_4$,

$$\mathcal{E}_i = \mathcal{E} + \frac{P}{\epsilon_0} - \frac{P}{\epsilon_0} + 0 + \frac{P}{3\epsilon_0}$$

or

$$\mathcal{E}_i = \mathcal{E} + \frac{P}{3\epsilon_0}. \tag{6.73}$$

Equation (6.73) shows that the internal field at an atom in a cubic structure is of the form of Eq. (6.66) with the value of $\gamma = \frac{1}{3}$.

Dielectric constants of solids. First consider solid insulators such as diamond, carbon, silicon, and germanium. Such materials are called *elemental dielectrics* and have cubic structure. In these materials there are no ions or permanent dipoles because all the atoms are alike. Hence $\alpha_i = 0 = \alpha_0$. Therefore,

$$P = P_e$$

$$P = N\alpha_e \mathcal{E}_i, \tag{6.74}$$

where N is the number of atoms per unit volume.

Substituting the expression for $\mathcal{E}_i$ from Eq. (6.73) in Eq. (6.74),

$$P = N\alpha_e \left(\mathcal{E} + \frac{P}{3\epsilon_0}\right). \tag{6.75}$$

Rearranging,

$$P = \frac{N\alpha_e \mathcal{E}}{1 - (N\alpha_e/3\epsilon_0)}. \tag{6.76}$$

According to Eq. (6.22),

$$P = \epsilon_0(\epsilon_r - 1)\mathcal{E}. \tag{6.22}$$

Equating the right sides of Eqs. (6.22) and (6.76),

$$\epsilon_0(\epsilon_r - 1) = \frac{N\alpha_e}{1 - (N\alpha_e/3\epsilon_0)}. \tag{6.77}$$

Rearranging the terms and adding 3 on both sides,

$$\frac{\epsilon_r - 1}{\epsilon_r + 2} = \frac{N\alpha_e}{3\epsilon_0}. \tag{6.78}$$

Equation (6.78) is called the *Clausius-Mosotti relationship*. It can be used to determine the electronic polarization in elemental dielectrics if the dielectric constant ϵ_r is known. It can also be used to predict the relative

dielectric constant if the electronic polarizability of the atom is known. For example,
$$\epsilon_r \text{ (for Ge)} = 16.$$

According to the Clausius-Mosotti relationship (6.78),
$$\frac{15}{18} = \frac{N\alpha_e}{3\epsilon_0}$$

or
$$N\alpha_e = \frac{5}{2}\epsilon_0$$

$$\epsilon_0 = 8.854 \times 10^{-12} \text{ F/m},$$

$$N = \frac{8}{(5.65)^3 \times 10^{-30}} = \frac{800}{(5.65)^3} \times 10^{28} = 4.45 \times 10^{28} \text{ m}^{-3}.$$

Therefore,
$$\alpha_e = \frac{5}{2} \times \frac{8.854 \times 10^{-12}}{4.45 \times 10^{28}} = 4.96 \times 10^{-40} \text{ F/m}^2.$$

In alkali halides the permanent dipole moment is zero ($\mu_P = 0$) and the total polarization is made up of electronic and ionic polarization:
$$P = P_e + P_i.$$

Therefore,
$$\epsilon_0(\epsilon_r - 1)\mathcal{E} = P_e + P_i. \qquad (6.79)$$

The examples of alkali halides are NaCl, KCl, KBr, KI, and many others.

6.3 Ferroelectricity

In 1921 Valasek discovered the phenomenon of ferroelectricity in Rochelle salt, which is sodium potassium tartrate tetrahydrate. Ferroelectricity is a result of dielectric hysteresis. Since then many other ferroelectric materials have been discovered. Typical examples are dihydrogen phosphates and arsenates of alkali metals such as KH_2PO_4 and oxygen octahedron groups such as barium titanate, $BaTiO_3$.

In these materials the electric flux density **D** is not determined uniquely by the applied field $\mathcal{E}$ but depends upon the previous history of the material. Theoretically speaking, if a virgin specimen (not subjected to an electric field previously) of a ferroelectric material is subjected to an electric field and the field is gradually increased, the polarization **P** increases along the

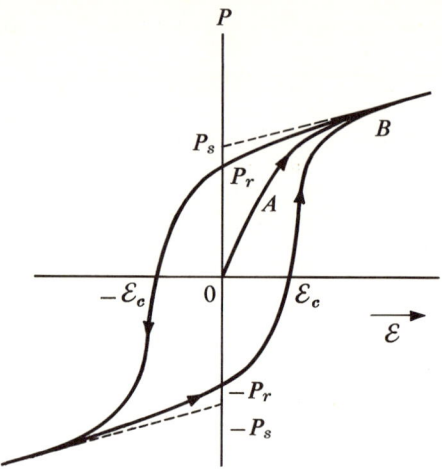

Fig. 6.15
Hysteresis loop for ferroelectric materials

curve OAB (Fig. 6.15). When the field is gradually decreased, it is found that at $\mathcal{E} = 0$ there is a polarization still left. This polarization value is called *residual polarization* $\mathbf{P}_r$. A negative field $-\mathcal{E}_c$ is required to reduce the polarization down to zero. $\mathcal{E}_c$ is called the *coercive field*. The existence of a dielectric hysteresis loop in a material implies that the substance possesses a spontaneous polarization, i.e., a polarization that persists when the applied field is zero. Figure 6.15 shows that $\mathbf{P}_s$ (depending upon the shape of the hysteresis loop obtained experimentally) depends upon a number of factors. It depends upon the dimensions of the specimen, the temperature, the humidity, the texture of the crystal, and the previous thermal and electrical history of the crystal.

The Curie point. The hysteresis loop of a ferroelectric material changes its shape as the temperature is increased. The height and width decrease with temperature. At a certain temperature known as the *ferroelectric Curie temperature*, the loop merges into a straight line Fig. 6.16 and the ferroelectric behavior of the material disappears.

Spontaneous polarization. Spontaneous polarization in a ferroelectric material disappears above the Curie temperature. If the temperature is below the Curie temperature, $\mathbf{P}$ does not change linearly with $\mathcal{E}$ and the relative dielectric constant is not constant. In such a case ϵ_r for a ferroelectric material may be defined for the virgin curve as a differential quantity,

$$\frac{dP}{d\mathcal{E}} = \epsilon_0(\epsilon_r - 1). \tag{6.80}$$

FERROELECTRICITY

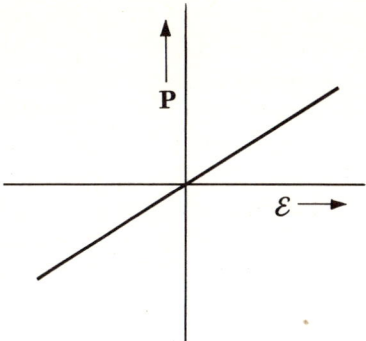

Fig. 6.16
P-$\mathcal{E}$ relationship above ferroelectric Curie temperature

We have already discussed that the internal field may be different from the applied field. The relationship between the two is given by Eq. (6.66):

$$\mathcal{E}_i = \mathcal{E} + \frac{\gamma \mathbf{P}}{\epsilon_0}. \tag{6.66}$$

Further, according to the theory discussed,

$$\mathbf{P} = N\alpha \mathcal{E}_i,$$

where $\alpha = \alpha_e + \alpha_i + \alpha_0$, the total polarizability. Therefore,

$$\mathbf{P} = N\alpha \left(\mathcal{E} + \frac{\gamma \mathbf{P}}{\epsilon_0} \right)$$

and, after simplifying,

$$\mathbf{P} = \frac{N\alpha \mathcal{E}}{1 - (N\alpha\gamma/\epsilon_0)}. \tag{6.81}$$

If the denominator of Eq. (6.81) equals zero, one obtains a nonvanishing solution for **P** and therefore there exists the possibility of spontaneous polarization. The condition for spontaneous polarization is

$$\frac{N\alpha\gamma}{\epsilon_0} = 1. \tag{6.82}$$

From Eq. (6.81),

$$\frac{\mathbf{P}}{\mathcal{E}} = \epsilon_0(\epsilon_r - 1) = \frac{N\alpha}{1 - (N\alpha\gamma/\epsilon_0)}. \tag{6.83}$$

We may not conclude from Eq. (6.83) that with $N\alpha\gamma/\epsilon_0 = 1$ the dielectric constant will become infinite but may conclude that the substance will become spontaneously polarized.

Dipole theory of ferroelectricity. Many different theories of different ferroelectric materials have been put forward, by Mason, Slater, Devonshire, and others. A comprehensive account of these is given in various textbooks, including the text by Megaw (see the Bibliography). We shall try to understand ferroelectricity by taking up a simplified version of the dipole theory of ferroelectricity. A ferroelectric material consists of permanent dipole moments in which the interaction between adjacent dipole moments is large and a dipole moment has a tendency to align itself in a direction parallel to that of its neighbor. Assuming that the permanent dipoles are the ones responsible for spontaneous polarization, we can write

$$\mathbf{P} = \mathbf{P}_0 = N\alpha_0 \mathcal{E}_i = L(\alpha) N \mu_P. \tag{6.51}$$

For high temperatures

$$L(\alpha) \simeq \alpha/3,$$

$$\mathbf{P} = \mathbf{P}_0 = \frac{N\mu_P^2}{3kT} \mathcal{E}_i. \tag{6.84}$$

The cause of spontaneous polarization is that the internal field $\mathcal{E}_i$ is different from the externally applied field because of a strong interaction between dipoles, and hence Eq. (6.84) is different from Eq. (6.53), in that the applied field has been replaced by the internal field.

According to Eq. (6.66),

$$\mathcal{E}_i = \mathcal{E} + \frac{\gamma \mathbf{P}}{\epsilon_0},$$

and hence substituting Eq. (6.66) in Eq. (6.84),

$$\mathbf{P} = \frac{N\mu_P^2}{3kT} \left(\mathcal{E} + \frac{\gamma \mathbf{P}}{\epsilon_0} \right),$$

or, simplifying,

$$\mathbf{P} = \frac{N\mu_P^2/3kT}{1 - (N\gamma\mu_P^2/3kT\epsilon_0)} \mathcal{E}$$

or

$$\frac{\mathbf{P}}{\mathcal{E}} = \frac{C\epsilon_0}{T - \theta}, \tag{6.85}$$

where $\theta = \gamma N \mu_P^2 / 3k\epsilon_0$ and $C = \theta/\gamma$. C in Eq. (6.85) is called the *Curie constant* and θ is the Curie temperature.

The left side of Eq. (6.85) may be written $\epsilon_0(\epsilon_r - 1)$, and therefore

$$\epsilon_r - 1 = \frac{C}{T - \theta}. \tag{6.86}$$

Equation (6.86) is known as the *Curie-Weiss law* and we have shown with the help of the simple dipolar theory that a ferroelectric material obeys

Eq. (6.86) at sufficiently large temperatures. We shall now try to show that spontaneous polarization is possible in a ferroelectric material below the Curie temperature. According to Eq. (6.51),

$$P = N\mu_P L(\alpha). \tag{6.51}$$

For very large values of α, low temperature, and very, very high electric fields, $L(\alpha) \to 1$, and the polarization is called the *saturation polarization* P_{sat}. Hence

$$P_{sat} = N\mu_P. \tag{6.87}$$

Equation (6.51) may therefore be rewritten

$$P/P_{sat} = L(\alpha). \tag{6.88}$$

If the internal field is different from the applied field,

$$L(\alpha) = L\left(\frac{\mu_P \mathcal{E}_i}{kT}\right),$$

and upon substitution of Eq. (6.66),

$$L(\alpha) = L\left[\frac{\mu_P}{kT}\left(\mathcal{E} + \frac{\gamma P}{\epsilon_0}\right)\right].$$

The expression for polarization is therefore

$$\frac{P}{P_{sat}} = L\left[\frac{\mu_P}{kT}\left(\mathcal{E} + \frac{\gamma P}{\epsilon_0}\right)\right]. \tag{6.89}$$

Spontaneous polarization exists if there is a nonvanishing solution for P in Eq. (6.89) with the applied electric field $\mathcal{E} = 0$. Hence for spontaneous polarization,

$$\frac{P}{P_{sat}} = L\left(\frac{\mu_P \gamma P}{kT\epsilon_0}\right). \tag{6.90}$$

If we put

$$\frac{\mu_P \gamma P}{kT\epsilon_0} = x, \tag{6.91}$$

$$\frac{P}{P_{sat}} = L(x). \tag{6.92}$$

Also, from Eq. (6.91),

$$\frac{P}{N\mu_P} = \left(\frac{kT\epsilon_0}{N\mu_P^2 \gamma}\right) x$$

or

$$\frac{P}{P_{sat}} = \left(\frac{T}{3\theta}\right) x \quad \text{(by incorporating the equation for } \theta\text{)}. \tag{6.93}$$

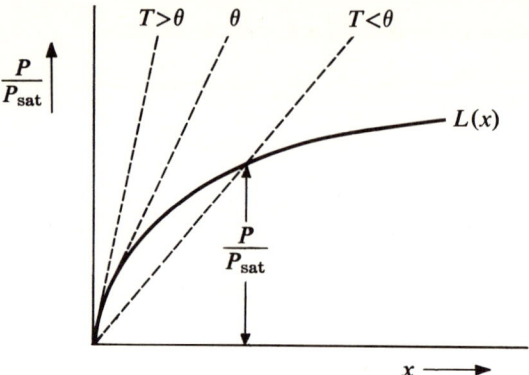

Fig. 6.17
Langevin function and expression (6.93)

P/P_{sat} satisfies Eqs. (6.92) and (6.93) simultaneously. Equation (6.92) has been represented in Fig. 6.17 by the curve $L(x)$. Equation (6.93) represents a set of straight lines which pass through the origin. A solution for a finite value would exist if there is an intersection of the straight line with the $L(x)$ curve. Temperatures $T = \theta$ result in spontaneous polarization. At very very high temperatures there is no spontaneous polarization, because the straight line does not intersect the $L(x)$ curve. Measured values of γ are: Rochelle salt, 2.1; Ba Ti O$_3$, 0.044; and KH$_2$PO$_4$, 0.37.

If the dipole theory based on

$$\mathcal{E}_i = \mathcal{E} + \frac{\gamma P}{\epsilon_0}$$

were correct, a large number of polar liquids would be ferroelectric. On the other hand, it is known that ferroelectric materials are rare. More exact calculations of the internal field in ferroelectric materials show that γ is not constant but depends upon the dielectric constant in such a manner that γ decreases with increasing ϵ_r, or

$$\mathcal{E}_i = \mathcal{E} + \frac{\gamma(\epsilon_r) P}{\epsilon_0}. \tag{6.94}$$

As such,

$$\frac{P}{\mathcal{E}} = \frac{(N\mu_P^2/3kT)}{1 - [N\gamma(\epsilon_r)\mu_P^2/3kT]}. \tag{6.95}$$

Because γ decreases with ϵ_r, the possibility of spontaneous polarization decreases.

In summary it may be said that research on ferroelectric is in progress to devise generalized theories of spontaneous polarization.

6.4 Complex dielectric constant and lossy dielectrics

So far we have discussed the effect of static fields on dielectrics. The behavior of dielectrics in alternating fields is a little more involved. With high-frequency applied fields the dielectrics absorb energy and behave as lossy dielectrics. To determine the effect of alternating fields on dielectrics, it will be helpful to consider the three polarizabilities — electronic, ionic, and orientation — one after another.

Frequency dependence of electronic polarization. While discussing the electronic polarizability with static fields two atomic models were employed. It was found that the model with uniform volume charge distribution of the electron over a sphere of radius R did not yield results very different from the quantum-mechanical model. We shall therefore employ the semiclassical model for a study of the frequency dependence of the electronic polarizability. Let us consider the situation that in the presence of an applied field $\mathcal{E}$, the electron cloud is moved in a direction opposite to the field and let us imagine that in equilibrium the center of the electron cloud with the applied field is at the point A separated from its field-free position $x = 0$. Now if at $t = 0$ the applied field is suddenly removed, the cloud experiences a Coulomb force of attraction. The magnitude of this force, calculated previously, is

$$F(\text{Coulomb}) = \frac{Z^2 e^2 x}{4\pi\epsilon_0 R^3}.$$

If the mass of the electron cloud is m, it experiences an acceleration and starts moving toward $x = 0$. The differential equation of motion of the electron cloud is

$$m\frac{d^2x}{dt^2} = -\frac{Z^2 e^2}{4\pi\epsilon_0 R^3} x.$$

If we put

$$\frac{Z^2 e^2}{4\pi\epsilon_0 R^3} = f \quad \text{(a constant)},$$

$$m\frac{d^2x}{dt^2} = -fx. \tag{6.96}$$

Putting $f/m = \omega_0^2$, Eq. (6.96) may be rewritten

$$\frac{d^2x}{dt^2} + \omega_0^2 x = 0. \tag{6.97}$$

Equation (6.97) is the equation of a single harmonic oscillator whose equation of motion for the boundary conditions stated is

$$x = -x_0 \sin \omega_0 t.$$

ω_0 is the natural angular frequency of oscillation of the electron cloud. It is shown below that $\omega_0 \simeq 10^{16}$ rad/sec and hence corresponds to the ultraviolet region:

$$\omega_0^2 = \frac{f}{m} = \frac{Z^2 e^2}{4\pi\epsilon_0 R^3 m}.$$

For helium $Z = 2$ and $R \simeq 10^{-10}$ m, so

$$\omega_0^2 = \frac{4(1.6 \times 10^{-19})^2}{4\pi \times 8.854 \times 10^{-12} \times 10^{-30} \times 9.1 \times 10^{-31}} \cong 10.1 \times 10^{32}.$$

Accordingly, the electron cloud would go on oscillating for an indefinite period of time about its field-free static position $x = 0$. This is not the case in practice. Equation (6.97) is incomplete in that it does not include the emission of radiation by an oscillating charge according to classical field theory. From field theory it is known that an accelerating charge radiates energy. The radiation of energy corresponds to the damping force of the electron cloud. Therefore Eq. (6.97) may be modified as

$$m\frac{d^2 x}{dt^2} = -fx - 2b\frac{dx}{dt}, \tag{6.98}$$

where $2b(dx/dt)$ is the damping force proportional to the velocity of the electron cloud and

$$2b = \frac{\mu_0 e^2 \omega_0^2}{6\pi mc},$$

where $\mu_0 = (4\pi/10) \times 10^{-7}$ H/m (permeability of vacuum), $c = 2.9979 \times 10^8$ m/sec (velocity of light), and $e = 1.6 \times 10^{-19}$ C. Equation (6.98) may be simplified as

$$\frac{d^2 x}{dt^2} + \omega_0^2 x + \frac{2b}{m}\frac{dx}{dt} = 0. \tag{6.99}$$

Equation (6.99) is the differential equation of motion of a damped oscillator and describes the motion of the electron cloud. The locus of the center of the electron cloud, plotted against time, is shown as a dashed line in Fig. 6.18.

Now if the applied field is $\mathcal{E}_0 \cos \omega t$, where ω is the angular frequency of the applied field, the Lorentz force on the electron cloud is $-Ze\mathcal{E}_0 \cos \omega t$ and the equation of motion of the electron may be written

$$m\frac{d^2 x}{dt^2} = -fx - 2b\frac{dx}{dt} - Ze\mathcal{E}_0 \cos \omega t$$

or

$$m\frac{d^2 x}{dt^2} + 2b\frac{dx}{dt} + fx + Ze\mathcal{E}_0 \cos \omega t = 0. \tag{6.100}$$

COMPLEX DIELECTRIC CONSTANT AND LOSSY DIELECTRICS

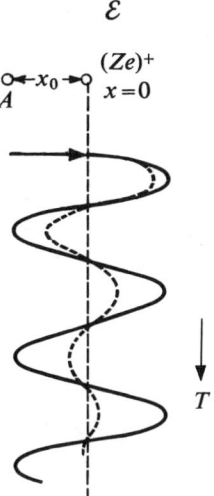

Fig. 6.18
Variation of electron displacement with time in an alternating electric field

It is now desired to solve Eq. (6.100) for $x(t)$. Because x would depend upon time once the displacement of the electron cloud has been determined, the induced dipole may be obtained by simply multiplying $-Ze$ with $x(t)$:

$$\mu_{ind} = Zex(t). \tag{6.101}$$

As a possible solution of Eq. (6.100) let us assume

$$x(t) = \text{Re}\,(B^*e^{j\omega t}), \tag{6.102}$$

where B^* is some complex quantity and Re is the real part of the quantity in parentheses.

It now remains to determine B^*, which may be obtained by substituting the assumed solution in Eq. (6.100):

$$\frac{dx}{dt} = \text{Re}(B^*j\omega e^{j\omega t}),$$

$$\frac{d^2x}{dt^2} = \text{Re}[B^*(-\omega^2)e^{j\omega t}],$$

$$Ze\mathcal{E}_0 \cos \omega t = \text{Re}(Ze\mathcal{E}_0 e^{j\omega t}),$$

and hence

$$\text{Re}\left[\left(-\omega^2 B^* + j\frac{2b\omega}{m}B^* + \omega_0^2 B^* + \frac{Ze}{m}\mathcal{E}_0\right)e^{j\omega t}\right] = 0. \tag{6.103a}$$

In Eq. (6.103a), because $e^{j\omega t} \neq 0$,

$$B^*\left[(\omega_0^2 - \omega^2) + j\frac{2b\omega}{m}\right] = -\frac{Ze}{m}\mathcal{E}_0$$

or

$$B^* = \frac{(-Ze/m)\mathcal{E}_0}{(\omega_0^2 - \omega^2) + j(2b\omega/m)}.$$

Hence

$$x(t) = \text{Re}\left[\frac{(-Ze/m)\mathcal{E}_0 e^{j\omega t}}{(\omega_0^2 - \omega^2) + j(2b\omega/m)}\right],$$

$$\mu_{\text{ind}}(t) = \text{Re}\left[\frac{(Z^2 e^2/m)\mathcal{E}_0 e^{j\omega t}}{(\omega_0^2 - \omega^2) + j(2b\omega/m)}\right]. \quad (6.103\text{b})$$

Equation (6.103b) is a relationship between the induced dipole moment due to motion of the electron cloud and the applied electric field. The definition of polarizabil'ty in alternating fields may be α_e^*, the complex electronic polarizability. Hence by definition

$$\mu_{\text{ind}}(t) = \text{Re}(\alpha_e^* \mathcal{E}_0 e^{j\omega t}). \quad (6.104)$$

Comparing the right sides of Eqs. (6.103) and (6.104),

$$\alpha_e^* = \frac{Z^2 e^2/m}{(\omega_0^2 - \omega^2) + j(2b\omega/m)}. \quad (6.105)$$

The complex polarizability α_e^* may be written as the sum of a real part α_e' and an imaginary part of α_e'':

$$\alpha_e^* = \alpha_e' - j\alpha_e''. \quad (6.106)$$

Equation (6.105) may be rewritten by multiplying the numerator and denominator of the right side by the conjugate complex of the denominator:

$$\alpha_e^* = \frac{Z^2 e^2}{m}\left[\frac{\omega_0^2 - \omega^2}{(\omega_0^2 - \omega^2)^2 + (4b^2\omega^2/m^2)} - j\frac{2b\omega}{m}\frac{1}{(\omega_0^2 - \omega^2)^2 + (4b^2\omega^2/m^2)}\right]. \quad (6.107)$$

Comparing the right sides of Eqs. (6.106) and (6.107),

$$\alpha_e' = \frac{Z^2 e^2}{m}\frac{\omega_0^2 - \omega^2}{(\omega_0^2 - \omega^2)^2 + (4b^2\omega^2/m^2)}, \quad (6.108)$$

$$\alpha_e'' = \frac{Z^2 e^2}{m}\frac{2b\omega}{m}\frac{1}{(\omega_0^2 - \omega^2)^2 + (4b^2\omega^2/m^2)}. \quad (6.109)$$

It can be seen easily that if d.c. fields are applied, i.e., if $\omega = 0$ and $\alpha_e'' = 0$, α_e' reduces to

$$\alpha_e' = \frac{Z^2 e^2}{m}\frac{1}{\omega_0^2} = \frac{Z^2 e^2}{f} = 4\pi\epsilon_0 R^3,$$

the value already obtained in Sec. 6.2 with d.c. fields.

The electronic polarization $P_e(t)$ is obtained by multiplying $\mu_{ind}(t)$ with N, the number of atoms per unit volume. Assuming we are considering gases, the interatomic distance is very large and therefore the internal field may be taken as equal to the applied field:

$$P_e(t) = \text{Re}(N\alpha_e^* \mathcal{E}_0 e^{j\omega t}) \tag{6.110}$$

or

$$P_e(t) = \text{Re}[N(\alpha_e' - j\alpha_e'')\mathcal{E}_0 e^{j\omega t}].$$

Simplifying the above equation,

$$P_e(t) = N\alpha_e' \mathcal{E}_0 \cos \omega t + N\alpha_e'' \mathcal{E}_0 \sin \omega t. \tag{6.111}$$

Substituting values of α_e' and α_e'' from Eqs. (6.108) and (6.109) in Eq. (6.111),

$$P_e(t) = N\frac{Z^2 e^2}{m} \frac{\omega_0^2 - \omega^2}{(\omega_0^2 - \omega^2)^2 + (4b^2\omega^2/m^2)} \mathcal{E}_0 \cos \omega t + N\frac{Z^2 e^2}{m} \frac{2b\omega}{m}$$

$$\times \frac{1}{(\omega_0^2 - \omega^2)^2 + (4b^2\omega^2/m^2)} \mathcal{E}_0 \sin \omega t. \tag{6.112}$$

The electronic polarization with an alternating field can be considered to consist of two parts, one in phase with the applied field $\mathcal{E}_0 \cos \omega t$ and the other phase shifted from the applied field by $\pi/2$ and lags. The frequency dependence of the in-phase and out-of-phase components is given in Fig. 6.19 and discussed below.

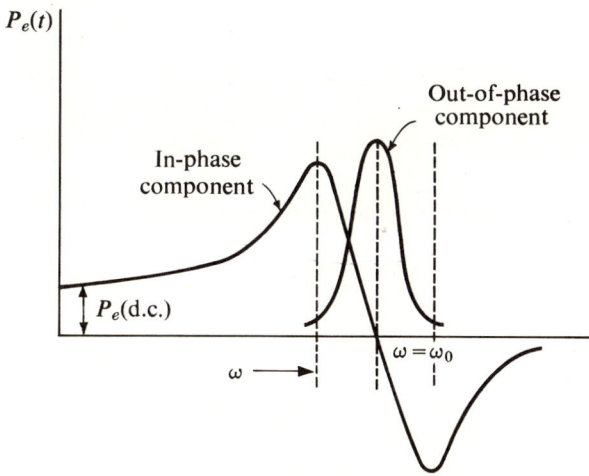

Fig. 6.19
Variation of electric polarization with frequency

The in-phase component of electronic polarization $P_e(t)_{in}$

$$P_e(t)_{in} = N\frac{Z^2 e^2}{m} \frac{\omega_0^2 - \omega^2}{(\omega_0^2 - \omega^2)^2 + (4b^2\omega^2/m^2)} \mathcal{E}_0 \cos \omega t.$$

If $\omega = 0$,

$$P_e(t)_{in} = P_e \quad \text{(d.c. value).}$$

Further, if $\omega < \omega_0$, the in-phase component is positive and negative if $\omega > \omega_0$. The in-phase component is zero if $\omega = \omega_0$ or the frequency of the applied field is equal to the natural frequency of oscillation of the electron cloud. It can be easily shown that for the electron cloud $2b \ll \omega_0$ and, as such, the in-phase component is practically constant from $\omega = 0$ to a value in the neighborhood of $\omega = \omega_0$. Therefore, changes in the in-phase component can be expressed in the neighborhood of ω_0 by taking a new variable: $\Delta\omega = \omega_0 - \omega$ and $\omega \simeq \omega_0$:

$$\omega_0^2 - \omega^2 = (\omega_0 + \omega)(\omega_0 - \omega) \cong 2\omega_0 \Delta\omega,$$

$$P_e(t)_{in} = \frac{NZ^2 e^2}{m} \frac{2\omega_0 \Delta\omega}{4\omega_0^2 (\Delta\omega)^2 + (4b^2\omega_0^2/m^2)} \mathcal{E}_0 \cos \omega t,$$

which may be simplified as

$$\frac{P_e(t)_{in}}{\mathcal{E}_0 \cos \omega t} = \frac{NZ^2 e^2}{m} \frac{\Delta\omega/2\omega_0}{(\Delta\omega)^2 + (b^2/m^2)}, \tag{6.113}$$

from which it can be seen that the expression has a maximum for $\Delta\omega = b/m$ and a minimum for $\Delta\omega = -b/m$.

The out-of-phase component of electronic polarization $P_e(t)_{out}$

$$P_e(t)_{out} = \frac{NZ^2 e^2}{m} \frac{2b\omega}{m} \frac{1}{(\omega_0^2 - \omega^2)^2 + (4b^2\omega^2/m^2)} \mathcal{E}_0 \sin \omega t.$$

If $\omega = 0$,

$$P_e(t)_{out} = 0.$$

The out-of-phase component of electronic polarization is a maximum at $\omega = \omega_0$:

$$\frac{P_e(t)_{out}}{\mathcal{E}_0 \sin \omega t} = \frac{NZ^2 e^2}{m} \frac{2b\omega}{m} \frac{1}{(\omega_0^2 - \omega^2)^2 + (4b^2\omega^2/m^2)}. \tag{6.114}$$

In terms of $\Delta\omega = \omega_0 - \omega$,

$$\frac{P_e(t)_{out}}{\mathcal{E}_0 \sin \omega t} \simeq \frac{NZ^2 e^2}{m} \frac{b}{2m\omega} \frac{1}{(\Delta\omega)^2 + (b^2/m^2)}. \tag{6.115}$$

Equation (6.115) shows that the maximum value of the out-of-phase component falls to $\frac{1}{2}$ at $\Delta\omega = \pm b/m$. The physical interpretation of the out-of-phase component is that the material absorbs energy in the region of frequency in the neighborhood of the natural frequency of oscillation of the electron cloud. The absorption of energy will be discussed after all types of polarization in alternating fields are discussed.

We may summarize by saying that electric fields of very high frequency affect the behavior of electronic polarizability. This frequency is in the ultraviolet region.

Frequency dependence of ionic polarization. By an analogous mathematical treatment it can be shown that if an alternating field is applied to a dielectric with ionic structure the ionic polarizability will be a complex quantity α_i^*, where

$$\alpha_i^* = \alpha_i' - j\alpha_i''.$$

The ionic polarization $P_i(t)$ will then be given by

$$P_i(t) = \mathrm{Re}(N\alpha_i^* \mathcal{E}_0 e^{j\omega t}),$$

or

$$P_i(t) = N\alpha_i' \mathcal{E}_0 \cos \omega t + N\alpha_i'' \mathcal{E}_0 \sin \omega t. \tag{6.116}$$

Equation (6.116) shows that ionic polarization with alternating fields also consists of two parts, one in phase with the applied field and the other lagging by $\pi/2$. The frequency response of the in-phase and out-of-phase components in ionic polarization is similar to the frequency response of the electronic polarization components. The only difference between the two is that the natural frequency of oscillation of an ion is much lower than that of an electron cloud. The natural frequency of oscillation of ions lies in the infrared region and is of the order of 10^{14} rad/sec.

In a material having both ionic and electronic polarization, the total in-phase component of the polarization varies as shown in Fig. 6.20. ω_{0_e} is

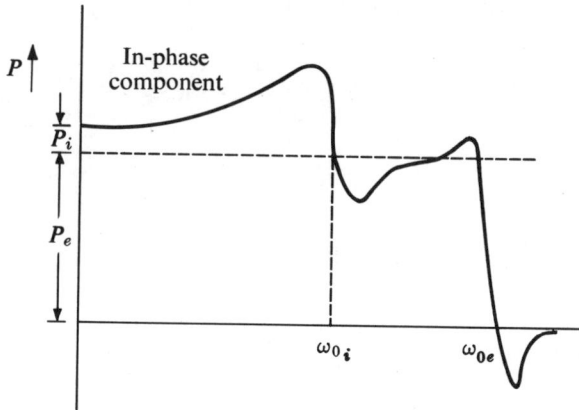

Fig. 6.20
Variation of the in-phase component of electric polarization with frequency for a material having both ionic and electronic polarization

the natural frequency of oscillation of the electron cloud and ω_{0_i} is the natural frequency of oscillation of the ions.

Solids and liquids without permanent dipole moments. It is now possible to discuss the effect of alternating fields on dielectrics without permanent dipole moments and, therefore, $\alpha_0 = 0$. We have already discussed that in such materials the internal field is different from the applied field. Therefore the total polarization, which consists of electronic and ionic polarizations, may be defined as

$$P(t) = \text{Re}[N(\alpha_e^* + \alpha_i^*)\mathcal{E}_{0_i}^* e^{j\omega t}], \qquad (6.117)$$

where $\mathcal{E}_{0_i}^*$ is the internal field, which may be complex.

From field theory it can be shown that the polarization in an alternating field has to be defined as

$$P(t) = \text{Re}[\epsilon_0(\epsilon_r^* - 1)\mathcal{E}_0 e^{j\omega t}], \qquad (6.118)$$

where ϵ_r^* is the relative dielectric constant, which may be complex. Equating the right sides of Eqs. (6.117) and (6.118) would help us express ϵ_r^* in terms of atomic properties α_e^* and α_i^*. There is, however, difficulty in expressing $\mathcal{E}_{0_i}^*$ in terms of $\mathcal{E}_0$, the applied field. Assuming that the crystal structure is cubic and the Lorentz expression is applicable,

$$\mathcal{E}_i(t) = \mathcal{E}(t) + \frac{P(t)}{3\epsilon_0}. \qquad (6.119)$$

Substituting Eq. (6.118) in Eq. (6.119),

$$\mathcal{E}_i(t) = \mathcal{E}_0 \cos \omega t + \text{Re}\left(\frac{\epsilon_r^* - 1}{3} \mathcal{E}_0 e^{j\omega t}\right).$$

Simplifying,

$$\mathcal{E}_i(t) = \text{Re}\left[\left(\frac{\epsilon_r^* + 2}{3}\right)\mathcal{E}_0 e^{j\omega t}\right]$$

or

$$\text{Re}[\mathcal{E}_{0_i}^* e^{j\omega t}] = \text{Re}\left[\frac{\epsilon_r^* + 2}{3} \mathcal{E}_0 e^{j\omega t}\right]. \qquad (6.120)$$

Equation (6.120) is a relationship between the internal field and the applied field. Hence

$$P(t) = \text{Re}\left[N(\alpha_e^* + \alpha_i^*)\frac{\epsilon_r^* + 2}{3}\mathcal{E}_0 e^{j\omega t}\right] = \text{Re}[\epsilon_0(\epsilon_r^* - 1)\mathcal{E}_0 e^{j\omega t}].$$

From the above equation for polarization it can be seen that

$$\frac{\epsilon_r^* - 1}{\epsilon_r^* + 2} = \frac{N}{3\epsilon_0}(\alpha_e^* + \alpha_i^*), \qquad (6.121)$$

which shows that if the polarizabilities are complex, the relative dielectric constant is complex. Equation (6.121) is the Clausius-Mosotti relationship for alternating fields. The complex relative dielectric constant may be expressed

$$\epsilon_r^* = \epsilon_r' - j\epsilon_r'', \qquad (6.122)$$

where ϵ_r' is the real part of the complex relative dielectric constant and ϵ_r'' is the imaginary part of the complex relative dielectric constant.

Frequency dependence of orientational polarization. We have already discussed that if permanent dipole moments μ_P are present in a substance, the dipole polarization in the absence of the field is zero (excluding ferroelectric materials), or

$$\sum \mu_P = 0.$$

When the substance is subjected to an electric field, the dipoles experience a torque and tend to align in the direction of the field; this produces a polarization which we have called *orientational polarization*. Now if the field is suddenly removed, the dipoles will again be randomly oriented. However, this would not occur simultaneously with the switching off of the field. A certain time is required for the dipoles to orient randomly. A time measure of the dipoles for random orientation is the dielectric relaxation time τ. The dipole relaxation time is defined as the time required for the dipoles to orient in such a way that the polarization reduces to $1/e$ or its original value in the presence of the field. In the definition of τ it is assumed that the polarization decays exponentially:

$$P_0(t) = P_{0(t=0)} e^{-t/\tau}. \qquad (6.123)$$

Such a treatment is called the *phenomenonological approach*. Equation (6.123) suggests that the differential equation for orientation polarization may be

$$\frac{d}{dt}[P_0(t)] = \frac{1}{\tau}[P_0 - P_{0(t\to\infty)}(t)], \qquad (6.124)$$

where $P_{0\,(t\to\infty)}$ is the final value of orientational polarization. With the differential equation of the orientational polarization in mind, the frequency dependence of orientational polarization can be discussed as follows.

Let us consider an alternating field $\mathcal{E}_0 \cos \omega t$ to be applied to a system of permanent dipoles:

$$P_{0(t\to\infty)} = \epsilon_0(\epsilon_{r_0} - 1)\mathcal{E}_0 \cos \omega t,$$

where ϵ_{r_0} is the contribution of orientational polarization to the relative dielectric constant. Hence Eq. (6.124) may be rewritten

$$\frac{d}{dt}[P_0(t)] = \frac{1}{\tau}[\epsilon_0(\epsilon_{r_0} - 1)\mathcal{E}_0 \cos \omega t - P_0(t)]. \qquad (6.125)$$

Let the solution of the differential equation (6.125) be

$$P_0(t) = \text{Re}[\epsilon_0(\varepsilon_{r_0}^* - 1)\mathscr{E}_0 e^{j\omega t}]. \tag{6.126}$$

The value of $(\varepsilon_{r_0}^* - 1)$ in the solution may be obtained by substituting the solution (6.126) in Eq. (6.125),

$$\frac{d}{dt}[P_0(t)] = \text{Re}[\epsilon_0 j\omega(\varepsilon_{r_0}^* - 1)\mathscr{E}_0 e^{j\omega t}],$$

and hence

$$\varepsilon_{r_0}^* - 1 = \frac{\epsilon_{r_0} - 1}{1 + j\tau\omega}. \tag{6.127}$$

Hence the introduction of the dipolar relaxation time leads us to the condition that the orientational polarization varies according to Eq. (6.128):

$$P_0(t) = \text{Re}\left[\frac{\epsilon_0(\epsilon_{r_0} - 1)}{1 + j\tau\omega} \mathscr{E}_0 e^{j\omega t}\right]. \tag{6.128}$$

Expanding the right side of Eq. (6.128),

$$P_0(t) = \frac{\epsilon_0(\epsilon_{r_0} - 1)}{1 + \tau^2\omega^2}\mathscr{E}_0 \cos \omega t + \frac{\epsilon_0(\epsilon_{r_0} - 1)\tau\omega}{1 + \tau^2\omega^2}\mathscr{E}_0 \sin \omega t. \tag{6.129}$$

Like electronic and ionic polarizations, the orientational polarization consists of two parts, one in phase with the applied field and the other lagging by $\pi/2$. The variation of both the in-phase and out-of-phase components with frequency depends upon the product $\tau\omega$. At $\tau\omega = 1$ the in-phase component reduces to one half of its d.c. value for $\tau\omega \gg 1$; it decreases

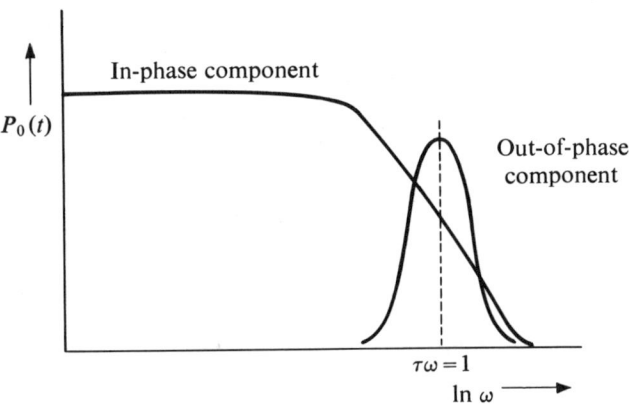

Fig. 6.21
Variation of in-phase and out-of-phase components of orientational polarization with frequency

down to very low values. The variation of the out-of-phase component depends upon $\tau\omega/(1 + \tau^2\omega^2)$, which has a maximum at $\tau\omega = 1$.

τ may vary from 10^{-6} to 10^{-10} sec and, as such, the frequency range affecting the behavior of orientational polarizations lies between 10^6 to 10^{10} Hz as changes occur in the neighborhood of $\tau\omega = 1$ (Fig. 6.21). The out-of-phase component of the orientational polarization causes an absorption of energy in the dielectric.

The decrease in the in-phase component with frequency may be interpreted as follows. At low frequencies the dipoles can respond to the variation of the electric field and move with the field direction. As the frequency increases, the dipoles can no longer cope with the changes in the direction of the field direction and hence the effective polarization falls to very low values. The frequencies at which the in-phase components decrease substantially lies in the radiofrequency region and hence is very important from the point of view of electrical engineers.

Absorption of energy and dielectric losses. With alternating fields the electric flux density has to be expressed as

$$D(t) = \text{Re}(\epsilon_0 \epsilon_r^* \mathcal{E}_0 e^{j\omega t}). \tag{6.130}$$

There is a variation of electric flux density with time, so there will be a flow of current and the current density will be

$$J(t) = \frac{d}{dt}(D),$$

and hence

$$J(t) = \text{Re}(\epsilon_0 \epsilon_r^* j\omega \mathcal{E}_0 e^{j\omega t}). \tag{6.131}$$

Taking

$$\epsilon_r^* = \epsilon_r' - j\epsilon_r'',$$

$$J(t) = \omega\epsilon_0 \mathcal{E}_0(\epsilon_r'' \cos \omega t - \epsilon_r' \sin \omega t). \tag{6.132}$$

Equation (6.132) shows that there are two components of the current density, one in phase with the applied field and the other out of phase with the field by 90°. The energy density, i.e., the energy per unit volume, is

$$W(t) = \frac{1}{2\pi} \int_0^{2\pi} J(t)\mathcal{E}(t)\, d(\omega t). \tag{6.133}$$

Substituting Eq. (6.132) in Eq. (6.133),

$$W(t) = \frac{1}{2\pi} \int_0^{2\pi} \omega\epsilon_0 \mathcal{E}_0^2(\epsilon_r'' \cos^2 \omega t - \epsilon_r' \sin \omega t \cos \omega t)\, d(\omega t)$$

or

$$W(t) = \frac{\omega}{2} \epsilon_0 \epsilon_r'' \mathcal{E}_0^2. \tag{6.134}$$

Equation (6.134) shows that the absorption of energy per unit volume depends upon ϵ_r'' and is independent of ϵ_r'. It can be seen that ϵ_r'' is the imaginary part of the relative dielectric constant and hence one could conclude that the absorption of energy depends upon the out-of-phase component of ionic, electronic, and orientational polarizations. The absorption of energy is possible in the radiofrequency range if orientational polarization or permanent dipole moments are present. The occurrence of absorption in the optical region is the source of the color of the substance. The energy density $W(t)$ is responsible for the losses in a dielectric, and these losses are referred to as *dielectric losses*.

To derive a circuit analogue of the phenomenon of dielectric losses we show in Fig. 6.22 a parallel-plate capacitor with unit area of cross section

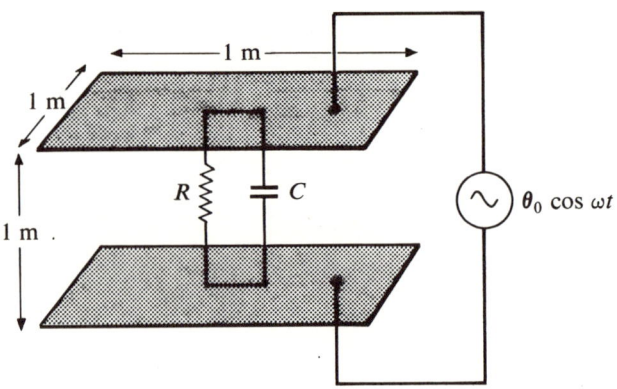

Fig. 6.22
Circuit to illustrate dielectric losses

of the plates and separated by a meter. If the applied voltage is $\theta_0 \cos \omega t$, according to Eq. (6.132),

$$J(t) = \omega \epsilon_0 \theta_0 \epsilon_r'' \cos \omega t - \omega \epsilon_0 \theta_0 \epsilon_r' \sin \omega t,$$

which shows that the dielectric may be considered to be made up of a parallel combination of resistance and capacitance.

If R is the resistance and C is the capacitance of the dielectric,

$$J(t) = \frac{\theta_0 \cos \omega t}{R} - \theta_0 C \omega \sin \omega t.$$

Hence

$$R = \frac{1}{\epsilon_0 \epsilon_r'' \omega},$$

$$C = \epsilon_0 \epsilon_r'.$$

As $\epsilon_r'' \to 0$, $R \to \infty$. The loss angle $\tan \delta$ is defined as ϵ_r''/ϵ_r' and is a measure of the dielectric losses:

$$\tan \delta = \epsilon_r''/\epsilon_r',$$

and hence

$$\tan \delta = 1/CR\omega.$$

Index of refraction and separation of ionic and electronic polarization. It is known from field theory that electromagnetic waves in a homogeneous isotropic medium travel with a phase velocity

$$v = \frac{c}{(\mu_r \epsilon_r)^{1/2}}, \tag{6.135}$$

where c is the velocity of light in a vacuum, μ_r the relative permeability, and ϵ_r the relative dielectric constant. The ratio of the phase velocity of light in vacuum to that in any other medium is known as the *index of refraction* of the medium η:

$$\eta = c/v = (\mu_r \epsilon_r)^{1/2}.$$

In dielectrics $\mu_r = 1$, and hence

$$\eta = (\epsilon_r)^{1/2}$$

or

$$\epsilon_r = \eta^2.$$

Let us consider an alkali halide such as NaCl. The polarization is made up of electronic and ionic polarization. A d.c. measurement of ϵ_r will give us $N\alpha_e + N\alpha_i$ by the relationship

$$\frac{\epsilon_r - 1}{\epsilon_r + 2} = \frac{N\alpha_e + N\alpha_i}{3\epsilon_0}. \tag{6.136}$$

An optical measurement of the index of refraction η will show that ϵ_r depends entirely upon $N\alpha_e$, because the ionic contribution to polarization is negligible (see Fig. 6.20). Hence

$$\frac{\eta^2 - 1}{\eta^2 + 2} = \frac{N\alpha_e}{3\epsilon_0}. \tag{6.137}$$

With the help of Eqs. (6.136) and (6.137) the unknown quantity α_i can be determined.

6.5 Ferroelectric energy conversion

Heat conversion into electrical energy can be accomplished with any charged capacitor in which the relative dielectric constant varies with temperature. The charge on a capacitor is equal to the product of capacitance and voltage:

$$Q = CV,$$

where for a parallel-plate capacitor

$$C = \frac{\epsilon_0 \epsilon_r A}{d}.$$

A decrease in relative dielectric constant ϵ_r with rising temperature will decrease the capacitance, which in turn requires a rise of voltage V if the charge is kept constant. The voltage rise is proportional to an energy rise. If ΔV is the voltage rise, the change in energy and the voltage rise are related by

$$\Delta W = \tfrac{1}{2} Q \, \Delta V,$$

where ΔW is the energy rise. Ferroelectric materials have strong temperature effects and are capable of high polarization and could thus store enough electric energy for conversion.

Above the Curie temperature, the relative dielectric constant is related to the temperature by the Curie-Weiss law equation (6.86),

$$\epsilon_r - 1 = \frac{C}{T - \theta} \tag{6.86}$$

or

$$\frac{1}{\epsilon_r - 1} = \frac{1}{C}(T - \theta). \tag{6.138}$$

Equation (6.138) is plotted in Fig. 6.23.

A study of the ferroelectric conversion of heat to electrical energy can be carried out with the help of Fig. 6.24(a) and the charge voltage diagram 6.24(b). In the first step, the capacitor made up of ferroelectric material is charged to a voltage V_1 of the battery with the material at a temperature T_1, preferably a little below the Curie point. In the charging process the energy delivered by the battery to the capacitor is $\tfrac{1}{2} QV_1$ and is the area

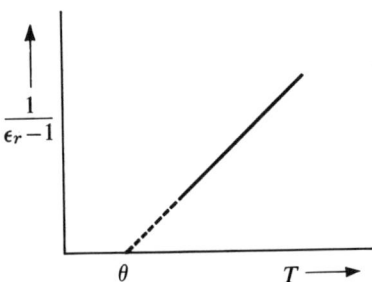

Fig. 6.23
Variation of $1/(\epsilon_r - 1)$ with temperature, Curie-Weiss law

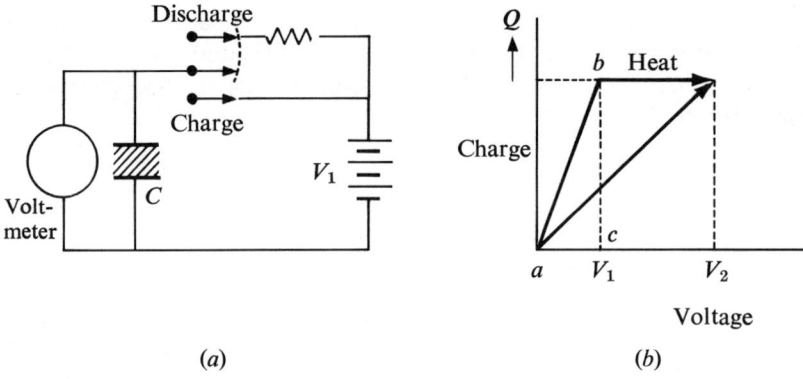

Fig. 6.24
(a) Ferroelectric energy converter, circuit. (b) Ferroelectric energy converter, charge-voltage diagram

of the triangle *abc* in Fig. 6.24'b). The capacitor is then isolated from the battery, and as a second step the capacitor is heated from the heat source. The dielectric constant decreases and charge on the capacitor remains the same, so the voltage across the capacitor increases. If C_1 is the capacitance at T_1 and C_2 is the capacitance at $T_2 > T_1$,

$$Q = V_1 C_1 = V_2 C_2,$$
$$V_2 > V_1 \quad C_2 < C_1,$$

or

$$V_2 = V_1 \frac{C_1}{C_2} = V_1 \frac{\epsilon_{r_1}}{\epsilon_{r_2}},$$

where ϵ_{r_1} is the relative dielectric constant at T_1 and ϵ_{r_2} is the relative dielectric constant at T_2. Using doped barium titanate Hoh reported a ratio $\epsilon_{r_1}/\epsilon_{r_2} = 5$ for a change in temperature of 15 to 30°C. The gain in energy in the capacitor $\Delta w = \frac{1}{2} Q(V_2 - V_1) = \frac{1}{2} Q \Delta V$. The capacitor is now discharged through a load resistor R and a rechargeable battery of voltage V_1. The capacitor has to be cooled to T_1 to complete the cycle.

Efficiency of ferroelectric energy conversion. Efficiency of ferroelectric energy conversion can be obtained by relating the electrical output ΔW to the total heat input $\Delta H + \Delta W$, where ΔH is the thermal energy required to heat the uncharged capacitor. Thus

$$\eta = \frac{\Delta W}{\Delta H + \Delta W}.$$

If A is the area of cross section of the plates of capacitor, ε is the field-intensity, d is the separation of plates $\Delta V = d\,\Delta\varepsilon$, and $\Delta W = \frac{1}{2}\,ADd\,\Delta\varepsilon$. Also $\Delta H = CAd\,\Delta T$, where C is the specific heat of the ferroelectric material. Hence

$$\eta = \frac{\frac{1}{2}ADd\,\Delta\varepsilon}{CAd\,\Delta T + \frac{1}{2}ADd\,\Delta\varepsilon} = \frac{D\,\Delta\varepsilon}{2C\,\Delta T + D\,\Delta\varepsilon}$$

or

$$\eta = \left(\frac{2C\,\Delta T}{D\,\Delta\varepsilon} + 1\right)^{-1}. \tag{6.139}$$

Equation (6.139) shows that for high efficiency the specific heat of the ferroelectric material should be low and D and $\Delta\varepsilon$ should be high. For

$$D = 1\ \text{C/m}^2,$$
$$C = 2.9 \times 10^6\ \text{m}^{-3}/\text{deg},$$
$$\Delta T = 30°\text{C},$$
$$\Delta\varepsilon = 10^7\ \text{V/m},$$

$$\eta = \left(\frac{5.8 \times 10^6 \times 30}{10^7} + 1\right)^{-1} = \frac{1}{18.4} = 5.6\%,$$

a theoretical value.

Problem 1

a. A field strength ε is applied to a dielectric. Show that the stored energy per unit volume in the medium due to polarization P is $\frac{1}{2}P\varepsilon$.

b. A potential of 5,000 V is applied to a capacitor of capacitance 1 μF. Compute the energy stored in the condenser. *Ans.* **b.** 12.5 J/m³

Problem 2

There are 1.6×10^{20} NaCl molecules per cubic meter in a vapor. Determine the orientational polarization of room temperature if the vapor is subjected to an electric field of 5,000 V/cm. Assume that the NaCl molecule consists of Na⁺ and Cl⁻ ions separated by 2.5 Å. *Ans.* 10^{-11} C/m²

Problem 3

An atom of polarizability α is placed in a homogeneous electric field ε. Show that the energy stored in the polarized atom is equal to $\frac{1}{2}\alpha\varepsilon^2$.

Problem 4

Consider a medium containing N free electrons per unit volume subjected to an alternating field of frequency ω. Assuming that the internal field seen by the electrons is the applied field, show that the relative dielectric constant of the medium is

$$\epsilon_r = 1 - \frac{Ne^2}{\epsilon_0 m \omega^2}.$$

Problem 5

Helium gas containing N atoms per unit volume is subjected to an alternating electric field of angular frequency ω. Derive an expression for the complex dielectric constant of the gas.

Problem 6

The equivalent circuit of a lossy capacitor may be represented as a parallel combination of a pure capacitance with resistance. If the complex relative dielectric constant $\epsilon_r^* = \epsilon_r' - j\epsilon_r''$, show that the value of the resistance element is given by

$$R = \frac{d}{\epsilon_0 \epsilon_r'' \omega A},$$

where ω is the frequency of the applied field, A the area of the parallel plates, and d the separation between plates.

Problem 7

Show that a parallel-plate capacitor made up of two parallel layers of material, one layer with relative dielectric constant ϵ_r, zero conductivity, and thickness t, and the other layer with $\epsilon_r = 0$, finite conductivity σ, and thickness bt, behaves as if the space between the capacitor plates was filled with a homogeneous dielectric constant ϵ_r^*, where

$$\epsilon_r^* = \frac{\epsilon_r(1+b)}{1 + (j\omega\epsilon_0\epsilon_r b/\sigma)},$$

where ω is the angular frequency.

Problem 8

When Lucite is subjected to an alternating field of frequency 4,000 M/sec, the real part of the complex relative dielectric constant is measured to be 2.57. The tangent of loss angle measures 0.0032.
a. Determine the imaginary part of the relative dielectric constant.
b. Determine the energy loss in the dielectric per unit volume if a field of $\mathcal{E} = 100 \cos \omega t$ V/m is applied.
Ans. a. 0.00824   b. 9.1 W/m³

Problem 9

The following data were obtained for KI: $\epsilon_r = 4.94$, $\eta^2 = 2.69$. Compute the ratio of electronic and ionic polarizabilities of the alkali halide.   Ans. 0.75

Problem 10

It is observed that in a solid solution of $SrTiO_3$ in $BaTiO_3$ the Curie temperature of the solid solution decreases linearly with increasing amounts of $SrTiO_3$. The relative dielectric constant of a solid solution of 9.15 percent of $SrTiO_3$ in $BaTiO_3$ was measured above the Curie temperature. The following data were obtained:
$\epsilon_r = 4{,}200$ at $100°C$, $\epsilon_r = 2{,}400$ at $120°C$.
a. Determine the Curie temperature of the solid solution.
b. Estimate the Curie temperature of pure $BaTiO_3$ if 1 percent of $SrTiO_3$ decreases the Curie temperature by $4.5°$.

    Ans. **a.** $73.3°C$ **b.** $114.8°C$

7

Magnetic Materials and Properties

7.1 Magnetic dipole moment and magnetization

In the construction of models which will aid in the understanding of magnetic phenomena, the concepts of magnetic dipole moment and magnetization are very fundamental. The concepts of magnetic dipole moment and magnetization will be derived from the basic concepts of magnetic field theory.

Magnetic forces on moving charges. Moving charges can experience forces other than the electrostatic forces associated with stationary charges. This force associated with the velocity of the charge is called a *magnetic force*. A moving charge experiencing such a force is said to be moving in a magnetic field of force. Experimentally a magnetic field vector **B** is defined in terms of the force **F** experienced by a charge q moving with a velocity **v** by the Lorentz relationship,

$$\mathbf{F} = q(\mathbf{v} \times \mathbf{B}). \tag{7.1}$$

B is called the *magnetic flux density* and is expressed in Wb/m² in the MKS system. **B** has a direction perpendicular to **F**. The magnetic forces acting on a charged particle are perpendicular to the velocity of the particle and hence tend to force the particle to move in a circular path.

A current-carrying conductor experiences a force in a magnetic field. This can be calculated with the help of Lorentz's force equation. It has already been pointed out in Sec. 4.4 that electrons in metals or conductors attain a drift velocity which is observed as a flow of current. The electrical current density,

$$\mathbf{J} = ne\mathbf{v}, \tag{7.2}$$

where n is the number of free electrons per unit volume, e the charge of an electron, and $\mathbf{v}$ the velocity (drift). The current density is uniform in a differential volume element of the conductor, and the force experienced by this element in the magnetic field is the sum of the forces on all moving charges within the volume. The force $d\mathbf{F}$ on this differential elemental volume is

$$d\mathbf{F} = (ne\, dA\, dL)\mathbf{v} \times \mathbf{B}. \tag{7.3}$$

dA is the cross-sectional area of the differential volume and dL is the length parallel to $\mathbf{v}$.

Substituting the value of $\mathbf{v}$ from Eq. (7.2) in terms of $\mathbf{J}$ in Eq. (7.3),

$$d\mathbf{F} = (dA\, dL)\mathbf{J} \times \mathbf{B}, \tag{7.4}$$

where $(dA\, dL)$ is the volume of the current element. Therefore, the force per unit volume in the conductor is

$$\frac{d\mathbf{F}}{dV} = \mathbf{J} \times \mathbf{B} \qquad \text{per unit volume.}$$

The total force on a current-carrying conductor is the vector sum

$$\mathbf{F} = \int \mathbf{J} \times \mathbf{B}\, dV, \tag{7.5}$$

where the integration is to be carried out throughout the volume of the conductor.

Equation (7.5) is the general relation for the force on a conductor in a magnetic field. If the conductor has a uniform cross-sectional area, the force on an element of length dL is, from Eq. (7.4),

$$d\mathbf{F} = dL\, \mathbf{I} \times \mathbf{B}, \tag{7.6}$$

where $\mathbf{I} = \mathbf{J}\, dA$. If $\mathbf{j}$ is a unit vector in the direction of $\mathbf{I}$,

$$d\mathbf{F} = \mathbf{j}\, dL\, I \times \mathbf{B}.$$

If $\mathbf{j}\, dL$ are combined into a vector $d\mathbf{L}$,

$$d\mathbf{F} = I(d\mathbf{L} \times \mathbf{B}). \tag{7.7}$$

The total force on a conductor of uniform cross section and current density is

$$\mathbf{F} = \oint I(d\mathbf{L} \times \mathbf{B}), \tag{7.8}$$

MAGNETIC DIPOLE MOMENT AND MAGNETIZATION

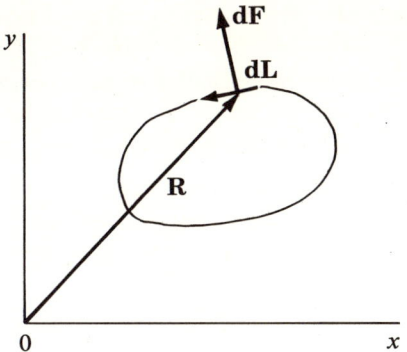

Fig. 7.1
Elementary current loop

in which the line integral is to be taken around the entire closed path of the current. It can be easily shown that the force on a closed circuit placed in a uniform magnetic field is zero.

Torque on a plane closed loop and magnetic dipole moment. Let us consider a loop in the xy plane with the x current I flowing counterclockwise (Fig. 7.1). Let us consider a segment $d\mathbf{L}$ on this loop located at $\mathbf{R}$ from the origin. Therefore,

$$d\mathbf{L} = \mathbf{i}\, dx + \mathbf{j}\, dy,$$
$$\mathbf{R} = \mathbf{i}x + \mathbf{j}y, \qquad (7.9)$$
$$\mathbf{B} = \mathbf{i}B_x + \mathbf{j}B_y + \mathbf{k}B_z,$$

and the force on the element $d\mathbf{L}$ is

$$d\mathbf{F} = I\, d\mathbf{L} \times \mathbf{B}.$$

The torque about the origin of this element is

$$d\mathbf{T} = \mathbf{R} \times d\mathbf{F}$$

or

$$d\mathbf{T} = \mathbf{R} \times (I\, d\mathbf{L} \times \mathbf{B}). \qquad (7.10)$$

Substituting equations (7.9) in Eq. (7.10) and simplifying,

$$d\mathbf{T} = [\mathbf{i}(B_y y\, dx - B_x y\, dy) + \mathbf{j}(B_x x\, dy - B_y x\, dx) - \mathbf{k}(B_z x\, dx + B_x y\, dy)]I. \qquad (7.11)$$

The total torque is obtained by the vector sum of Eq. (7.11),

$$\mathbf{T} = \oint d\mathbf{T} = (-\mathbf{i}B_y + \mathbf{j}B_x)AI, \qquad (7.12)$$

where A is the area of the loop.

Equation (7.12) is independent of the choice of the coordinate system because the moment of the couple is independent of the choice of axis. Equation (7.12) can be made independent of the coordinate system by expressing the area of the current loop as a vector quantity whose magnitude is equal to the area of the loop and direction given by a normal to the loop according to the right-hand rule:

$$\mathbf{A} = A\mathbf{k}. \tag{7.13}$$

Taking the cross product of **A** and **B**:

$$\mathbf{A} \times \mathbf{B} = A\mathbf{k} \times (B_x\mathbf{i} + B_y\mathbf{j} + B_z\mathbf{k}).$$

Simplifying,

$$\mathbf{A} \times \mathbf{B} = (-B_y\mathbf{i} + B_x\mathbf{j})A. \tag{7.14}$$

Substituting Eq. (7.14) in Eq. (7.12),

$$\mathbf{T} = I\mathbf{A} \times \mathbf{B}. \tag{7.15}$$

If **n** is defined as a unit vector normal to the loop given by the right-hand rule, $I\mathbf{A}$ may be written

$$I\mathbf{A} = IA\mathbf{n} = \boldsymbol{\mu}_m, \tag{7.16}$$

and hence

$$\mathbf{T} = IA\,\mathbf{n} \times \mathbf{B} = \boldsymbol{\mu}_m \times \mathbf{B}. \tag{7.16a}$$

Equations (7.15), (7.16), and (7.16a) show that for a given current in a planar loop and a given uniform field, the torque developed on a plane closed loop in a magnetic field depends upon the area of the loop and its orientation. The quantity μ_m is defined as the magnetic dipole moment of the loop of wire.

Equation (7.16a) is similar to Eq. (6.39), where the torque developed in an electric dipole moment placed in an electric field $\mathcal{E}$ was calculated:

$$T = \boldsymbol{\mu}_P \times \mathcal{E} \tag{6.39}$$

Magnetization. *Magnetization* is defined as the magnetic dipole moment of a material per unit volume. If a volume ΔV contains various dipole moments with the ith dipole moment μ_{m_i} the magnetization **M** is given by

$$\mathbf{M} = \lim_{\Delta V \to 0} \frac{\sum_i \boldsymbol{\mu}_{m_i}}{\Delta V}, \tag{7.17}$$

where ΔV is a volume element. Of course $\Delta V \to 0$ implies dimensions exceedingly small compared to the sample but still large compared to the separation of individual dipole moments.

MAGNETIC DIPOLE MOMENT AND MAGNETIZATION

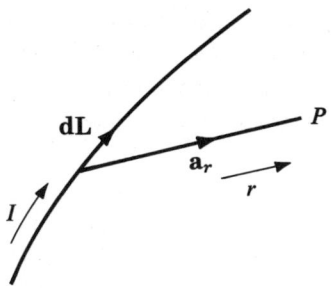

Fig. 7.2
Calculation of magnetic field by
Biot-Savart law

Biot-Savart law. Referring to Fig. 7.2 let us assume that a current I is flowing through a differential vector length $d\mathbf{L}$ of a conductor. The Biot-Savart law states that the magnitude of the magnetic field intensity at any point P produced by the current I flowing through the differential element is given in the MKS system by

$$d\mathbf{H} = \frac{I\,d\mathbf{L} \times \mathbf{a}_r}{4\pi r^2}, \tag{7.18}$$

where $\mathbf{a}_r$ is a unit vector in the direction of P and $\mathbf{H}$ is the magnetic field intensity measured in A/m. The Biot-Savart law is not restricted in its application to any particular type of material and is true for all materials.

In free space the relation between $\mathbf{B}$ and $\mathbf{H}$ is

$$\mathbf{B} = \mu_0 \mathbf{H}, \tag{7.19}$$

where μ_0 is called the *permeability of free space*:

$$\mu_0 = 4\pi \times 10^{-7} \text{ H/m}.$$

The relationship in a magnetic material is often expressed as

$$\mathbf{B} = \mu_0 \mu_r \mathbf{H} \tag{7.20}$$

in which μ_r is called the *relative permeability*. In many practical cases, the relationship between $\mathbf{B}$ and $\mathbf{H}$ is highly nonlinear multivalued, and anisotropic. In such cases Eq. (7.20) is useful only to indicate that $\mathbf{B}$ and $\mathbf{H}$ are related. The Biot-Savart law is often stated in terms of flux density,

$$d\mathbf{B} = \mu_0 \mu_r \frac{I\,d\mathbf{L} \times \mathbf{a}_r}{4\pi r^2}, \tag{7.21}$$

with these important qualifications in mind.

Ampère's law. On the basis of experimentation, Ampère proposed the following relationship between steady currents and the magnetic field in thd space surrounding them. He stated that the magnetomotive force is equal to the total current surrounded by the path of integration:

$$\oint \mathbf{H} \cdot d\mathbf{L} = I, \qquad (7.22)$$

where I represents the current enclosed by the curve chosen.

Either the Biot-Savart law or Ampère's law may be appropriate in the determination of the magnitude of the magnetic field intensity due to current flowing through a conductor. The equivalence of the two laws can be illustrated for the simple case of the determination of the magnetic field due to an infinitely long straight conductor carrying a current I.

Applying Eq. (7.18), the magnetic field intensity H at a point located at a distance r from the conductor may be computed. The result is

$$H = \frac{I}{2\pi r}.$$

The same result may be obtained by applying Ampère's law [Eq. (7.22)].

Relationship between M, H, and μ_r. The laws briefly stated above are helpful in establishing a relationship between **M**, **H**, and μ_r. Such a relationship is very useful in discussing the macroscopic properties of magnetic materials in terms of microscopic properties.

Consider a solenoid whose length is large compared with its radius. We assume it to be connected to a battery or other source of steady current through an ammeter. If the solenoid is in vacuum, the magnitude of the magnetic flux density is given by

$$\mathbf{B} = \mu_0 \mathbf{H}. \qquad (7.23)$$

From the Biot-Savart law or Ampère's law it can be shown for a solenoid whose length is large compared with its radius that the magnetic induction at an interior point far from the ends is

$$B = \mu_0 n i \quad \text{in vacuum}, \qquad (7.23a)$$

where n is the number of turns per unit length and i is the current flowing in the circuit. If the solenoid is filled with a homogeneous, isotropic medium of permeability μ_r,

$$B = \mu_0 \mu_r n i = \mu_0 \mu_r H. \qquad (7.24)$$

Subtracting equations (7.23) from Eq. (7.24),

$$(\mu_0 \mu_r - \mu_0) n i = (\mu_0 \mu_r - \mu_0) H = \mu_0 n i_M,$$

where $i_M = (\mu_r - 1)i$ may be considered as an equivalent magnetization current. The current through the solenoid remains the same while the magnetic induction changes, so one is led to assume that when the medium of relative permeability μ_r is placed inside the solenoid, the result is the same as a solenoidal current ni_M per unit length of its surface adjacent to the solenoid and, consequently, a magnetic moment per unit length of magnitude is

$$ni_M A = (\mu_r - 1)HA.$$

The magnetic moment per unit volume is

$$\mathbf{M} = (\mu_r - 1)\mathbf{H} \tag{7.25}$$

or

$$\mathbf{M} = \chi\mathbf{H}, \tag{7.26}$$

where

$$\chi = \mu_r - 1. \tag{7.27}$$

is known as the magnetic susceptibility.

Equations (7.26) and (7.27) form a link between macroscopic and microscopic properties. Multiplying Eq. (7:25) by μ_0,

$$\mu_0 \mathbf{M} = \mu_0(\mu_r - 1)\mathbf{H}$$

or

$$\mathbf{B} = \mu_0(\mathbf{M} + \mathbf{H}). \tag{7.28}$$

Elementary magnet. The magnetization in magnetic materials can be explained by considering the behavior of electrons in atoms. Before discussing electrons in atoms it may be helpful to conceive of an electron as a particle of charge $-e$ and mass m orbiting around a fixed point 0 in a circular path of radius r. Such an electron may be said to constitute an elementary magnet (Fig. 7.3).

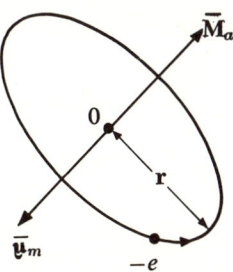

Fig. 7.3
Elementary magnet

Let ω_0 be the angular velocity of the electron in radians per second. This corresponds to a frequency $\omega_0/2\pi$. In 1 sec the electron crosses any point on the orbit $\omega_0/2\pi$ times. This corresponds to a flow of current $-e\omega_0/2\pi$ in the loop, which is the orbit of the electron.

The area of cross section of the loop $= \pi r^2$. The magnitude of the magnetic dipole moment associated with the electron motion is

$$\mu_m = -\frac{e\omega_0}{2\pi}\pi r^2 = -\frac{e\omega_0 r^2}{2}. \tag{7.29}$$

The dipole moment is directed downward, because an orbiting of negative charge in an anticlockwise direction corresponds to a conventional current in the clockwise direction. The angular momentum of the orbiting electron is

$$M_a = m\omega_0 r^2. \tag{7.30}$$

The angular momentum is directed upward for a mass m orbiting anticlockwise. Hence from Eqs. (7.29) and (7.30),

$$\frac{\mu_m}{M_a} = -\frac{e}{2m}. \tag{7.31}$$

Equation (7.31) shows that the ratio of the magnetic dipole moment and the angular momentum of an orbiting electron is independent of the angular velocity or the radius of the orbit.

Effect of an externally applied magnetic field on an elementary magnet. Let us consider that the elementary magnet is subjected to a magnetic field **B** at some angle to the plane of the orbit (Fig. 7.4). It will be

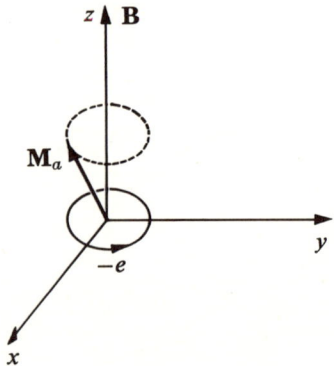

Fig. 7.4
Effect of an externally applied magnetic field on an elementary magnet

shown that the orbit as a whole precesses with the direction of **B** as an axis. This was first shown by Larmor, and the angular velocity of precession is called the *Larmor angular velocity* ω_L.

It was shown earlier that a magnetic dipole moment $\pmb{\mu}_m$ placed in a magnetic field **B** experiences a torque:

$$\mathbf{T} = \pmb{\mu}_m \times \mathbf{B}. \tag{7.16a}$$

For the elementary magnets,

$$\frac{\pmb{\mu}_m}{\mathbf{M}_a} = -\frac{e}{2m}. \tag{7.31}$$

The torque causes a rate of change of angular momentum

$$\mathbf{T} = \frac{d\mathbf{M}_a}{dt} = \pmb{\mu}_m \times \mathbf{B}. \tag{7.32}$$

Multiplying both sides by $(-e/2m)$,

$$-\frac{e}{2m}\frac{d\mathbf{M}_a}{dt} = -\frac{e}{2m}(\pmb{\mu}_m \times \mathbf{B})$$

or

$$\frac{d\pmb{\mu}_m}{dt} = -\frac{e}{2m}(\pmb{\mu}_m \times \mathbf{B}). \tag{7.33}$$

Three components of $(d\mu_m/dt)$ along x, y, and z are

$$\frac{d\mu_{mx}}{dt} = -\frac{e}{2m}(\mu_{my}B_z - \mu_{mz}B_y), \tag{7.34}$$

$$\frac{d\mu_{my}}{dt} = -\frac{e}{2m}(\mu_{mz}B_x - \mu_{mx}B_z), \tag{7.35}$$

$$\frac{d\mu_{mz}}{dt} = -\frac{e}{2m}(\mu_{mx}B_y - \mu_{my}B_x). \tag{7.36}$$

If the magnetic field is acting in the z direction,

$$\mathbf{B} = B_z\mathbf{K}.$$

Therefore,

$$B_x = B_y = 0,$$

and hence

$$\frac{d\mu_{mx}}{dt} = -\frac{e}{2m}\mu_{my}B_z, \tag{7.37}$$

$$\frac{d\mu_{my}}{dt} = \frac{e}{2m}\mu_{mx}B_z. \tag{7.38}$$

$$\frac{d\mu_{mz}}{dt} = 0$$

To solve simultaneous equations (7.37) and (7.38), differentiate both sides of each equation with respect to t,

$$\frac{d^2\mu_{mx}}{dt^2} = -\frac{e}{2m} B_z \frac{d\mu_{my}}{dt}, \qquad (7.39)$$

$$\frac{d^2\mu_{my}}{dt^2} = \frac{e}{2m} B_z \frac{d\mu_{mx}}{dt}. \qquad (7.40)$$

Substituting Eqs. (7.37) and (7.38) in Eqs. (7.39) and (7.40),

$$\frac{d^2\mu_{mx}}{dt^2} = -\left(\frac{e}{2m}\right)^2 B_z^2 \mu_{mx}, \qquad (7.41)$$

$$\frac{d^2\mu_{my}}{dt^2} = -\left(\frac{e}{2m}\right)^2 B_z^2 \mu_{my}. \qquad (7.42)$$

Equations (7.41) and (7.42) may be rewritten by calling

$$\frac{e}{2m} B_z = \omega_L.$$

Therefore,

$$\frac{d^2\mu_{mx}}{dt^2} = -\omega_L^2 \mu_{mx}, \qquad (7.43)$$

$$\frac{d^2\mu_{my}}{dt^2} = -\omega_L^2 \mu_{my}. \qquad (7.44)$$

Equations (7.43) and (7.44) show that changes in μ_m occur only in the xy plane at right angles to the direction of the applied magnetic field. The two components of μ_m can be compared to a linear oscillator whose frequency is ω_L, which is called the *Larmor frequency*. Equations (7.37) and (7.38) indicate that μ_{mx} and μ_{my} are 90° out of phase with each other in space and time. It may be concluded that by application of a magnetic field the origin of the elementary magnet remains the same; however, the dipole moment precesses about the z axis, the direction of the applied magnetic field, with Larmor frequency ω_L.

This precessional frequency results in an equivalent net magnetic moment parallel to the magnetic field. This change in the dipole moment is called the *induce dipole moment*:

$$\mu_{mind} = -\tfrac{1}{2}er^2\omega_L = -\frac{e^2}{4m}r^2 \mathbf{B}. \qquad (7.45)$$

The induced dipole moment has a direction opposite to the applied magnetic field. The induced dipole moment is proportional to the applied magnetic field **B**. This phenomenon results in a negative susceptibility in diamagnetic materials, as discussed in more detail in sec. 7.2.

MAGNETIC DIPOLE MOMENT AND MAGNETIZATION

Permanent magnetic dipole moments in materials. The presence of elementary magnets, called *permanent dipole moments*, determines some of the properties of magnetic materials. We shall discuss different types of magnetic materials later, but now we would like to study the presence of permanent magnetic dipoles in materials. It was shown earlier that if a charged particle has an angular momentum it represents an elementary magnet. Looking at an atom, we find the following contributions of permanent dipole moments:

1. Orbital angular momentum of electrons in the atom.
2. Spin angular momentum of electrons.
3. Nuclear spin angular momentum.

The total dipole moment of an atom is the sum of all three contributions. We shall discuss each one of them briefly.

1. Orbital angular momentum of electrons in atoms. It has already been shown in the discussion on the elementary magnet that an electron having an angular momentum will have a permanent dipole moment associated with it. While discussing the hydrogen atom in Sec. 1.6 we introduced the orbital angular momentum quantum number l for electrons in atoms. The values of l depend upon n, the principal quantum number. For a given value of n the possible values of l are

$$l = 0, 1, 2, 3, \ldots, n - 1.$$

The angular momentum associated with a given value of l was shown to be

$$\hbar[l(l + 1)]^{1/2}.$$

If $n = 1$, $l = 0$, and hence $M_a = 0$.

It has also been shown that there is a magnetic moment associated with the rotation of a charge. The magnetic field set up is perpendicular to the plane of the orbit. The angular momentum vector M_a in general will have a component parallel to the external field. Because of the reaction of the two magnetic fields, it was shown earlier in this section that the angular momentum vector would precess about the external magnetic field with the Larmor frequency ω_L. The possible components of the angular momentum along the direction of the applied magnetic field are determined by the magnetic quantum number m_l, which has the following possible values for a given value of l:

$$m_l = 0, \pm 1, \pm 2, \pm 3, \ldots, \pm l.$$

For $l = 1$,

$$m_l = 0, +1, \text{ aad } -1.$$

MAGNETIC MATERIALS AND PROPERTIES

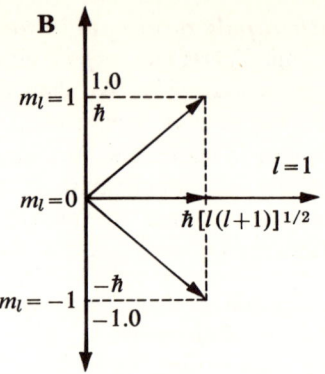

Fig. 7.5
Three possible orientations of angular momentum in an external magnetic field for $l = 1$

Hence the components of the angular momentum along the direction of the field for $l = 1$ are $\hbar$, 0, and $-\hbar$ (Fig. 7.5). Now

$$\frac{\mathfrak{p}_m}{\mathbf{M}_a} = -\frac{e}{2m},$$

and hence the possible magnetic dipoles are

$$-\frac{e\hbar}{2m}, \ 0, \ \frac{e\hbar}{2m}$$

or

$$-\frac{e\hbar}{4\pi m}, \ 0, \ \frac{e\hbar}{4\pi m}.$$

The quantity $\mu_B = e\hbar/4\pi m = 9.27 \times 10^{-24}$ A-m² and is called a *Bohr magneton*. If there are three electrons in the state $l = 1$, the vector sum of the three magnetic dipole moments corresponding to three values of m_l, 0, +1, and −1, add up to zero, and hence the resultant magnetic dipole moment is zero.

One therefore expects a resultant orbital magnetic momentum in atoms which contain incompletely filled shells. Even then the resultant may be zero. Elements of the iron group are interesting from the point of view of electrical engineers. The last two shells in the iron group are: Fe, $3d^2$ $4s^2$; Co, $3d^7$ $4s^2$; and Ni, $3d^8$ $4s^2$. The d shell is incomplete, so that in elements of the iron group the incomplete shell lies near the outer portion of the atom. Because of the incompleteness of the shell, the free atoms have a permanent orbital dipole moment. However, in a microscopic specimen of material these orbital moments are evidently quenched and hence make negligible contribution to the magnetic properties of the group. Experimental evidence in ferroresonance studies indicate the dominance of the spin contribution.

2. *Spin angular momentum of electrons.* The electron spin was discussed in Sec. 1.7. It was postulated that an electron has a quantum-mechanical property called *spin*. The possible angular momentum components of the spin along an externally applied field are only two, $\hbar/2$ and $-\hbar/2$. This resulted in two spin quantum numbers, $m_s = \pm\frac{1}{2}$. The magnetic dipole moment in general is given by

$$\mu_m = g\left(\frac{e}{2m}\right)\frac{\hbar}{2}, \tag{7.46}$$

where g is called the *spectroscopic splitting factor*. As we have seen, $g = 1$ for orbital magnetic moments alone. For electron spin $g = 2.0023$ (without any contribution from orbital angular momentum) and hence

$$\mu_{m_s} \simeq \mu_B.$$

The electron spin gives rise to a magnetic dipole moment approximately equal to a Bohr magneton in the direction of an externally applied field.

In case both orbital angular momentum and spin angular momentum are present, the orbital and spin angular momentums may be combined together vectorially to give a total angular momentum. If we represent the total angular momentum by a quantum number j, for a given value of l and $m_s = \pm\frac{1}{2}$, j can have values $l \pm \frac{1}{2}$. In a multielectron atom the same procedure may be used to describe the atom itself. The orbital moment of the atom is obtained by vector addition of the individual orbital moments of the electrons to give $\hbar[L(L+1)]^{1/2}$. A vector addition of individual spin moments of the electrons gives $\hbar[S(S+1)]^{1/2}$, in which S is an integer or half-integer obtained from a combination of m_s vectors. The total momentum of the atom is given by $\hbar[J(J+1)]^{1/2}$, in which J is an integer or half-integer obtained from a combination of L and S vectors. This is known as *LS* coupling, or Russell-Saunders coupling. When adding magnetic moments due to spin and orbital motions in order to obtain the resultant magnetic moment, the factor g may take into account the fact that orbital momentum and spin momentum are in different ratios to the respective magnetic moments. The general expression for g in this case is given by Lande's equation as

$$g = 1 + \frac{J(J+1) + S(S+1) - L(L+1)}{2J(J+1)}, \tag{7.47}$$

which may be derived from appropriate momentum vector algebra.

Hund's rule for incomplete shells. Hund's rule for incomplete shells determines J, L, and S, the quantities necessary to determine g, the spectroscopic splitting factor, for the electronic ground state of an atom.

1. The electron spins add to give a maximum value of S. This should be consistent with Pauli's principle; e.g., assume that the $d(l = 2)$ shell has three electrons. With $l = 2$ the possible values of m_l are (for the entire shell) 2, 1, 0, -1, -2. If the shell is filled, the spins associate themselves with 10 electrons and are $+\frac{1}{2}$, $+\frac{1}{2}$, $+\frac{1}{2}$, $+\frac{1}{2}$, $+\frac{1}{2}$, $-\frac{1}{2}$, $-\frac{1}{2}$, $-\frac{1}{2}$, $-\frac{1}{2}$, $-\frac{1}{2}$. With 3 electrons in the d shell, adding the first three positive values,

$$S = \tfrac{3}{2}.$$

In another example let us consider 6 electrons in the d shell:

$$S = \tfrac{1}{2} + \tfrac{1}{2} + \tfrac{1}{2} + \tfrac{1}{2} + \tfrac{1}{2} - \tfrac{1}{2} = 2.$$

2. Consistent with rule 1, the orbital momentum quantum numbers add up to give the maximum value of L; e.g., assume that the d shell has 3 electrons. The possible values of m_l are (for the entire shell) 2, 1, 0, -1, -2:

$$L = 2 + 1 + 0 = 3.$$

As another example, let us consider 9 electrons in the d shell. One electron is uncompensated and hence

$$L = 2.$$

3. The J value is equal to $|L - S|$ when the shell is less than half full and $|L + S|$ when more than half full.

As an example we shall calculate the g values for two iron-group ions.

g value for Cr^{3+}. Looking at the periodic table, the electronic configuration in Cr atomic number 24 is

$$1s^2 2s^2 2p^6 3s^2 3p^6 3d^5 4s^1.$$

If 3 electrons are taken away, Cr results in Cr^{3+}, with electronic configuration

$$1s^2 2s^2 2p^6 3s^2 3p^6 3d^3.$$

All the shells except $3d$ are complete. Hence the contribution to the permanent dipole moment is only from the $3d$ shell. It has 3 electrons:

$$S = \tfrac{3}{2},$$
$$L = 2 + 1 + 0 = 3.$$

The shell can take 10 electrons but has only 3 electrons, so it is less than half filled; hence

$$J = L - S = 3 - \tfrac{3}{2} = \tfrac{3}{2}.$$

Therefore,
$$g = 1 + \frac{\frac{3}{2} \times \frac{5}{2} + \frac{3}{2} \times \frac{5}{2} - 4 \times 3}{\frac{15}{4} \times 2} = \frac{2}{5} = 0.4.$$

g value of Cu^{2+}. Looking at the periodic table, the electronic configuration in Cu, atomic number 29, is

$$1s^2 2s^2 2p^6 3s^2 3p^6 3d^{10} 4s^2.$$

If 2 electrons are taken away, Cu results in Cu^{2+}, with electronic configuration

$$1s^2 2s^2 2p^6 3s^2 3p^6 3d^9.$$

1. $S = \frac{1}{2} + \frac{1}{2} + \frac{1}{2} + \frac{1}{2} + \frac{1}{2} - \frac{1}{2} - \frac{1}{2} - \frac{1}{2} - \frac{1}{2} = \frac{1}{2}.$
2. $L = 2.$
3. The shell is short one electron, so it is filled more than half:

$$J = L + S = 2 + \tfrac{1}{2} = 2\tfrac{1}{2}.$$

Hence
$$g = 1 + \frac{\frac{5}{2} \times \frac{7}{2} + \frac{1}{2} \times \frac{3}{2} - 2 \times 3}{2 \times \frac{5}{2} \times \frac{7}{2}} = 1.2.$$

3. *Nuclear magnetic moments.* Another contribution to the permanent dipole moment may be due to the nucleus having an angular momentum. If μ_{m_n} is the magnetic dipole moment due to the spinning of the nucleus

$$\mu_{m_n} = \frac{e\hbar}{2m_p}, \qquad (7.48)$$

where m_P is the mass of the proton. The nucleus is approximately 10^3 times heavier than the electron, so the magnetic moments of nuclei are smaller than the magnetic moments of an electron by a factor 10^{-3}.

7.2 Diamagnetism

The magnetization **M** and the applied magnetic field strength **H** are related to each other by Eq. (7.26):

$$\mathbf{M} = \chi \mathbf{H}. \qquad (7.26)$$

In diamagnetic materials, the magnetization is proportional to the applied field but the magnetic susceptibility χ is negative. All materials exhibit the diamagnetic effect, diamagnetic materials being those in which there are no other magnetic effects. The origin of diamagnetic susceptibility can be understood by considering electrons in an atom. The outer electrons shield the rest of the atomic structure from an applied field. The electrons have angular momentum and the electron motion may be considered as electric current in a manner similar to elementary magnets. Let us consider a single electron in the outermost shell such as an electron in a copper atom executing a circular orbit around the nucleus at a radius r. Let the angular velocity of the electron be ω_0 with no applied field. The motion of the electron corresponds to a magnetic dipole moment in the absence of a magnetic field:

$$\mathbf{\mu}_{m_0} = -\tfrac{1}{2}e^2\omega_0.$$

When the flux through the atom is changed, an induced current is set up through the atom which opposes the change in flux. An induced current corresponds to a change in the frequency of the electronic motion. The change in frequency is ω_L and the induced current persists without decaying as long as the magnetic field is present. The magnetic moment associated with the induced current is called the *diamagnetic moment*.

With no applied field the balance of forces in classical mechanics require that the centrifugal force equals Coloumb's force of attraction:

$$\frac{mv^2}{r} = \frac{e^2}{4\pi\epsilon_0 r^2}$$

or

$$mr\omega_0^2 = \frac{e^2}{4\pi\epsilon_0 r^2}, \tag{7.49}$$

where ω_0 is the angular velocity.

In a magnetic field **B** perpendicular to the plane of the orbit an additional Lorentz force acts on the electron. The stability of the electron orbit requires a balance between the three forces acting on the electron:

1. Coulomb's force of attraction $e^2/4\pi\epsilon_0 r^2$.
2. Lorentz force $evB = er\omega B$.
3. Centrifugal force $mr\omega^2$,

$$mr\omega^2 = \frac{e^2}{4\pi\epsilon_0 r^2} \pm er\omega B,$$

(where the $+$ or $-$ sign depends upon the direction of applied field), or

$$\omega^2 = \frac{e^2}{4\pi\epsilon_0 mr^3} \pm \frac{eB}{m}\omega$$

(where ω is the angular velocity in the presence of the applied external field).

If $B = 0$, $\omega = \omega_0$. Therefore,

$$\omega_0^2 = \frac{e^2}{4\pi\epsilon_0 mr^3},$$

and hence

$$\omega^2 = \omega_0^2 \pm \frac{eB}{m}\omega. \tag{7.50}$$

DIAMAGNETISM

Since
$$\frac{eB}{m} \ll \omega_0,$$

$$\omega \cong \omega_0 \pm \frac{eB}{2m}.$$

Because $eB/2m = \omega_L$ as previously defined,

$$\omega \cong \omega_0 \pm \omega_L.$$

It may be recalled that ω_L is such that the induced dipole moment opposes the applied magnetic field. Therefore,

$$\mathbf{\mu}_{mind} = -\frac{e^2}{4m} r^2 \mathbf{B}. \tag{7.51}$$

Equation (7.51) gives the induced dipole moment of an electron when the atom is subjected to a magnetic field.

The above procedure can be carried over to the case of an atom having Z electrons orbiting around the nucleus. If r_i^2 is the mean square radius of the ith electron, projected on a plane perpendicular to the direction of the applied magnetic field the total induced moment in an atom is

$$\mathbf{\mu}_{mind} = -\frac{e^2}{4m} \mathbf{B} \sum_{i=1}^{Z} r_i^2. \tag{7.52}$$

If there are N atoms per unit volume, the magnetization $\mathbf{M}$ is given by

$$\mathbf{M} = -\frac{e^2 N}{4m} \mathbf{B} \sum_{i=1}^{Z} r_i^2. \tag{7.53}$$

Now $\mathbf{B} = \mu_0 \mu_r \mathbf{H}$, and hence

$$\mathbf{M} = -\frac{e^2 N \mu_0 \mu_r \mathbf{H}}{4m} \sum_{i=1}^{Z} r_i^2. \tag{7.54}$$

Therefore the susceptibility χ, which is given by $\mathbf{M}/\mathbf{H}$, is

$$\chi = \frac{\mathbf{M}}{\mathbf{H}} = -\frac{e^2 N \mu_0 \mu_r}{4m} \sum_{i=1}^{Z} r_i^2 \tag{7.55}$$

Equation (7.55) shows that the susceptibility is negative. Assuming one electron and taking $m = 9.1 \times 10^{-31}$ Kg, $r_1^2 = 10^{-20}$ m, $N \sim 5 \times 10^{28}$ m^{-3}, and $e = 1.6 \times 10^{-19}$ C,

$$\chi = -\frac{2.56 \times 10^{-38} \times 5 \times 10^{28} \times 4\pi \times 10^{-7} \times 10^{-20}}{4 \times 9.1 \times 10^{-31}} \mu_r,$$

If we assume that $\mu_r \cong 1$ for this material then we find $\chi \cong -5 \times 10^{-6}$ which verifies the assumption.) The experimental value for copper is -0.9×10^{-5}

at room temperature. The electronic structure is independent of temperature, so the diamagnetic susceptibility is independent of temperature.

The effective radius, r_i, may be related to the average orbital radius of spherical charge distribution ρ_i as follows. In the case

$$\rho_i^2 = x_i^2 + y_i^2 + z_i^2,$$

where x_i, y_i, and z_i are the rectangular coordinates of the ith electron referred to the nucleus. Since $r_i^2 = x_i^2 + y_i^2$

$$r_i^2 \text{ equals } \tfrac{2}{3}\rho_i^2,$$

where ρ_i^2 is the mean-square distance of the ith electron from the nucleus. Then

$$\chi_D = -\frac{\mu_0 N e^2}{6m} \sum_{i=1}^{Z} \rho_i^2, \tag{7.56}$$

where χ_D is the diamagnetic susceptibility. Equation (7.56) is known as Langevin's equation and is valid when quantum mechanics is used to calculate ρ_i^2 from the electronic wave function. Because χ_D is proportional to ρ_i^2, the outer electrons make the largest contribution to χ_D.

7.3 Paramagnetism

Many salts of iron and the rare-earth families have permeabilities slightly greater than 1, usually lying between 1 and $1 + 10^{-3}$. These materials do not show hysteresis, and their permeabilities are independent of field strength. Such materials are called *paramagnetic*. There are two kinds of paramagnetism — strong and weak. Weak paramagnetism in many metallic elements is caused by conduction electrons, and the susceptibility for such materials is practically independent of temperature. Strong paramagnetism is caused by a permanent magnetic moment of the component atom or molecule. The origin of the two kinds of paramagnetism is explained by the electronic shells in an atom. As already discussed, electrons in the outermost shell are loosely bound to the nucleus. In a metal they are conduction electrons, and the spins of a portion of these electrons can be changed by an applied magnetic field in a way that can be explained by quantum mechanics. This gives rise to weak paramagnetism. If the other shells are completely filled, the magnetic moments due to their orbital and spin motions are self-neutralizing, so that the only contribution is diamagnetic. In the paramagnetic salts of iron and the rare-earth groups of elements some of the shells are incomplete. The electrons in the incomplete shell have a resultant moment that is large compared with the spin of the conduction electrons or diamagnetic moment of the closed shells. Such materials with permanent magnetic dipole moments show strong paramagnetism. The following discussion pertains to strong paramagnetism.

Classical theory of paramagnetism. Consider the unit volume of a medium containing N atoms. Let μ be the maximum value of the dipole contribution of each atom capable of being aligned with an external field. Therefore, there are N magnetic dipoles, each of magnitude μ. In general $\mu = gJ\mu_B$. Suppose the medium is subjected to a magnetic field **H**. If the interaction between the dipoles is weak, the field seen by the dipoles is the applied field. In the classical theory it will be assumed that the dipoles may have all possible directions in the medium with no externally applied field. When the medium is placed in a magnetic field, the precessional motion is damped and the dipoles tend to be aligned in the direction of the magnetic field. The dipoles have all possible orientations, so the magnetization **M** per unit volume may be computed in the same way as the polarization **P** for a dipolar gas (Sec. 6.2). The magnetostatic energy of a dipole of dipole moment **µ** in the presence of a magnetic field **H** is $-\boldsymbol{\mu}\cdot\mathbf{H}$. The magnetization M may be found as

$$M = N\mu\left(\coth\frac{\mu_0\mu H}{kT} - \frac{kT}{\mu_0\mu H}\right). \tag{7.57}$$

Equation (7.57) is analogous to Eq. (6.50).
Introducing

$$x = \frac{\mu_0\mu H}{kT},$$

$$M = N\mu\left(\coth x - \frac{1}{x}\right). \tag{7.57a}$$

The function $(\coth x - 1/x)$ is called the *Langevin function* $L(x)$. Hence

$$M = N\mu L(x). \tag{7.57b}$$

The Langevin function was plotted in Fig. 6.1(a). For larger values of x, which means for larger values of H and smaller values of T, $L(x)$ approaches unity and

$$M = M_s = N\mu,$$

where M_s is the saturation magnetization. It can be easily seen that to achieve the condition of absolute dipole moment alignment in the direction of the applied field, the magnetic field must be impractically high or the temperature almost zero. For all practical cases, as long as

$$\mu_0\mu H \ll kT,$$

the Langevin function $L(x)$ may be approximated by $x/3$ and hence

$$M = \frac{N\mu_0\mu^2 H}{3kT}. \qquad (7.58)$$

If we substitute

$$C = \frac{N\mu_0\mu^2}{3k}$$

in Eq. (7.58),

$$M = \frac{C}{T} H$$

or

$$\chi_P = C/T. \qquad (7.59)$$

This is known as the *Curie-Weiss law*.

Equation (7.59) shows that the paramagnetic susceptibility is inversely proportional to the absolute temperature. An estimation of the value of χ_P at room temperature can be made as follows: μ is of the order of a Bohr magneton μ_B; $\mu = \mu_B$; $N = 5 \times 10^{28}$ m^{-3}:

$$C = \frac{5 \times 10^{28} \times 1.257 \times 10^{-6} \times (9.27 \times 10^{-24})^2}{3 \times 1.38 \times 10^{-23}} \cong 0.1.$$

Hence $\chi_P \simeq 10^{-3}$, and the assumption that the dipoles are non-interacting to give $H = H$ app is a good one. The paramagnetic susceptibility is two orders of magnitude greater than the diamagnetic susceptibility discussed earlier. Examples are:

$$\text{Fe}_2\text{O}_3, \chi_P = 1.4 \times 10^{-3}; \quad \text{CrCl}_3, \chi_P = 1.5 \times 10^{-3}.$$

Two-level quantum-mechanical theory of paramagnetism. In the classical theory of paramagnetism it was assumed hat the dipoles are freely rotating. It was shown earlier that according to quantum mechanics the permanent magnetic moment of an atom or ion is not freely rotating but is restricted to a finite set of orientations relative to the applied magnetic field. To simplify matters let us first consider that when a medium is subjected to a magnetic field, the dipole moments associated with the atoms or ions in the medium can either align themselves parallel or antiparallel to the applied field. The theory based on this assumption is called the *two-level quantum-mechanical theory of paramagnetism*.

Consider again a medium containing N dipole moments corresponding to N atoms per unit volume. Let μ be the dipole moment associated with each atom. When the medium is subjected to a magnetic field, the magnetic

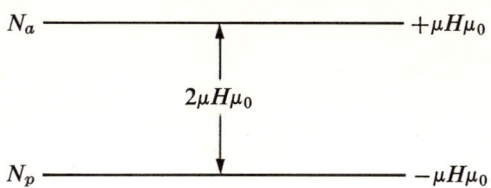

Fig. 7.6
Two-level systems with populations N_a and N_p

dipole moments either align themselves parallel or antiparallel to the applied field. Let us assume that the dipole moments do not interact with each other.

In the presence of an applied magnetic field let N_P dipole moments be aligned parallel to the field and N_a be aligned antiparallel to the field. The total magnetization is

$$M = (N_p - N_a)\mu. \tag{7.60}$$

It must be remembered that

$$N = N_p + N_a. \tag{7.61}$$

The values of N_p and N_a depend upon the temperature of the medium and the magnetic field strength H of the applied field. They can be calculated with the help of Boltzmann's statistics.

In the presence of a magnetic field H, the magnetostatic energy of a dipole of dipole moment μ is $-\boldsymbol{\mu}\cdot\mu_0\mathbf{H}$. This is in accordance with the electrostatic energy analysis in Sec. 6.2. Hence the minimum energy would be contained by those dipoles aligned in the direction of the field (Fig. 7.6). The antiparallel dipoles would have an energy $\mu_0\mu H$. The difference of energy between antiparallel and parallel dipole moments is $2\mu_0\mu H$. It may be recalled that

$$\mu = gJ\mu_B.$$

At absolute zero temperature, the state of minmum energy represents the most stable configuration, and therefore all the dipoles or elementary magnets would be aligned parallel to the magnetic field. At any finite temperature, the equilibrium state must be that of a less perfect alignment. The classical statistical mechanics of Maxwell and Boltzmann, discussed in Sec. 2.2, prescribes the method to calculate the distribution of dipoles between aligned and nonaligned states. According to the statistics,

$$\frac{N_a}{N_p} = \exp\left(-\frac{2\mu_0\mu H}{kT}\right). \tag{7.62}$$

In Eqs. (7.61) and (7.62) the only two unknowns are N_a and N_p. Substituting Eq. (7.62) in Eq. (7.61),

$$N_p + N_p \exp\left(-\frac{2\mu_0\mu H}{kT}\right) = N$$

or

$$N_p = \frac{N}{1 + e^{-2\mu_0\mu H/kT}} = \frac{Ne^{\mu_0\mu H/kT}}{e^{\mu_0\mu H/kT} + e^{-\mu_0\mu H/kT}}.$$

Similarly,

$$N_a = \frac{N}{1 + e^{2\mu_0\mu H/kT}} = \frac{Ne^{-\mu_0\mu H/kT}}{e^{\mu_0\mu H/kT} + e^{-\mu_0\mu H/kT}}.$$

Therefore,

$$N_p - N_a = N \frac{e^{\mu_0\mu H/kT} - e^{-\mu_0\mu H/kT}}{e^{\mu_0\mu H/kT} + e^{-\mu_0\mu H/kT}}$$

or

$$N_p - N_a = N \tanh \frac{\mu_0\mu H}{kT}.$$

Therefore the total magnetization according to Eq. (7.57) is

$$M = N\mu \tanh \frac{\mu_0\mu H}{kT}. \tag{7.63}$$

If $T \to 0$, $\tanh \to 1$, and hence

$$M_{T\to 0} = N\mu.$$

It was mentioned earlier and can be seen now that in this state all dipoles N are aligned. The value of magnetization at $T = 0$ may be called *saturation magnetization*:

$$M_{T\to 0} = M_{\text{sat}}.$$

Therefore Eq. (7.63) may be rewritten

$$\frac{M}{M_{\text{sat}}} = \tanh \frac{\mu_0\mu H}{kT}. \tag{7.64}$$

In Fig. 7.7 M/M_{sat} has been plotted against the variable $\mu_0\mu H/kT$. For very low values of $\mu_0\mu H/kT$,

$$\tanh \frac{\mu_0\mu H}{kT} \simeq \frac{\mu_0\mu H}{kT}.$$

And in this condition,

$$\frac{M}{M_{\text{sat}}} = \frac{\mu_0\mu H}{kT} \tag{7.65}$$

or

$$M = \frac{N\mu_0\mu^2 H}{kT}. \tag{7.66}$$

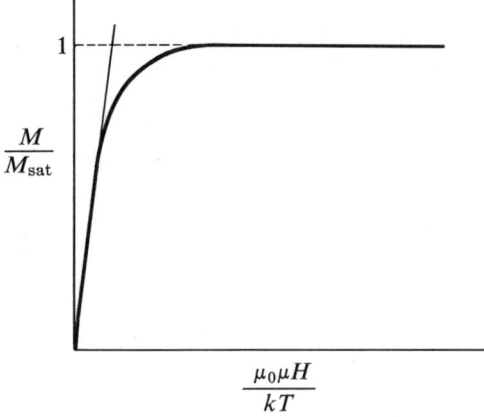

Fig. 7.7
Variation of normalized magnetization M/M_{sat} with $\mu_0\mu H/kT$

In practice the condition $\mu_0\mu H \ll kT$ is satisfied for practical values of H and not too low temperatures.

Equation (7.62) shows that

$$\frac{M}{H} = \chi_P = \frac{N\mu_0\mu^2}{kT}, \tag{7.67}$$

where χ_P, the paramagnetic susceptibility, is inversely proportional to the absolute temperature T. The alignment of dipoles results from the applied field and therefore the magnetization disappears when the applied field is removed. The temperature causing thermal motion of the dipoles is responsible for bringing randomness back to the system.

Equations (7.67) and (7.58) show that the paramagnetic susceptibility χ_P in a two-level quantum mechanical model differs by a factor 3 from the classical model. The susceptibility χ_P may also be expressed as

$$\chi_P = \frac{Ng^2\mu_0\mu_B^2 J^2}{kT}, \tag{7.68}$$

which reduces to

$$\chi_P = \frac{N\mu_0\mu_B^2}{kT} \tag{7.68a}$$

for $g = 2, J = \frac{1}{2}$.

Magnetic quantum number M_J and Brillouin theory. We shall now discuss paramagnetic materials considering that the permanent magnetic moment of a given atom or ion is not freely rotating but restricted to a finite set of orientations relative to the applied field.

The magnetic quantum numbers M_J associated with a given value of J, the total angular momentum quantum number, are

$$M_J = J, J-1, J-2, \ldots, 0, \ldots, -(J-1), -J.$$

Therefore an atom with angular momentum quantum number J has $2J+1$ equally spaced energy levels in a magnetic field. The magnetic dipole moment for an atom for a given value of M_J is $M_J g \mu_B$. Its potential energy in a magnetic field H is $-M_J g \mu_0 \mu_B H$.

According to classical mechanics, the total magnetization in a medium of N atoms per unit volume is

$$M = N \left[\frac{\sum\limits_{-J}^{J} M_J g \mu_B \mu_0 \exp(M_J g \mu_B H \mu_0/kT)}{\sum\limits_{-J}^{J} \exp(M_J g \mu_B H \mu_0/kT)} \right], \qquad (7.69)$$

where in Eq. (7.69) the term in brackets is the statistical average of the magnetic moment component per atom in the direction of the applied field H.

When $M_J g \mu_B H \mu_0 / kT \ll 1$,

$$\exp\left(\frac{M_J g \mu_B H \mu_0}{kT}\right) \cong 1 + \frac{M_J g \mu_B H \mu_0}{kT},$$

$$\frac{\sum\limits_{-J}^{J} \mu_0 M_J g \mu_B \exp(M_J g \mu_B H \mu_0/kT)}{\sum\limits_{-J}^{J} \exp(M_J g \mu_B H \mu_0/kT)} = \frac{\sum\limits_{-J}^{J} M_J \mu_0 g \mu_B [1 + (M_J g \mu_0 \mu_B H/kT)]}{\sum\limits_{-J}^{J} [1 + (M_J g \mu_B H \mu_0/kT)]}$$

$$= \frac{\sum\limits_{-J}^{J} M_J g \mu_B \mu_0 + (M_J^2 g^2 \mu_{0B}^2 H/kT)}{\sum\limits_{-J}^{J} [1 + (M_J g \mu_B H \mu_0/kT)]}. \qquad (7.70)$$

Now if $\sum\limits_{-J}^{J} M_J = 0$,

$$\sum\limits_{-J}^{J} 1 + \frac{M_J g \mu_B H \mu_0}{kT} = 2J + 1 \qquad (7.71\text{a})$$

$$\sum\limits_{-J}^{J} M_J^2 = (\tfrac{1}{3})(2J+1)J(J+1). \qquad (7.71\text{b})$$

Hence

$$M = Ng^2 \frac{J(J+1)}{3} \frac{\mu_0 \mu_B^2 H}{kT}, \qquad (7.72)$$

and therefore

$$\chi_P = Ng^2 \mu_0 J(J+1) \frac{\mu_B^2}{2kT}. \qquad (7.73)$$

Equation (7.73) is the quantum-mechanical analogue of Eq. (7.68).

In general, Eq. (7.69) can be used to show that
$$M = NgJ\mu_B B_J(x), \qquad (7.74)$$
where
$$x = \frac{gJ\mu_B H}{kT}$$
and $B_J(x)$ is the Brillouin function (after Brillouin, who first developed the theory). It is defined by the equation
$$B_J(x) = \frac{2J+1}{2J}\coth\left[\frac{(2J+1)x}{2J}\right] - \frac{1}{2J}\coth\left(\frac{x}{2J}\right). \qquad (7.75)$$

With $J \to \infty$, i.e., an infinite number of possible orientations, the Brillouin function reduces to the Langevin function:
$$B_J(x) = L(x) = \coth x - \frac{1}{x},$$
and hence
$$M = Ng\mu_B J\left(\coth x - \frac{1}{x}\right).$$
At low temperatures
$$B_J(x) \to 1,$$
$$M \to Ng\mu_B J.$$

Corresponding to perfect alignment is a condition of saturation, discussed earlier. We shall close our discussion of paramagnetic materials at this point referring the reader to the Bibliography for further details.

7.4 Ferromagnetism; Domain-wall motion

It was shown, both in the classical and quantum-mechanical approaches, that saturation magnetization could be achieved in paramagnetic material at very low temperatures and high applied magnetic field H. In ferromagnetic materials saturation magnetization occurs at ordinary temperatures and at ordinary and sometimes even at small values of applied field. Ferromagnetic materials are those which possess spontaneous magnetization. Spontaneous magnetization is magnetization in the material in the absence of an externally applied magnetic field. Ferromagnetic materials have a characteristic temperature below which they exhibit the ferromagnetic behavior of spontaneous magnetization. This temperature is called the *ferromagnetic Curie temperature* θ_f. Above this temperature the behavior of ferromagnetic materials is paramagnetic.

From Eq. (7.28),
$$\mathbf{B} = \mu_0(\mathbf{M} + \mathbf{H}). \qquad (7.28)$$
Let $M = M_r$ and $B = B_r$ at $H = 0$. Therefore, $B_r = \mu_0 M_r$. $\qquad (7.76)$

In carbon steel a measured value of $B_r = 1$ Wb/m². Let us assume that we have 10^{29} atomic dipole moments per m³ and let us assume that each atomic dipole moment is 1 Bohr magneton:

$$1 \text{ Bohr magneton} = 10^{-23} \text{ A-m}^2$$

$$\mu_0 = 4\pi \times 10^{-7} \text{ H/m}.$$

With $B_r = 1$ Wb/m²,

$$M_r = \frac{1 \text{ Wb/m}^2}{4\pi \times 10^{-7} \text{ H/m}} \cong 10^6 \text{ A/m}.$$

This suggests that in carbon steel the dipole moments are 10^{29}, in number per unit volume, each one of them with a dipole moment 10^{-23} A/m² are all in the same direction and may be said to be lined up. In an attempt to get an answer to spontaneous magnetization, Weiss in 1907 postulated the existence of an internal field proportional to M, which acts on each atomic dipole due to the effect of neighboring dipoles and which is superimposed on the externally applied field. The effective magnetic field H_{eff} on an atomic dipole is, therefore,

$$\mathbf{H}_{\text{eff}} = \mathbf{H} + \gamma \mathbf{M}, \tag{7.77}$$

where $\mathbf{H}$ is the externally applied field and γ is a constant of proportionality between the internal field and the magnetization and is known as the *Weiss molecular-field coefficient*. The Weiss field $\gamma \mathbf{M}$ has a tendency to align the dipoles and is opposed by the motion of the thermal agitation of the elementary moments. We may now replace the magnetic field strength $\mathbf{H}$ by $\mathbf{H}_{\text{eff}}$ in the classical and quantum-mechanical theories of paramagnetism and look at the possibility of spontaneous magnetization.

Classical theory. According to the classical theory of paramagnetism, the magnetization M was found to be given, by Eq. (7.57a):

$$M = N\mu\left(\coth x - \frac{1}{x}\right), \tag{7.57a}$$

where

$$x = \frac{\mu_0 \mu H}{kT}.$$

For ferromagnetic materials replacing H by H_{eff} would change x:

$$x = \frac{\mu_0 \mu (H + \gamma M)}{kT}. \tag{7.78}$$

Calling $N\mu = M_{\text{sat}}$, the saturation magnetization

$$\frac{M}{M_{\text{sat}}} = \coth x - \frac{1}{x}. \tag{7.79}$$

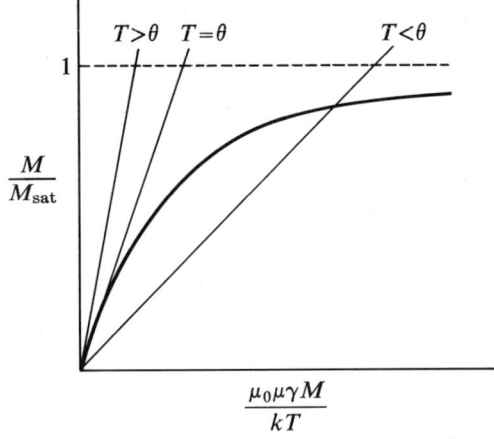

Fig. 7.8
Method of finding the spontaneous magnetization at a temperature T

Spontaneous magnetization is defined as the magnetization for $H = 0$, i.e., no applied field. Under this condition

$$x = \frac{\mu_0 \mu \gamma M}{kT}. \tag{7.80}$$

Figure 7.8 shows a plot of M/M_{sat} versus $x = \mu_0 \mu \gamma M/kT$, given by Eq. (7.79). With the help of Eq. (7.80),

$$\frac{M}{M_{sat}} = \frac{M}{N\mu} = \frac{kTx}{N\gamma\mu_0\mu^2}. \tag{7.81}$$

If a constant θ is defined such that

$$\theta = \frac{\gamma N \mu_0 \mu^2}{3k}, \tag{7.81a}$$

$$\frac{M}{M_{sat}} = \frac{T}{3\theta} x. \tag{7.82}$$

Spontaneous magnetization would occur for a simultaneous solution of the equations (7.79) and (7.82). Equation (7.82) represents a straight line with slope $T/3\theta$.

For $T < \theta$, a solution, $M \neq 0$, for the simultaneous equations (7.79) and (7.82) exists and hence spontaneous magnetization does occur.

When $T = \theta$, the slope of the line (7.82) is the same as that of (7.79) at the point $x = 0$, because

$$\frac{M}{M_{sat}} = \coth x - \frac{1}{x}$$

at $x \to 0$ is

$$\frac{M}{M_{sat}} = \frac{x}{3} = \frac{\mu_0\mu\gamma M}{3kT}$$

and

$$\frac{M/M_{sat}}{\mu_0\mu\gamma M/kT} = \frac{1}{3},$$

which is the same as the slope of the line (7.82),

$$\frac{M/M_{sat}}{x} = \frac{1}{3} \quad \text{for } T = \theta.$$

This shows that the spontaneous magnetization which exists for $T < \theta$ vanishes at $T = \theta$.

For $T > \theta$, no nontrivial solution exists for the simultaneous equations (7.79) and (7.82), and hence there is no possibility of spontaneous magnetization.

Behavior of ferromagnetic material for $T > \theta$ (classical theory). It has already been said that there exists no possibility of spontaneous magnetization for $T > \theta$. It will be of interest to study the behavior of the ferromagnetic materials for these temperatures in the presence of an applied magnetic field H.

Referring back to Eq. (7.57a),

$$M = N\mu\left(\coth x - \frac{1}{x}\right), \tag{7.57a}$$

where **H** is replaced by $\mathbf{H}_{eff} = \mathbf{H} + \gamma \mathbf{M}$ and hence

$$x = \frac{\mu_0\mu(H + \gamma M)}{kT}.$$

If T is large, x may be very small and hence $\coth x - 1/x$ for $x \ll 1$ approaches $x/3$. Therefore,

$$M = N\mu\frac{x}{3};$$

hence

$$M = \frac{N\mu_0\mu^2}{3kT}(H + \gamma M). \tag{7.83}$$

Simplifying,

$$M\left(1 - \frac{N\mu_0\mu^2\gamma}{3kT}\right) = \frac{N\mu_0\mu^2}{3kT} H$$

or

$$\chi = \frac{M}{H} = \frac{N\mu_0\mu^2/3kT}{1 - (N\mu_0\mu^2\gamma/3kT)}.$$

Using the definition of θ from Eq. (7.81a),

$$\chi = \frac{\theta/\gamma}{T - \theta}.$$

Putting

$$\theta/\gamma = C, \tag{7.84}$$

$$\chi = \frac{C}{T - \theta}. \tag{7.85}$$

C is called the *Curie constant*. θ is called the paramagnetic Curie temperature, and Eq. (7.85), showing the variation of susceptibility with temperature, is the Curie-Weiss law. Comparing Eq. (7.85) with Eq. (7.68) we find that ferromagnetic materials behave in the same fashion as paramagnetic material for $T \gg \theta$. Figure 7.9 shows the variation of $1/\chi$ versus temperature for paramagnetic materials and ferromagnetic materials in the temperature range in which they behave as paramagnetic materials. The slope of the curve determines $1/C$. θ can also be measured. The experimentally obtained value of $\gamma = 1000$ is a thousand times more than one would

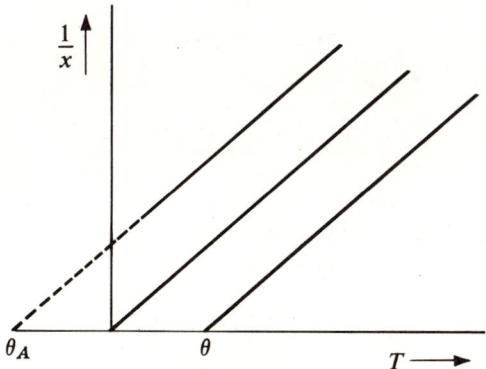

Fig. 7.9
Reciprocal susceptibility versus temperature for para-, ferro-, and antiferromagnetic materials above the critical temperature

expect if one assumed that the internal field is due to interatomic classical dipole interaction.

Quantum-mechanical theory. The magnetization M according to the quantum-mechanical theory is expressed in terms of the total angular momentum quantum number J by Eq. (7.74),

$$M = NgJ\mu_B B_J(x), \tag{7.74}$$

where for ferromagnetic materials H has to be replaced by H_{eff} in the definition of x:

$$x = \frac{gJ\mu_B\mu_0}{kT}(H + \gamma M). \tag{7.86}$$

$B_J(x)$, the Brillouin function, is given by Eq. (7.75),

$$B_J(x) = \frac{2J+1}{2J}\coth\left[\frac{(2J+1)x}{2J}\right] - \frac{1}{2J}\coth\frac{x}{2J}, \tag{7.75}$$

$$\frac{M}{M_{\text{sat}}} = \frac{M}{NgJ\mu_B} = B_J(x). \tag{7.74a}$$

In the absence of an applied magnetic field $H = 0$ and

$$x = \frac{gJ\mu_B\mu_0}{kT}\gamma M. \tag{7.87}$$

From Eq. (7.87),

$$M = \frac{xkT}{\gamma gJ\mu_B\mu_0}, \tag{7.88}$$

$$M_{\text{sat}} = NgJ\mu_B.$$

Therefore,

$$\frac{M}{M_{\text{sat}}} = \frac{xkT}{N\gamma(gJ\mu_B)^2\mu_0}. \tag{7.88a}$$

Spontaneous magnetization will occur if there exists a solution of the simultaneous equations (7.74a) and (7.88a).

If a constant θ is defined such that

$$\theta = \frac{N\gamma(gJ\mu_B)^2\mu_0}{k}\cdot\frac{(J+1)}{3J}, \tag{7.88b}$$

$$\frac{M}{M_{\text{sat}}} = \frac{T}{\theta}\frac{J+1}{3J}(x). \tag{7.89}$$

Since $\mu = gJ\mu_B$, θ given by Eqs. (7.88b) and (7.81a) are the same as $J \to \infty$.

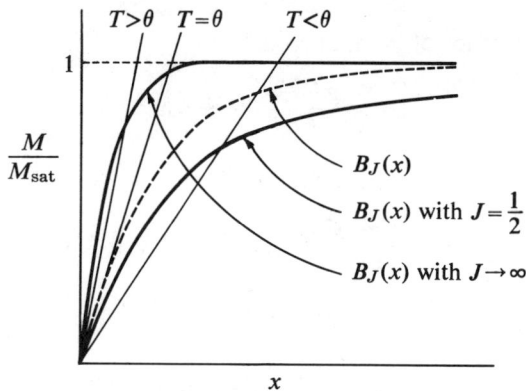

Fig. 7.10
M/M_{sat} versus x from Eq. (7.89) for $T > \theta_f$, $T < \theta_f$, and $T < \theta_f$

Figure 7.10 shows a plot of (M/M_{sat}) versus x from Eq. (7.89) for three values of temperatures, $T > \theta$, $T = \theta$, and $T < \theta$. It also shows a plot of (M/M_{sat}) versus x as given by Eqs. (7.75) with the value of x corresponding to Eq. (7.87) for $J \to \infty$ and $J = \frac{1}{2}$. The curve $B_J(x)$ lies somewhere between the two curves depending upon the value of J and is shown dashed in Fig. (7.10). The dashed horizontal line is the saturation magnetization $M = M_{sat} = NgJ\mu_B$. With no applied magnetic field one obtains a nonvanishing solution for M for $T < \theta$. This suggests the possibility of spontaneous magnetization.

When $T = \theta$, the slope of the line is the same as that of the curve given by Eq. (7.74) at the origin. This shows that spontaneous magnetization which occurs for $T < \theta$ vanishes at $T = \theta$.

For $T > \theta$, the simultaneous equations (7.74) and (7.89) have no nonvanishing solutions and hence no spontaneous magnetization.

Behavior of ferromagnetic materials for $T > \theta$; Derivation of the Curie-Weiss law. For high temperatures $x \ll 1$ and

$$B_J(x) \cong \frac{(J+1)x}{3J}.$$

Hence magnetization according to Eq. (7.74) is

$$M = NgJ\mu_B \frac{(J+1)x}{3J}$$

or

$$M = \frac{Ng\mu_B}{3}(J+1)x, \tag{7.90}$$

where in the presence of an applied field

$$x = \frac{gJ\mu_B\mu_0}{kT}(H + \gamma M). \tag{7.86}$$

Substituting Eq. (7.86) in Eq. (7.90),

$$M = \frac{Ng^2\mu_B^2}{3}\frac{(J+1)\mu_0 J}{kT}(H + \gamma M), \tag{7.91}$$

and replacing

$$g^2\mu_B^2 J^2 \text{ by } \mu^2$$

$$M = \frac{N\mu^2\mu_0}{kT} \cdot \frac{J+1}{3J}(H + \gamma M) \tag{7.92}$$

or

$$M = \frac{N\mu^2\mu_0 \dfrac{J+1}{3J} \dfrac{H}{kT}}{1 - \dfrac{N\mu^2\mu_0\gamma}{kT}\dfrac{J+1}{3J}} \tag{7.93}$$

Defining

$$\theta = \frac{N\mu_0\mu^2\gamma}{k}\frac{J+1}{3J} \tag{7.94}$$

$$M = \frac{CH}{T - \theta} \tag{7.95}$$

where $C = \theta/\gamma$.

Hence

$$\chi = \frac{C}{T - \theta},$$

which is Eq. (7.85) derived from the classical approach.

Weiss's theory was a great success in explaining spontaneous magnetization and the behavior of ferromagnetic material at temperatures greater than the Curie temperature θ.

It may be mentioned that the actual transition from ferromagnetic behavior to paramagnetic behavior occurs at a temperature slightly lower than θ. It is θ_f, the ferromagnetic Curie temperature. This relatively simple theory is not capable of describing the behavior of ferromagnetic materials in the vicinity of the Curie temperature.

The M-H curve of a virgin specimen of a ferromagnetic material below the ferromagnetic Curie temperature is shown in Fig. 7.11. For a virgin specimen $M = 0$ at $H = 0$. To explain this fact, Weiss postulated the the existence of domains within the material. Within these domains the

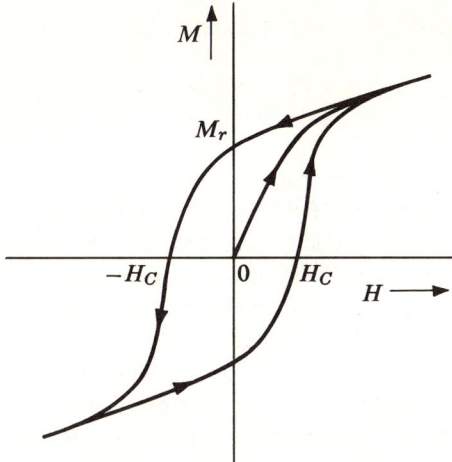

Fig. 7.11
Hysteresis loop of ferromagnetic material

material is magnetized to saturation but the direction of magnetization differs from domain to domain, resulting in a zero resultant magnetization of the specimen. We shall take up the domain-wall theory of ferromagnetic materials a little later. With an increase in the value of H, the magnetization of the specimen increases from 0 to M_{sat}. Along the virgin curve one can speak of a differential permeability as $1 + dM/dH$, which is a function of H. The differential permeability may reach values of the order of 10^3 to 10^6 in high-permeability materials such as iron, nickel, cobalt, and appropriate alloys. Upon reducing the value of H it is found that there is a finite value of $M = M_r$ at $H = 0$. The presence of M_r has already been discussed by the Weiss theory in terms of the internal field.

The value of the internal field γM, according to the Weiss theory, can be calculated as follows:

From Eq. (7.81a) the internal field in a saturated state is $\gamma M_{\text{sat}} = 3k\theta/\mu_0\mu$. In iron, $\theta \cong 1000°K$, $\mu = 2.2$ Bohr magnetons, 1 Bohr magneton $= 9.27 \times 10^{-24}$ A-m^2, and $k = 1.38 \times 10^{-23}$ J/deg. Therefore,

$$\gamma M_{\text{sat}} = \frac{3k\theta}{\mu_0\mu} = \frac{3 \times 1.38 \times 10^{-23} \times 10^3}{2.2 \times 9.27 \times 10^{-24} \times 4\pi \times 10^{-7}} = 1.56 \times 10^9 \text{A/m}.$$

Heisenberg, Bloch, and others were finally able to explain these high internal fields with the help of quantum-mechanics.

Exchange interactions. The flux density due to the magnetic dipole moment of 1 Bohr magneton μ_B at an atom A would induce a magnetic field at an atom B separated by a distance r of $\mu_0\mu_B/4\pi r^3$.

Assuming $r \cong 10^{-10}$ m, $\mu_0 \gamma M$ at atom B due to atom A is

$$\frac{\mu_0 \mu_B}{4\pi r^3} = \frac{4\pi \times 10^{-7} \times 9.27 \times 10^{-24}}{4\pi \times 10^{-30}} \simeq 1 \text{Wb/m}^2.$$

Hence the dipole-dipole interaction predicts internal fields which are lower than the actual internal fields by 10^3 and is therefore not in a position to justify a large ferromagnetic effect. The first theoretical explanation of the large Weiss field (internal field) in ferromagnetic materials was proposed by Heisenberg in 1928. He explained that the large internal field may be described in terms of quantum exchange interactions between the electrons. These interactions are essentially electrostatic in nature. Considering exchange interactions between nearest neighbors, one could show that there is a tendency for net spin alignment in transitions. A detailed discussion of the theory is not within the scope of this book and the reader is referred to the Bibliography.

Domains. Weiss introduced the idea of domain to explain the magnetization of a ferromagnetic material. In virgin specimens of ferromagnetic material the net magnetization may be zero. Weiss hypothesized the existence of small regions in which there exists spontaneous magnetization; i.e., the atomic dipole moments are all aligned parallel to one another at a temperature far below the Curie point. The over-all magnetization is the sum of domain magnetization vectors. The domains are randomly oriented and therefore their vector sum may be zero. Figure 7.12 shows a possible domain configuration in a virgin specimen. Magnetization of a given material occurs by the growth of a domain favorably oriented in the direction of the applied field at the expense of the others or by the coherent rotation of the magnetic moment in various individual domains.

Origin of domains. The origin of domains may be sought in magnetostatic energy. Let us consider a ferromagnetic material (Fig. 7.3) saturated in the direction shown. Energy is associated with the spatial distribution of the magnetic field. The form of the energy may be expressed as an integral over space,

$$E_p = \frac{1}{2} \int \mathbf{H} \cdot \mathbf{M} \, dv = \frac{\mu_0}{2} \int H^2 \, dv.$$

If a transition region is introduced between the two halves of the sample of Fig. 7.13, the spatial field distribution is considerably reduced and the associated magnetostatic energy is also reduced. However, the transition region or wall now involves exchange and other energies, normally com-

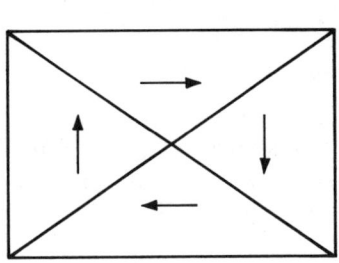

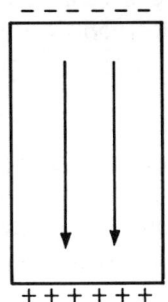

Fig. 7.12
Domain structure in a virgin specimen

Fig. 7.13
Origin of domain

bined in the concept of wall energy. The final configuration is such as to result in a minimum of the total energy. Walls are created and domains formed until the creation of additional walls increases the energy more than the magnetostatic energy is decreased.

In Fig. 7.14 is shown a possible equilibrium state with two domains separated by a domain wall.

Planer domain-wall motion. A practical group of ferromagnetic materials is composed of nickel-iron alloys. Figure 7.15 shows a typical

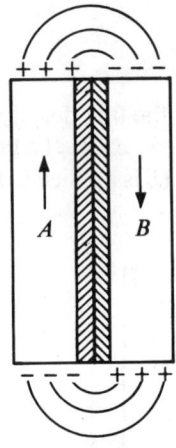

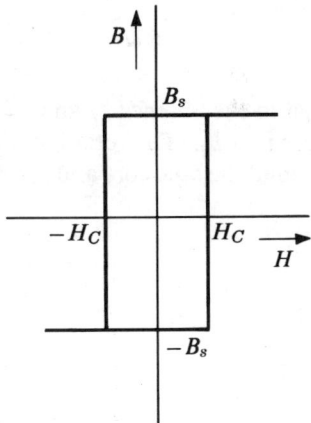

Fig. 7.14
Transition region created between the two halves of the sample of Fig. 7.13 to reduce the magnetostatic energy

Fig. 7.15
Typical *B-H* curve for commercially available ferromagnetic material

B-H curve of such materials taken with d.c. fields. Typical values of B_s and H_C for 50-50 Ni-Fe are

$$B_s = 1.5 \text{ Wb/m}^2,$$
$$H_C = 120 \text{ AT/m}.$$

A common use of these materials is in toroidal-shaped tape cores. Strips of the alloy are rolled to form a core called a tape core (Fig. 7.16). The tapes are thin and broad. Let *l* be the circumference of core. Referring to Fig. 7.17, which shows a cross section of the tape, let $W \gg d$. Let us suppose

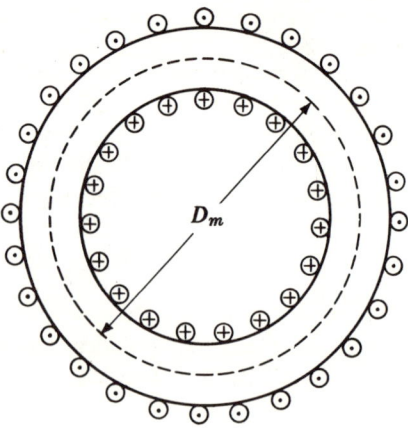

Fig. 7.16
Planar domain-wall motion

that in the absence of an externally applied field the flux density in the tape core is $-B_s$. The total flux through the tape is $-B_s d.W$. If a coil is wound around the tape core and if a magnetic field $H_a > H_c$ is applied to the material

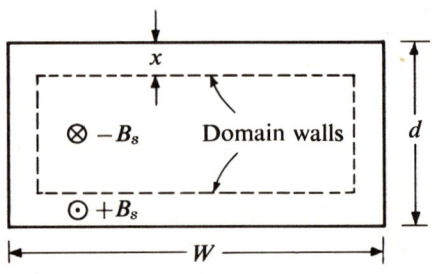

Fig. 7.17
Tape-wound core

in a direction out of the paper, flux reversal can occur. The specimen has a single domain with magnetization in to the paper before the application of the field. Domain walls for the reversal of the flux in an applied field may be created at the surface of the material, because the surfaces represent discontinuities which may be nucleating centers for domain walls. The domain walls so formed may propagate toward the center of the strip. In this very idealized model, complete reversal of flux will have taken place when the domain wall has moved by a distance $d/2$. Of course, other mechanisms may contribute to flux reversal, but are neglected in this model.

Let us consider the situation when the domain wall has moved by a distance x. The total flux during this motion is the difference of flux out of the paper minus the flux in the paper,

$$\Phi = 2xWB_s - W(d - 2x)B_s.$$

Therefore, the rate of change of flux is

$$\frac{d\Phi}{dt} = 4B_s W \frac{dx}{dt}. \tag{7.96}$$

A rate of change of flux induces a voltage in the tape core in accordance with the Faraday law of electromagnetic induction:

$$e = \frac{d\Phi}{dt} = 4B_s W \frac{dx}{dt}, \tag{7.96a}$$

in which dx/dt is the velocity of the domain wall.

Depending upon the resistance of the tape core, the induced voltage results in the flow of a current i is

$$i = e/R,$$

where $R = \rho 2W/lx$ and ρ is the resistivity. Hence

$$i = \frac{2B_s}{\rho} lx \frac{dx}{dt}. \tag{7.97}$$

This current opposes the applied ampere turns and is shown in Fig. 7.16. It is called an *eddy current*. According to Ampère's law,

$$\oint \mathbf{H} \cdot d\mathbf{l} = \text{total current enclosed by the path.}$$

From the hysteresis loop of the material, the change in flux must occur at $H = H_C$ and hence

$$\oint \mathbf{H}_C \cdot d\mathbf{l} = \text{applied ampère turns} - \text{opposing ampère turns.}$$

Considering a mean diameter of core D_m and assuming N turns of the coil wound around the tape with a current I flowing through them,

$$\oint \mathbf{H}_C \cdot dl = IN - i$$

or

$$H_C \pi D_m = H_a \pi D_m - \frac{2B_s}{\rho} \pi D_m x \frac{dx}{dt} \qquad (7.98)$$

where H_a is the applied field,

$$H_a = \frac{IN}{\pi D_m}.$$

Simplifying Eq. (7.98),

$$H_a - H_C = \frac{2B_s}{\rho} x \frac{dx}{dt}. \qquad (7.99)$$

$2B_s/\rho$ may be called the *eddy-current damping constant* and is denoted

$$H_a - H_C = \beta_e x \frac{dx}{dt}. \qquad (7.100)$$

Hence $H_a - H_c$ is the effective field H_e. Therefore,

$$H_e = \beta_e x \frac{dx}{dt}. \qquad (7.101)$$

Integrating Eq. (7.101) and assuming that the applied field is a step function applied at $t = 0$,

$$\beta_e \left(\frac{x^2}{2} + C \right) = H_e t,$$

where C is the constant of integration at $t = 0$, $x = 0$, and hence $C = 0$. Therefore,

$$\beta_e \frac{x^2}{2} = H_e t$$

or

$$x = \left(\frac{2 H_e t}{\beta_e} \right)^{1/2}. \qquad (7.102)$$

If the domain walls penetrate from both sides of the material, the time required for the flux to reverse from $-B_s$ to B_s is t_s, which is given by Eq. (7.102) when $x = d/2$:

$$t_s = \frac{\beta_e d^2}{8 H_e}. \qquad (7.103)$$

Therefore $t_s H_e = S_w = \beta_e (d^2/8)$, where S_w is sometimes called the *switching constant*. Equation (7.103) shows that the time for flux reversal is depen-

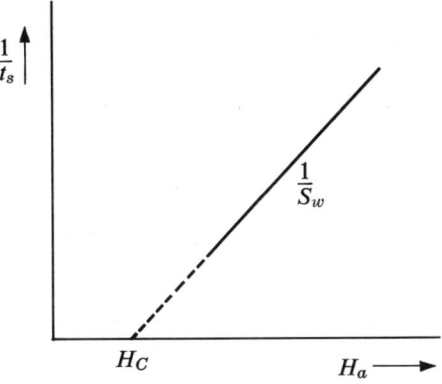

Fig. 7.18
Reciprocal switching time versus applied magnetic field

dent upon β_e, d, and $1/H_e$. The time required for flux reversal t_s is small if β_e is small, i.e., if the resistivity of the material is high. The thinner the tape, the faster the flux reversal. The switching constant can be determined as the slope of the $1/t_s$ versus H_a curve (Fig. 7.18). Fast reversal times are advantageous in computer applications.

The induced voltage as a function of time is on a per-turn-per-wrap basis:

$$e(t) = \frac{d\Phi}{dt} = 4B_s W \frac{dx}{dt}. \tag{7.104}$$

Differentiating Eq. (7.102) with respect to time,

$$\frac{dx}{dt} = \frac{1}{2}\left(\frac{2H_e}{\beta_e}\right)^{1/2} t^{-1/2}$$

and substituting in Eq. (7.104),

$$e(t) = 2B_S W \left(\frac{2H_e}{\beta_e t}\right)^{1/2} \tag{7.105}$$

Figure 7.19 shows a plot of $e(t)$ against time. At $t = t_s$ the flux is reversed and the induced voltage drops to zero. By winding another coil on the core and recording the induced voltage as a function of time, the total flux change in the core can be measured. It should be emphasized that the actual flux reversal may be a much more complicated process than the one assumed in this example.

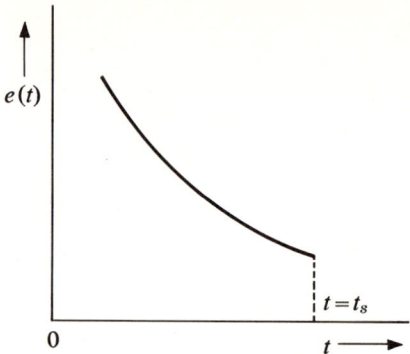

Fig. 7.19
Induced voltage as a function of time

7.5 Antiferromagnetism

The explanation for ferromagnetism is based on the fact that quantum-mechanical exchange forces produce large internal fields and the neighboring dipoles tend to align in the same direction. It can be shown quantum mechanically that when the distance between interacting atoms is very small, the exchange forces produce a tendency to antiparallel alignment of the neighboring spin dipole moments. Such materials are called *antiferromagnetic materials*. Neel and Bitter were the first to study such materials. Antiferromagnetic materials show spontaneous magnetization below a certain temperature, known as the *Neel temperature* θ_N. Above the Neel temperature antiferromagnetic materials behave as paramagnetic material and the susceptibility is given by

$$x = \frac{C}{T + \theta_A} \tag{7.106}$$

for $T > \theta_N$. Below the Neel temperature the system becomes organized. As $T \to 0$ half of the dipoles align in one direction and the other half in the opposite direction. This means that the susceptibility when plotted against temperature shows a maximum at the Neel temperature. The following model explains antiferromagnetism. We follow the same procedure as in ferromagnetism with the difference that in antiferromagnetism we consider a crystal containing two types of atoms A and B distributed over interlocking lattices as shown in Fig. 7.20. Atoms B occupy the lattice points and atoms A occupy the center of the body.

Classical theory. Referring to Eq. (7.67),

$$M = N\mu \left(\coth x - \frac{1}{x} \right),$$

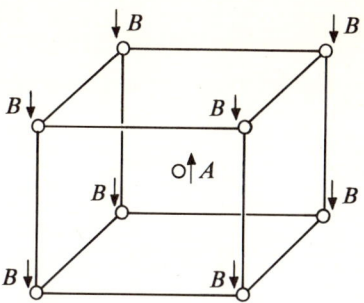

Fig. 7.20
Crystal containing two types of atoms A and B, distributed over interlocking lattices

where

$$x = \frac{\mu_0\mu(H - \gamma M)}{kT}.$$

Here the interaction factor or Weiss molecular field constant is negative, because the dipoles tend to align antiparallel.

For $T > \theta_N$, $x \ll 1$

$$M = N\mu \frac{x}{3} = \frac{N\mu_0\mu^2}{3kT}(H - \gamma M). \tag{7.107}$$

If M_A is the total magnetization of A-type atoms; M_B is the total magnetization of B-type atoms; γ_1 is the Weiss molecular field constant for like atoms AA or BB and γ_2 is the Weiss molecular field constant for unlike atoms AB or BA. The magnetization M_A is given by

$$M_A = \frac{N\mu_0\mu^2}{2 \times 3kT}(H - \gamma_1 M_A - \gamma_2 M_B) \tag{7.108}$$

and the magnetization M_B by

$$M_B = \frac{N\mu_0\mu^2}{2 \times 3kT}(H - \gamma_1 M_B - \gamma_2 M_A). \tag{7.109}$$

In Eqs. (7.108) and (7.109), N of Eq. (7.107) has been replaced by $N/2$, because half the total number of atoms are of type A and the other half of type B.

The total magnetization is the sum of partial magnetizations M_A and M_B:

$$M = M_A + M_B.$$

Adding Eqs. (7.108) and (7.109),

$$M = \frac{N\mu_0\mu^2}{6kT}[2H - \gamma_1(M_A + M_B) - \gamma_2(M_A + M_B)],$$

$$M = \frac{N\mu_0\mu^2}{6kT}[2H - (\gamma_1 + \gamma_2)M],$$

or

$$M = \frac{N\mu_0\mu^2}{3kT}\left(H - \frac{\gamma_1 + \gamma_2}{2}M\right).$$

Simplifying,

$$M = \frac{(N\mu_0\mu^2/3kT)H}{1 + (N\mu_0\mu^2/3kT)(\gamma_1 + \gamma_2/2)}$$

or

$$\chi = \frac{M}{H} = \frac{C}{T + (C/2)(\gamma_1 + \gamma_2)}, \qquad (7.110)$$

where $C = N\mu_0\mu^2/3k$, the definition of C used earlier in the ferromagnetic case.

Calling $(C/2)(\gamma_1 + \gamma_2) = \theta_A$,

$$\chi = \frac{C}{T + \theta_A}, \qquad (7.110a)$$

which is the same relation as Eq. (7.106). θ_A is called the *asymptotic Curie temperature*. Figure 7.9 shows a plot of $1/\chi$ versus T for antiferromagnetic materials also. At low temperatures M_A and M_B are very large but oppositely directed. As the temperature starting from absolute zero is increased, M_A and M_B of the two sublattices in the absence of an externally applied field vary as in ferromagnetic materials. Spontaneous magnetization disappears at a certain temperature. This can be shown as follows.

Rewriting Eqs. (7.108) and (7.109) for $H = 0$,

$$M_A = \frac{N\mu_0\mu^2}{2 \times 3kT}(-\gamma_1 M_A - \gamma_2 M_B), \qquad (7.111)$$

$$M_B = \frac{N\mu_0\mu^2}{2 \times 3kT}(-\gamma_1 M_B - \gamma_2 M_A). \qquad (7.112)$$

Introducing C in Eqs. (7.111) and (7.112),

$$M_A = \frac{C}{2T}(-\gamma_1 M_A - \gamma_2 M_B)$$

or

$$M_A\left(1 + \frac{C\gamma_1}{2T}\right) + \frac{C\gamma_2}{2T}M_B = 0. \qquad (7.113)$$

ANTIFERROMAGNETISM

Similarly,

$$M_B = \frac{C}{2T}(-\gamma_1 M_B) - \frac{C}{2T}\gamma_2 M_A$$

or

$$M_B\left(1 + \frac{C\gamma_1}{2T}\right) + \frac{C\gamma_2}{2T}M_A = 0. \qquad (7.114)$$

One solution of simultaneous equations (7.113) and (7.114) is that

$$M_A = M_B = 0.$$

Spontaneous magnetization would take place if there existed a solution of the simultaneous equations for a finite value of M_A and M_B. This is possible if

$$\begin{vmatrix} 1 + \dfrac{C\gamma_1}{2T} & \dfrac{C\gamma_2}{2T} \\ \dfrac{C\gamma_2}{2T} & 1 + \dfrac{C\gamma_1}{2T} \end{vmatrix} = 0,$$

$$\left(1 + \frac{C\gamma_1}{2T}\right)^2 - \left(\frac{C\gamma_2}{2T}\right)^2 = 0,$$

or

$$1 + \left(\frac{C\gamma_1}{2T}\right)^2 + \frac{C\gamma_1}{T} - \left(\frac{C\gamma_2}{2T}\right)^2 = 0. \qquad (7.115)$$

The positive roots of Eq. (7.115) give the transition temperature T. Therefore, for positive roots,

$$1 + \frac{C\gamma_1}{2T} = \frac{C\gamma_2}{2T},$$

$$1 = \frac{C}{2T}(\gamma_2 - \gamma_1),$$

or

$$T = \frac{C(\gamma_2 - \gamma_1)}{2} = \theta_N. \qquad (7.116)$$

Equation (7.116) gives the Neel temperature. It is at this temperature that a kink is observed in the susceptibility versus temperature curve.

Quantum-mechanical theory. We have already seen that at high temperatures the quantum-mechanical theory only defines the value of μ^2 in terms of the total angular momentum quantum number J:

$$\mu^2 = \mu_B^2 g^2 J^2.$$

The expression for Neel temperature θ_N turns out to be the same as that given by Eq. (7.116).

It should be noted that the Neel temperatures increase as the value of γ_2 increases, and decrease as the value of γ_1 increases.

From Eqs. (7.116) and (7.110a),

$$\frac{\theta_N}{\theta_A} = \frac{\gamma_2 - \gamma_1}{\gamma_2 + \gamma_1}. \tag{7.117}$$

Typical examples of antiferromagnetic materials are: $MnF_2, \theta_N = 72°K$, $\theta_A = 113°K$; $FeF_2, \theta_N = 79°K, \theta_A = 117°K$; $CoF_2, \theta_N = 38°K, \theta_A = 53°K$. Measured values of θ_N and θ_A can be obtained. They indicate that γ_1 must be positive, indicating that there is antiferromagnetic interaction.

7.6 Ferrimagnetism and ferrites

Ferrimagnetic materials are those materials in which neighboring dipoles have a tendency to orient themselves antiparallel, but the adjacent dipoles are of unequal magnitude. Let us consider four atoms comprising a molecule (Fig. 7.21). The net moment of the molecule is $2(\mu_B - \mu_A)$, where μ_B is

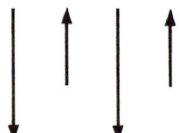

Fig. 7.21
Four atoms comprising a molecule and the associated dipole moments

the atomic magnetic dipole moment of atoms B and μ_A of atoms A. Ferrites are nonmetallic ferrimagnetic materials and are very important from the point of view of electrical engineering. Magnetite is a ferrimagnetic material with the chemical formula $Fe^{2+}Fe_2^{3+}O_4$. If the divalent ferrous ion in magnetite is replaced by another divalent metal, such as Mg, Cu, Zn, etc., a ferrite is obtained. The chemical formula for simple ferrites may therefore be written $Me^{2+}Fe_2^{3+}O_4$, where Me^{2+} represents the divalent metallic ion.

Let us look at the molecule $Fe^{2+}Fe_2^{3+}O_4$. It consists of one ion of Fe^{2+} and two of Fe^{3+}. Fe^{2+} has an atomic structure $1s^2 2s^2 2p^6 3s^2 3p^6 3d^6$. The d shell can take 10 electrons. The total spin quantum number S, therefore, equals 4. If we assume 1 Bohr magneton as the contribution of 1 spin, the Fe^{2+} ion corresponds to 4 Bohr magnetons. Similarly, Fe^{3+} corresponds to 5 Bohr magnetons, and two ions Fe_2^{3+} would give rise to 10 magnetons.

The oxygen atom has an even number of electrons in the outermost shell $1s^2 2s^2 2p^6 3s^2 3p^4$, and hence has no spin dipole moment contribution. If the spin of the ions Fe^{2+} and Fe_2^{3+} are all lined up parallel, one expects $(4 + 2 \times 5) = 14$ Bohr magnetons per molecule. However, experiments show only 4 Bohr magnetons per molecule. Neel in 1948 gave an explanation of the discrepancy. He showed that in ferrimagnetic materials the exchange interactions are of the antiferromagnetic type.

The crystal structure of the material plays an important role in determining the physical properties of the materials. Ferrites have a structure similar to $MgAl_2O_4$, a spinel structure. It is not our purpose here to go into the details of the structure. It may be mentioned that the arrangement of ions is such that the net spin dipole moments of Fe_2^{3+} ions turn out to be zero, because half of them are at tetrahedral sites and the other half at octrahedral sites, with antiferromagnetic exchange interaction between them. The net spin dipole moment is then only due to Fe^{2+} ions and is 4 Bohr magnetons, in conformity with experimental results.

Being nonmetallic, the dc resistivity of ferrites is many orders of magnitudes higher than that of para-, dia-, and ferromagnetic materials. This makes ferrites useful for high-frequency work, because the flow of eddy currents are checked by the increased resistance and the losses in iron are reduced. Ferrites have already been used in high-frequency transformers and high-frequency generators.

7.7 Cooling by adiabatic demagnetization

Temperatures below 1°K can be obtained with the help of paramagnetic salts. This is done by utilizing the thermodynamic properties of paramagnetic materials. Figure 7.22 gives the entropy of the salt as a function of

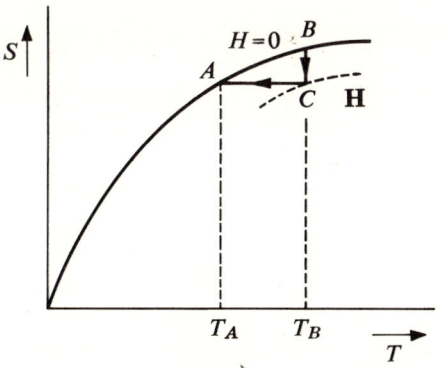

Fig. 7.22
Entropy of a paramagnetic salt as a function of temperature

temperature (OAB). The entropy, which is the degree of disorder, is a minimum at $T = 0$. The first step is to cool the crystal, as much as possible, by conventional low-temperature techniques. Cooling to 4°K, which is the temperature of liquid helium, is no problem. Let this state be represented by a point B on the entropy S versus T curve OAB. The next step is to increase the intensity of the externally applied magnetic field from $H = 0$ to some value of H with the temperature maintained at T_B. During this process, which is isothermal, the applied magnetic field introduces an alignment of the atomic magnetic dipoles and hence a decrease in entropy of the crystal. Let C be the new entropy of the crystal lying in the S versus T curve for the applied value of H. The specimen is now adiabatically demagnetized with change in entropy $dS = 0$. The entropy now corresponds to point A. Successive repetition of these processes result in very, very low temperatures, even of the order of 10^{-3}°K.

Quantitatively it may be explained as follows. According to the second law of thermodynamics,

$$dE = dQ - dW. \tag{5.1}$$

Now

$$dE = \left(\frac{\partial E}{\partial T}\right)_H dT + \left(\frac{\partial E}{\partial H}\right)_T dH \quad \text{because } E(T, H),$$

$$dW = M \, dH,$$

where H is the magnetic field and M is the magnetization. Also, $dQ = T \, dS$, where dS is the change of entropy. Therefore,

$$T \, dS = \left(\frac{\partial E}{\partial T}\right)_H dT + \left[\left(\frac{\partial E}{\partial H}\right)_T + M\right] dH.$$

dS is a total differential, so from Maxwell's relations

$$\frac{1}{T}\left[\left(\frac{\partial E}{\partial H}\right)_T + M\right] = \left(\frac{\partial S}{\partial H}\right)_T = \left(\frac{\partial M}{\partial T}\right)_H$$

$$dS = \left(\frac{\partial M}{\partial T}\right)_H dH \tag{7.118}$$

for an isothermal case. Integrating Eq. (7.118),

$$S = S_{H=0} + \int_0^H \left(\frac{\partial M}{\partial T}\right)_H dH. \tag{7.119}$$

From Eq. (7.71),

$$M = Ng^2\mu_0 \frac{J(J+1)\mu_B^2}{3kT} H. \tag{7.71}$$

Differentiating Eq. (7.71) with respect to temperature,

$$\left(\frac{\partial M}{\partial T}\right)_H = -\frac{N\mu_0 g^2 \cdot J(J+1)\mu_B^2}{3kT^2} H. \qquad (7.120)$$

Substituting Eq. (7.120) in Eq. (7.119),

$$S = S_{H=0} + \int_0^H -\frac{N\mu_0 g^2 \cdot J(J+1)\mu_B^2}{3kT^2} H\, dH.$$

Hence

$$S = S_{H=0} + \frac{-N\mu_0 g^2 \cdot J(J+1)\mu_B^2}{6kT^2} H^2. \qquad (7.121)$$

The change in entropy at a given temperature T_B for a given value of H is

$$\Delta S = -\frac{N\mu_0 g^2 \cdot J(J+1)\mu_B^2}{6kT_B^2} H^2. \qquad (7.122)$$

The distance BC in Fig. 7.22 is ΔS.

The magnetocaloric effect described above has been used by many workers for the production of very low temperatures. The first experimental test of the phenomenon was carried out at temperatures down to 1.3°K. The results confirmed the theory in a satisfactory manner. Temperatures as low as 0.0015°K have already been obtained by the effect. Typical paramagnetic salts used in attaining low temperatures by magnetocaloric effect are $FeNH_4(SO_4)_2 \cdot 12H_2O$ and CeF_3.

Problem 1

Using the Biot-Savart law, show that the magnetic flux density B at a distance r due to current I flowing in an infinitely long straight conductor is given by

$$B = \mu_0 \mu_r \frac{I}{2\pi r}.$$

Problem 2

Consider a spinning spherical shell of charge e and mass m, both uniformly distributed over its surface. Show that the ratio of the magnetic moment to angular momentum is $e/2m$.

Problem 3

Using Fermi-Dirac statistics, show that the weak paramagnetic susceptibility for conduction electrons in a metal for small values of H is nearly independent of temperature.

MAGNETIC MATERIALS AND PROPERTIES

Problem 4

a. In the quantum-mechanical theory of paramagnetism using the statistical average of the magnetic moment component per atom [Eq. (7.69)], derive the Brillouin function

$$B_J(x) = \frac{2J+1}{2J} \coth \frac{2J+1}{2J} x - \frac{1}{2J} \coth \frac{x}{2J},$$

where $x = gJ\mu_0\mu_B H/kT$.

b. Show that for $J \to \infty$, the Brillouin function $B_J(x)$ reduces to the Langevin function $L(x)$.

c. Show that for $J = \frac{1}{2}$, the Brillouin function $B_J(x) = \tanh x$.

Problem 5

a. Using the quantum-mechanical theory of paramagnetism, show that if $gJ\mu_B H \ll kT$ the differential permeability

$$\frac{dM}{dH} = \frac{Ng^2\mu_B^2\mu_0 J(J+1)}{3kT}$$

as a first-order approximation of the Brillouin function.

b. Show that if the first two terms in the series expansion of Brillouin functions are used, the differential permeability is given by

$$\frac{dM}{dH} = \frac{Ng^2\mu_0\mu_B^2 J(J+1)}{3kT} - \frac{N(gJ\mu_B)^4\mu_0^3}{15(kT)^3} H^2 \frac{(2J+1)^4 - 1}{2J^4}.$$

c. Show that in the classical case, the above equation reduces to

$$\frac{dM}{dH} = \frac{N\mu_0\mu^2}{3kT} - \frac{N\mu}{15}\left(\frac{\mu_0\mu}{kT}\right)^3 H^2.$$

Problem 6

A magnetic alloy tape core has the following properties: $\rho = 50 \times 10^{-6}\ \Omega$ cm, $B_s = 0.5$ Wb/m², $H_c = 4$ AT/m, and $W = 3$ mm (width of the tape). It is required to design a tape core with a switching time of 1 μsec such that the emf induced across the sensing coil per turn per wrap of the core is not less than 3 mV. Find the required thickness of the magnetic tape and the magnitude of the applied field. Assume perfect coupling.   Ans. $d = 2\ \mu$; $H_a = 5$ At/m

Problem 7

From Lande's equation for the spectroscopic splitting factor g, it is shown that

a. $g = 2$ for spin moments alone

b. $g = 1$ for orbital moments alone.

8

Transport Properties

8.1 Irreversible processes; conjugate forces and fluxes

Ordinary thermodynamics deals with equilibrium processes. A study of the nonequilibrium processes is called *irreversible thermodynamics*. Examples of irreversible processes are Fourier's law between heat flow and temperature gradient, Fick's law between flow of matter of a component in a mixture and its concentration gradient, Ohm's law between electrical current and potential gradient, and many others. When two or more of these irreversible processes occur simultaneously, they interfere and give rise to new effects. An example of cross effects is the *Peltier effect*, which deals with the rejection or absorption of heat at junctions of metals and semiconductors when an electric current flows through them. The thermodynamics of irreversible processes is based on Onsager's theory.

Onsager's theorem. We have mentioned that temperature gradient, potential gradient, concentration gradients, etc., give rise to the occurrence of irreversible phenomena. Onsager called these quantities *forces* and denoted them by X_i ($i = 1, 2, 3, \ldots, n$). These are not forces in the Newtonian sense and are often called *affinities*. The forces cause certain irreversible phenomena such as electric current, heat flow, etc. These are called *fluxes* or *flows of currents* and are denoted by J_i ($i = 1, 2, 3, \ldots, n$).

In general, any force can give rise to any flux. For example, a concentration gradient and a temperature gradient could both cause diffusion flow and also heat flow. Irreversible phenomena may therefore be expressed

$$J_i = \sum_{K=1}^{n} L_{iK} X_K \quad \text{where } i = 1, 2, 3, \ldots, n, \tag{8.1}$$

which states that any flow is caused by a contribution of all forces. The coefficients L_{iK} with $i, K = 1, 2, \ldots, n$ are called *phenomenological coefficients*. The coefficients L_{ii} represent the coefficients between the same force and flux only. Examples of this are electrical conductivity and the constant of proportionality between electric potential gradient and electric current. The coefficients L_{iK}, where $i \neq K$, are connected to interference phenomenon.

Onsager's fundamental theorem of irreversible processes states that provided a proper choice is made for the fluxes J_i and forces X_i, the matrix of phenomenological coefficients L_{iK} is symmetric, or

$$L_{iK} = L_{Ki}. \tag{8.2}$$

These identities are called *Onsager reciprocal relations* and express a connection between two reciprocal phenomena.

Proper choice of forces and fluxes. We have already seen in the statement of Onsager's theorem that a proper choice of forces and fluxes has to be made if the relationship (8.1) is to be used. To explain this we take the following example. Referring to Fig. 8.1, consider a bar in which

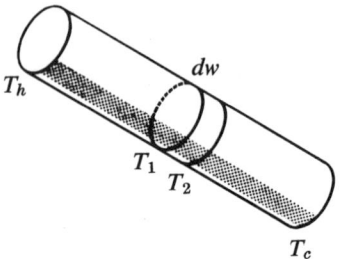

Fig. 8.1
Bar with each end maintained at different temperatures

T_h is the temperature of the hot end and T_c the temperature of the cold end. Consider an infinitesimal width dw of the bar with T_1 the hot temperature and T_2 the temperature of the cold side of this width. For this infinitesimal width,

$$dE = T\, dS + \bar{\mu}\, dN. \tag{8.3}$$

This is Eq. (5.12), with the assumption that the volume of the bar remains constant. $\bar{\mu}$ is the electrochemical potential, dS the change in entropy, and dE the change in internal energy. Equation (8.3) is valid for equilibrium conditions and may therefore be applied for the region of width dw.

Because of the temperature difference between the two ends of the width dw, fluxes flow through this width and are

$$\mathbf{J}_S = \left(\frac{dS}{dt} dA\right)\mathbf{n},$$

$$\mathbf{J}_E = \left(\frac{dE}{dt} dA\right)\mathbf{n},$$

$$\mathbf{J}_N = \left(\frac{dN}{dt} dA\right)\mathbf{n},$$

where dA is the element of area; $\mathbf{n}$ is a unit vector normal to dA; and $\mathbf{J}_S$ $\mathbf{J}_E$, $\mathbf{J}_N$ are entropy, energy, and particle current densities, respectively. From Eq. (8.3), differentiating with respect to time,

$$\left(\frac{dE}{dt} dA\right)\mathbf{n} = T\left(\frac{dS}{dt} dA\right)\mathbf{n} + \bar{\mu}\left(\frac{dN}{dt} dA\right)\mathbf{n}$$

or

$$\mathbf{J}_E = T\mathbf{J}_S + \bar{\mu}\mathbf{J}_N. \tag{8.4}$$

Taking the divergence of both sides,

$$\text{div } \mathbf{J}_E = \text{div } T\mathbf{J}_S + \text{div } \bar{\mu}\mathbf{J}_N.$$

Expanding the right side,

$$\text{div } \mathbf{J}_E = T \text{ div } \mathbf{J}_S + \mathbf{J}_S \cdot \text{grad } T + \bar{\mu} \text{ div } \mathbf{J}_N + \mathbf{J}_N \cdot \text{grad } \bar{\mu}. \tag{8.5}$$

Because of the conservation of energy and conservation of particles,

$$\text{div } \mathbf{J}_E = 0,$$
$$\text{div } \mathbf{J}_N = 0.$$

Hence

$$0 = T \text{ div } \mathbf{J}_S + \mathbf{J}_S \cdot \text{grad } T + \mathbf{J}_N \cdot \text{grad } \bar{\mu},$$

$$\text{div } \mathbf{J}_S = -\frac{\mathbf{J}_S}{T} \cdot \text{grad } T - \frac{\mathbf{J}_N}{T} \cdot \text{grad } \bar{\mu},$$

or

$$\text{div } \mathbf{J}_S = + T\mathbf{J}_S \cdot \text{grad } \frac{1}{T} - \mathbf{J}_N \cdot \frac{\text{grad }(\bar{\mu})}{T}. \tag{8.6}$$

If we define the heat current density $\mathbf{J}_Q$ as $T\mathbf{J}_S$, in analogy with $dQ = T \, dS$,

$$\text{div } \mathbf{J}_S = \mathbf{J}_Q \cdot \text{grad } \frac{1}{T} - \mathbf{J}_N \cdot \frac{\text{grad } \bar{\mu}}{T}. \tag{8.7}$$

The electric current density $\mathbf{J} = e\mathbf{J}_N$, where e is the electronic charge and hence

$$\text{div } \mathbf{J}_S = \mathbf{J}_Q \cdot \text{grad } \frac{1}{T} - \frac{\mathbf{J}}{e} \cdot \frac{\text{grad } \bar{\mu}}{T}, \tag{8.8}$$

where div $\mathbf{J}_S$ is the generation rate of entropy per unit volume in the elemental volume of width dw.

Therefore, $dS_v/dt = \text{div } \mathbf{J}_S$. Hence

$$\frac{dS_v}{dt} = \mathbf{J}_Q \cdot \text{grad} \frac{1}{T} - \frac{\mathbf{J}}{e} \cdot \frac{\text{grad } \bar{\mu}}{T}. \tag{8.9}$$

It has been shown by de Groot (see the Bibliography) that the time derivative of the entropy during an irreversible process is

$$\frac{dS_v}{dt} = \sum_i J_i X_i. \tag{8.10}$$

Comparing Eqs. (8.9) and (8.10) it may be concluded that $\mathbf{J}_Q$ and grad $(1/T)$, $-\mathbf{J}/e$ and grad $\bar{\mu}/T$, form conjugate pairs of fluxes and forces. Hence we may write

$$-\frac{\mathbf{J}}{e} = L_{11} \frac{\text{grad } \bar{\mu}}{T} + L_{12} \text{ grad } \frac{1}{T}, \tag{8.11}$$

$$\mathbf{J}_Q = L_{21} \frac{\text{grad } \bar{\mu}}{T} + L_{22} \text{ grad } \frac{1}{T}. \tag{8.12}$$

Equations (8.11) and (8.12) represent the proper choices of fluxes and forces and hence the Onsager relationship, $L_{12} = L_{21}$, is satisfied. It now remains to determine the constants L_{11}, L_{12}, and L_{22} in terms of known quantities.

Electrical conductivity. σ is given by Ohm's law,

$$\mathbf{J} = \sigma \mathcal{E}$$

or

$$\sigma = \frac{\mathbf{J}}{\mathcal{E}}.$$

Because $\mathcal{E}$, the electric field, is $\mathcal{E} = -\text{grad } \vartheta$,

$$\sigma = -\frac{\mathbf{J}}{\text{grad } \vartheta}.$$

If there exist no temperature and concentration gradients, $\bar{\mu} = K + e\vartheta$ would be simply $\bar{\mu} = e\vartheta$, because $K = 0$. K is the chemical potential. For positive charge carriers, $e = 1.6 \times 10^{-19}$ C, and for negative charge carriers $e = -1.6 \times 10^{-19}$ C. Hence

$$\sigma = -\frac{\mathbf{J}}{1/e \text{ grad } \bar{\mu}}\bigg)_{\text{grad } T=0}. \tag{8.13}$$

With the definition of σ given by Eq. (8.13), L_{11} can be expressed in terms of σ from Eq. (8.11).

If grad $T = 0$, Eq. (8.11) reduces to

$$-\frac{\mathbf{J}}{e} = L_{11} \frac{\operatorname{grad} \bar{\mu}}{T}$$

or

$$-\frac{\mathbf{J}}{1/e \operatorname{grad} \bar{\mu}}\bigg)_{\operatorname{grad} T=0} = \sigma = \frac{e^2 L_{11}}{T}. \tag{8.14}$$

Hence

$$L_{11} = \frac{\sigma T}{e^2}. \tag{8.15}$$

Constant L_{11} is therefore expressed in terms of the constants σ, T, and e by Eq. (8.15).

Thermal conductivity. $\varkappa$ is defined by Fourier's law,

$$\mathbf{J}_Q = -\varkappa \operatorname{grad} T)_{\mathbf{J}=0}. \tag{8.16}$$

From Eqs. (8.11) and (8.12) the thermal conductivity $\varkappa$ may be expressed in terms of the constants L_{11}, L_{22}, and L_{12}.

If $\mathbf{J} = 0$,

$$0 = L_{11} \frac{\operatorname{grad} \bar{\mu}}{T} + L_{12} \operatorname{grad} \frac{1}{T}$$

or

$$\frac{\operatorname{grad} \bar{\mu}}{T} = -\frac{L_{12}}{L_{11}} \operatorname{grad} \frac{1}{T}. \tag{8.17}$$

Substituting Eq. (8.17) in Eq. (8.12),

$$\mathbf{J}_Q)_{\mathbf{J}=0} = -\frac{L_{12} L_{21}}{L_{11}} \operatorname{grad} \frac{1}{T} + L_{22} \operatorname{grad} \frac{1}{T},$$

or simplifying with the fact that $L_{12} = L_{21}$,

$$\mathbf{J}_Q)_{\mathbf{J}=0} = \left(L_{22} - \frac{L_{12}^2}{L_{11}}\right) \operatorname{grad} \frac{1}{T}. \tag{8.18}$$

Because

$$\varkappa = -\frac{\mathbf{J}_Q}{\operatorname{grad} T}\bigg)_{\mathbf{J}=0},$$

Eq. (8.18) gives us

$$\varkappa = \frac{L_{11} L_{22} - L_{12}^2}{L_{11} T^2}. \tag{8.19}$$

Equation (8.19) expresses constants L_{11}, L_{22}, L_{12} in terms of thermal conductivity. It will be shown below that L_{12} could be expressed in terms of the so-called *Seebeck coefficient*.

The Seebeck coefficient or thermoelectric power. The Seebeck effect was discovered in 1822. Whenever junctions of two different materials are at temperatures T_0 and a slightly higher temperature $T_0 + \Delta T$, a voltage or, more accurately, in the language of electrical engineering, an electromotive force, is developed in the circuit. The emf is known as a *Seebeck emf* and the effect is known as the *Seebeck effect*. For small temperature differences the emf developed is proportional to the temperature difference and depends upon the materials used to form the junctions. Referring to Fig. 8.2, let the two materials be A and B and the difference

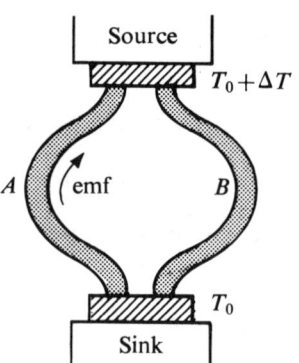

Fig. 8.2
Calculation of the Seebeck coefficient for two different materials, A and B

of temperature between the hot and cold ends be ΔT.

If V_{AB} is the voltage in the circuit,

$$V_{AB} = S_{AB}\Delta T, \tag{8.20}$$

where S_{AB} is defined as the relative Seebeck coefficient between materials A and B. If, for one pair of materials A and B, $V_{AB} = S_{AB}\Delta T$, and for a second pair of materials B and C, $V_{BC} = S_{BC}\Delta T$, the voltage in a circuit comprising A and C is

$$V_{AC} = (S_{AB}\Delta T + S_{BC}\Delta T) = (S_{AB} + S_{BC})\Delta T.$$

This additive property of the relative Seebeck coefficients is called the *law of intermediate conductors* and it leads us to the inference that

$$S_{AB} = S_A - S_B,$$

where S_A and S_B are the absolute Seebeck coefficients of materials A and B. Looking at material A, electrons near the hot junction will acquire kinetic energy and move away. The motion of electrons sets up an electrical potential gradient due to the temperature graduate. A more general definition of the Seebeck coefficient or thermoelectric power would therefore be

$$S = -\frac{\text{grad } \bar{\mu}}{e \text{ grad } T}\bigg)_{\mathbf{J}=0}. \tag{8.21}$$

Referring back to Eq. (8.11), if $\mathbf{J} = 0$,

$$0 = L_{11}\frac{\text{grad } \bar{\mu}}{T} + L_{12} \text{ grad } \frac{1}{T},$$

$$L_{11}\frac{\text{grad } \bar{\mu}}{T} = +\frac{L_{12}}{T^2} \text{ grad } T,$$

or

$$\frac{\text{grad } \bar{\mu}}{\text{grad } T} = \frac{L_{12}}{L_{11}T}.$$

Using the definition of S from Eq. (8.21),

$$S = -\frac{L_{12}}{eTL_{11}}. \tag{8.22}$$

Now, from Eq. (8.15),

$$L_{11} = \frac{\sigma T}{e^2}. \tag{8.15}$$

Therefore,

$$S = -\frac{L_{12}e}{\sigma T^2}$$

or

$$L_{12} = -\frac{\sigma T^2 S}{e}. \tag{8.23}$$

Eq. (8.11) may now be rewritten in terms of constants σ and S as

$$-\frac{\mathbf{J}}{e} = \frac{\sigma T}{e^2}\frac{\text{grad } \bar{\mu}}{T} - \frac{\sigma T^2 S}{e} \text{ grad } \frac{1}{T}, \tag{8.24}$$

or simplifying,

$$-\mathbf{J} = \frac{\sigma}{e} \text{ grad } \bar{\mu} + \sigma S \text{ grad } T. \tag{8.25}$$

Equations (8.15) and (8.22) help us determine L_{22} in terms of σ, S, and $\varkappa$. From Eq. (8.19),

$$L_{22} = \frac{L_{11}T^2\varkappa + L_{12}^2}{L_{11}},$$

$$L_{22} = T^2\varkappa + \frac{L_{12}^2}{L_{11}},$$

$$L_{22} = T^2\varkappa + \frac{\sigma^2 T^4 S^2}{e^2 \sigma T} e^2,$$

or

$$L_{22} = T^2\varkappa + \sigma T^3 S^2. \tag{8.26}$$

Incorporating Eqs. (8.23) and (8.26) in Eq. (8.12) and taking $L_{12} = L_{21}$,

$$\mathbf{J}_Q = -\frac{\sigma T^2 S}{e}\frac{\operatorname{grad}\bar{\mu}}{T} + (T^2\varkappa + \sigma T^3 S^2)\operatorname{grad}\frac{1}{T} \qquad (8.27)$$

or

$$\mathbf{J}_Q = -\frac{\sigma TS}{e}\operatorname{grad}\bar{\mu} - (\varkappa + T\sigma S^2)\operatorname{grad} T. \qquad (8.28)$$

Equations (8.24) and (8.27) represent equations of electric and thermal current densities in terms of parameters of the material.

Equation (8.25) reduces to the basic Ohm's law if grad $T = 0$ and if in the definition of $\bar{\mu}$ the chemical potential $K = 0$; i.e., there is no concentration gradient:

$$\mathbf{J} = \sigma\mathbf{E}.$$

Peltier and Thompson coefficients. In 1834 Peltier discovered what is called the *Peltier effect.* Whenever a circuit consisting of two dissimilar materials joined together carries an electric current, heat is evolved at one junction and absorbed at the other. This process is thermodynamically reversible. The rate at which heat is absorbed is proportional to the current flowing through the materials and depends upon the nature of the two materials:

$$Q_{AB} = \pi_{AB} I, \qquad (8.29)$$

where Q_{AB} is the rate at which Peltier heat is absorbed at the junction, I the current flowing through the junction, and π_{AB} the relative Peltier coefficient between materials A and B. The law of intermediate conductors holds and hence

$$Q_{AC} = Q_{AB} + Q_{BC} = (\pi_{AB} + \pi_{BC})I.$$

Hence
$$\pi_{AB} = \pi_A - \pi_B, \qquad (8.30)$$

where π_A and π_B are the absolute Peltier coefficients of materials A and B, respectively.

Thompson effect. If there is a temperature gradient along a conducting material, heat is absorbed or released in proportion to the current passing through it. This is known as the *Thompson effect.* The effect may be defined by

$$q_A = \tau_A I \frac{dT}{dx}, \qquad (8.31)$$

where q_A is the rate of heat absorption per unit length of the conductor, τ_A is the Thompson coefficient of the material, and dT/dx is the temperature gradient. Thompson's coefficient τ_A is positive if heat is absorbed when the conventional current I and dT/dx are in the same direction. The Thompson coefficient is unique in that it refers to the properties of only one material.

Kelvin's thermodynamics relationship. From the laws of thermodynamics, a relationship follows between the Seebeck coefficient S, Peltier coefficient π, and Thompson coefficient τ. Let us consider a circuit consisting of two homogeneous conductors A and B with a temperature difference dT between the two junctions. According to the first law of thermodynamics, the algebraic sum of all types of energy generated in the circuit in a unit time in steady state is zero. Let us consider a current dI flowing through the conductors. Peltier heat generated at one junction per unit time $= \pi_{ab}\, dI$. Peltier heat absorbed at the other junction per unit time $= \pi_{ba}\, dI$:

$$\pi_{ba} = \pi_{ab} + \frac{d\pi_{ab}}{dT}\, dT.$$

The total Thompson heat generated per unit time in one conductor is $\tau_a\, dT\, dI$, whereas the Thompson heat absorbed per unit time at conductor B is $\tau_b\, dT\, dI$. The Seebeck voltage is V_{ab}, and hence the power consumed in the circuit is equal to $S_a\, dT\, dI$. From the first law of thermodynamics,

$$\frac{d\pi_{ab}}{dT}\, dT\, dI + (\tau_a - \tau_b)\, dT\, dI = S_{ab}\, dT\, dI$$

or

$$\frac{d\pi_{ab}}{dT} + (\tau_a - \tau_b) = S_{ab}. \tag{8.32}$$

Equation (8.32) expresses a relationship between the coefficients α, π, and τ. Another relationship between the three coefficients may be derived from the second law of thermodynamics, because the three thermoelectric phenomena — the Seebeck, Peltier, and Thompson effects — may be regarded as reversible. Referring to Fig. 8.3, let the hot junction be at a temperature T_1 and the cold junction at a temperature T_0. Neglecting the

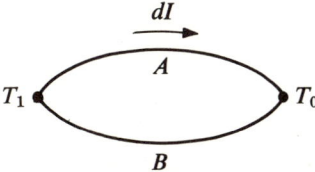

Fig. 8.3
Two materials joined together at both ends and with a temperature gradient along each material

irreversible processes, such as Joule heat, by assuming small currents, the reversibility of the process requires that the total change of entropy of all processes is zero. The hot junction receives an amount of heat $\pi_{ab_1} dI$ and the cold junction gives up an amount of heat $\pi_{ab_0} dI$. At the same time a temperature difference dT generates a heat $\tau\, dI\, dT$ in each conductor. Therefore,

$$\frac{\pi_{ab_1}}{T_1} dI - \frac{\pi_{ab_0}}{T_0} dI - dI \int_{T_0}^{T_1} \frac{\tau_b - \tau_a}{T} dT = 0. \tag{8.33}$$

Dividing by dI and differentiating Eq. (8.33) with respect to T,

$$\frac{d(\pi_{ab}/T)}{dT} - \frac{\tau_b - \tau_a}{T} = 0,$$

$$\frac{T(d\pi_{ab}/dT) - \pi_{ab}}{T^2} - \frac{\tau_b - \tau_a}{T} = 0,$$

or

$$\frac{d\pi_{ab}}{dT} - \frac{\pi_{ab}}{T} - (\tau_a - \tau_b) = 0. \tag{8.34}$$

Combining Eqs. (8.32) and (8.34),

$$S_{ab} = \frac{\pi_{ab}}{T}. \tag{8.35}$$

This is the first of Kelvin's relationships and relates the Seebeck and Peltier coefficients. All three quantities of Eq. (8.35) can be measured separately. For homogeneous material the first Kelvin relationship checks within experimental error. Differentiating Eq. (8.35) with respect to T,

$$\frac{dS_{ab}}{dT} = \frac{T(d\pi_{ab}/dT) - \pi_{ab}}{T^2} = \frac{1}{T}\left(\frac{d\pi_{ab}}{dT} - \frac{\pi_{ab}}{T}\right).$$

Replacing the quantity in parentheses by the right side of Eq. (8.34),

$$\frac{dS_{ab}}{dT} = \frac{\tau_a + \tau_b}{T}. \tag{8.36}$$

Equation (8.36) is the second Kelvin relationship. Because S_{ab} may be regarded as $S_a - S_b$,

$$\frac{d}{dT}(S_a - S_b) = -\frac{\tau_a}{T} + \frac{\tau_b}{T}$$

or

$$S_a = \int_0^T \frac{\tau_a}{T} dT \tag{8.37}$$

and

$$S_b = \int_0^T \frac{\tau_b}{T} dT. \tag{8.38}$$

IRREVERSIBLE PROCESSES; CONJUGATE FORCES AND FLUXES

The absolute Seebeck coefficient and the absolute Thompson coefficient of a material are therefore related by

$$\frac{dS}{dT} = \frac{\tau}{T}. \tag{8.39}$$

Relations (8.35) and (8.39) can be derived with the help of Eqs. (8.24) and (8.27) of conjugate forces and fluxes.

Kelvin's relationship with the Onsager theorem. The Seebeck coefficient has already been determined in terms of the constant L_{12} of Eq. (8.23). The Peltier coefficient may be more generally defined as

$$\pi = \frac{\mathbf{J}_Q}{\mathbf{J}}\bigg)_{\text{grad } T=0}. \tag{8.40}$$

Hence from Eqs. (8.24) and (8.27) with grad $T = 0$,

$$-\frac{\mathbf{J}}{e} = \frac{\sigma}{e^2} \text{ grad } \bar{\mu},$$

$$\mathbf{J}_Q = -\frac{\sigma TS}{e} \text{ grad } \bar{\mu},$$

or

$$\frac{\mathbf{J}}{\mathbf{J}_Q}\bigg)_{\text{grad } T=0} = \pi = ST,$$

which is the first Kelvin relation.

Kelvin's second relationship. Kelvin's second relationship, according to Eq. (8.39), correlates the Thompson coefficient with the Seebeck coefficient. Consider now the Thompson effect in two steps, with the help of Onsager's relation. In the first step let there be a temperature gradient along a bar (Fig. 8.1) with no current flowing in it; i.e., $\mathbf{J} = 0$. According to Eq. (8.4),

$$\mathbf{J}_E = T\mathbf{J}_S + \bar{\mu}\mathbf{J}_N. \tag{8.4}$$

For the law of conservation of energy and particles to hold,

$$\text{div } \mathbf{J}_E = 0,$$

$$\text{div } \mathbf{J}_N = 0.$$

The above conditions would be true only if

$$\text{div grad } T = 0.$$

If an electric current is made to pass through the bar as a second steps the absorption or generation of heat would change the temperature distribution. In such a case div $J_E \neq 0$ but div $\mathbf{J} = 0$ still, because of the conservation of particles. So div grad $T = 0$ still holds.

According to the second law of thermodynamics, Eq. (8.4) still holds:

$$\mathbf{J}_E = T\mathbf{J}_S + \bar{\mu}\mathbf{J}_N. \tag{8.4}$$

$\mathbf{J}_S$ may be determined in terms of $\mathbf{J}$ as follows. According to Eq. (8.28),

$$\mathbf{J}_Q = -\frac{\sigma TS}{e} \operatorname{grad} \bar{\mu} - (\varkappa + T\sigma S^2) \operatorname{grad} T. \tag{8.28}$$

Rearranging terms on the right side,

$$\mathbf{J}_Q = -TS\left(\frac{\sigma}{e} \operatorname{grad} \bar{\mu} + \sigma S \operatorname{grad} T\right) - \varkappa \operatorname{grad} T.$$

The term in parentheses is the right side of Eq. (8.25). Hence

$$\mathbf{J}_Q = TS\mathbf{J} - \varkappa \operatorname{grad} T \tag{8.41}$$

but $T\mathbf{J}_S = \mathbf{J}_Q$. Therefore, substituting Eq. (8.41) in Eq. (8.4),

$$\mathbf{J}_E = TS\mathbf{J} - \varkappa \operatorname{grad} T + \frac{\bar{\mu}}{e} \mathbf{J} \tag{8.42}$$

because $e\mathbf{J}_N = \mathbf{J}$. Taking divergence of both sides of Eq. (8.42),

$$\operatorname{div} \mathbf{J}_E = \operatorname{div} TS\mathbf{J} - \operatorname{div} \varkappa \operatorname{grad} T + \operatorname{div} \frac{\bar{\mu}}{e} \mathbf{J}. \tag{8.43}$$

Expanding,

$$\operatorname{div} \mathbf{J}_E = TS \operatorname{div} \mathbf{J} + T\mathbf{J} \cdot \operatorname{grad} S - \varkappa \operatorname{div} \operatorname{grad} T - \operatorname{grad} T \cdot \operatorname{grad} \varkappa$$
$$+ \frac{\bar{\mu}}{e} \operatorname{div} \mathbf{J} + \mathbf{J} \cdot \operatorname{grad} \frac{\bar{\mu}}{e}.$$

Because

$$\operatorname{div} \mathbf{J} = 0 \quad \text{and} \quad \operatorname{div} \operatorname{grad} T = 0,$$

$$\operatorname{div} \mathbf{J}_E = T\mathbf{J} \cdot \operatorname{grad} S + \mathbf{J} \cdot \operatorname{grad} \frac{\bar{\mu}}{e}.$$

Also, $\nabla T \cdot \nabla \varkappa = 0$, the gradient of $\varkappa$ being assumed zero.

Using the definition of conductivity,

$$\mathbf{J} \cdot \operatorname{grad} \frac{\bar{\mu}}{e} = -\frac{J^2}{\sigma},$$

and hence

$$\operatorname{div} \mathbf{J}_E = T\mathbf{J} \cdot \operatorname{grad} S - \frac{J^2}{\sigma}. \tag{8.44}$$

The second term on the right side of Eq. (8.44) is the Joule heat, and if it is small and hence neglected,

$$\operatorname{div} \mathbf{J}_E = T\mathbf{J} \cdot \operatorname{grad} S. \tag{8.45}$$

div $\mathbf{J}_E$ represents the heat absorption or rejection current density.

By definition the Thompson coefficient is defined as the Thompson heat current density absorbed or generated per unit electric current density and per unit temperature gradient. Hence

$$\tau = \frac{\text{div } \mathbf{J}_E}{\mathbf{J} \cdot \text{grad } T}.$$

Therefore, from Eq. (8.45),

$$\tau = \frac{T\mathbf{J} \cdot \text{grad } S}{\mathbf{J} \cdot \text{grad } T}.$$

In case we use grad $S = (dS/dT)$ grad T,

$$\tau = \frac{T(dS/dT)\mathbf{J} \cdot \text{grad } T}{\mathbf{J} \cdot \text{grad } T}$$

or

$$\tau = T \frac{dS}{dT},$$

which is the second Kelvin relationship.

8.2 Thermoelectric energy conversion

The thermoelectric effects described earlier may be utilized in a thermoelectric circuit to convert thermal energy to electrical energy (power generation) or to convert electrical energy to useful heating or cooling. Efficient operation of thermoelectric devices depends upon optimization of the circuit parameters and materials used in the devices.

Operation of a thermoelectric generator. Figure 8.4 shows a basic circuit for a thermoelectric power generator. There 1 and 2 are dissimilar elements, T_1 is the hot-junction temperature, and T_0 is the temperature of

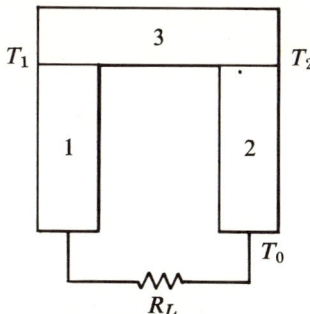

Fig. 8.4
Basic circuit for a thermoelectric power generator

the heat sink. The bridge 3 is assumed to be an electrical link with a thermoelectric coefficient much lower than that of elements 1 and 2. Let the resistance of 3 be negligible compared to 1 and 2. R_L is the load resistance. In the steady state heat is supplied at a temperature T_1, and a temperature difference $T_1 - T_0$ is maintained across elements 1 and 2. If S_{12} is the resultant Seebeck coefficient of the two elements, a voltage $S_{12}(T_1 - T_0)$ is developed across the load resulting in current I in the load resistance and the circuit. To predict the operation of the device, we have to write a heat-balance equation for the system. Let us denote the internal resistance of the two elements r_1 and r_2 and the thermal conductivities $\varkappa_1$ and $\varkappa_2$. Denoting the resistivity ρ and assuming the lengths of the two elements to be equal to l_1 and l_2 and the areas of cross section A_1 and A_2, we have

$$r = r_1 + r_2 = \left(\frac{\rho_1}{A_1} l_1 + \frac{\rho_2}{A_2} l_2\right),$$

and if $l_1 = l_2 = l$,

$$r = \left(\frac{\rho_1}{A_1} + \frac{\rho_2}{A_2}\right) l,$$

$$K' = \left(\frac{\varkappa_1 A_1}{l_1} + \frac{\varkappa_2 A_2}{l_2}\right),$$

and if $l_1 = l_2 = l$,

$$K' = (\varkappa_1 A_1 + \varkappa_2 A_2) \frac{1}{l},$$

where K' is the amount of heat conducted per second per degree difference in temperature by both arms. Taking a heat balance at T_1, we find that the heat absorbed from the source Q_a is balanced by conduction heat Q_x flowing in the arms from T_1 to T_0, and the Peltier heat Q_P developed due to the generated current flowing through the junction. Besides these values of heat accounted for, there is another heat coming to the junction due to Joule heat Q_j developed in the thermoelements. We shall assume that of half total Joule heat $I^2 r$ generated in the thermoelement passes to the hot junction and the other half to the cold junction.

The equation for heat balance is, therefore

$$Q_a + \tfrac{1}{2} Q_j = Q_p + Q_x, \qquad (8.46)$$

where

Q_a = heat absorbed from the source,

$Q_p = \pi_{12} I = S_{12} I T_1$,

$Q_j = I^2(R_1 + R_2) = I^2 \left(\frac{\rho_1 l_1}{A_1} + \frac{\rho_2 l_2}{A_2}\right) = I^2 l \left(\frac{\rho_1}{A_1} + \frac{\rho_2}{A_2}\right),$

$\dot{Q}_x = K'(T_1 - T_0) = (\varkappa_1 A_1 + \varkappa_2 A_2) \frac{1}{l} \Delta T.$

Hence

$$Q_a = S_{12}IT_1 - \frac{I^2}{2}\left(\frac{\rho_1}{A_1} + \frac{\rho_2}{A_2}\right)l + (\varkappa_1 A_1 + \varkappa_2 A_2)\frac{1}{l}\Delta T. \qquad (8.47)$$

In the above treatment, the Thompson heat is neglected. This is justified if dS/dT in the temperature range T_0 to T_1 is negligible.

The efficiency η of the heat conversion to electrical energy is defined as the ratio of the useful electrical energy $I^2 R_L$ delivered to the load resistance R_L and the heat absorbed from the source:

$$\eta = \frac{I^2 R_L}{Q_a}. \qquad (8.48)$$

The current in the circuit is

$$I = \frac{V_{12}}{r + R_L} = \frac{S_{12}\Delta T}{r + R_L}. \qquad (8.49)$$

Substituting Eqs. (8.49) and (8.47) in Eq. (8.48),

$$\eta = \frac{S_{12}^2(\Delta T)^2 R_L}{(r + R_L)^2} \bigg/ \frac{S_{12}^2 T_1 \Delta T}{r + R_L} + K'\Delta T - \frac{1}{2}\frac{S_{12}^2(\Delta T)^2 r}{(r + R_L)^2} \qquad (8.50)$$

If we substitute $m = R_L/r$ and simplify Eq. (8.50),

$$\eta = S_{12}^2 (\Delta T)^2 \frac{1}{r}\frac{m}{(m+1)^2} \bigg/ S_{12}^2 T_1 \Delta T \frac{1}{r(m+1)} + K'\Delta T - \frac{1}{2}\frac{S_{12}^2(\Delta T)^2}{r(m+1)^2}$$

$$\eta = \frac{\Delta T}{T_1}\frac{m/m+1}{1 + (K'r/S_{12}^2)(m+1/T_1) - \frac{1}{2}(\Delta T/T_1)(1/m+1)}. \qquad (8.51)$$

Equation (8.51) shows that the efficiency depends upon $\Delta T/T_1$, which is the Carnot efficiency; $K'r/S_{12}^2$, a term depending upon the properties of the material; and $m = R_L/r$.

Calling $S_{12}^2/K'r = Z$, a figure of merit of the material for thermoelectric power generation, and choosing an arbitrary value of S_{12} and m, the efficiency in Eq. (8.51) depends upon the product $K'r$. For a high value of efficiency, $K'r$ should be very small, and for optimum efficiency, $K'r$ should be minimum:

$$K'r = (\varkappa_1 A_1 + \varkappa_2 A_2)\left(\frac{\rho_1}{A_1} + \frac{\rho_2}{A_2}\right) = \varkappa_1\rho_1 + \varkappa_2\rho_2 + \varkappa_1\rho_2\frac{A_1}{A_2} + \varkappa_2\rho_1\frac{A_2}{A_1}. \qquad (8.52)$$

To find a minimum value for $K'r$, the derivative of Eq. (8.52) with respect to A_1/A_2 should equal zero, or

$$\varkappa_1\rho_2 = \varkappa_2\rho_1\left(\frac{A_2}{A_1}\right)^2$$

or

$$\frac{\rho_1}{\varkappa_1}\frac{\varkappa_2}{\rho_2} = \left(\frac{A_1}{A_2}\right)^2,$$

and therefore the minimum value of $K'r = [(\varkappa_1\rho_1)^{1/2} + (\varkappa_2\rho_2)^{1/2}]^2$.

Hence the figure of merit for optimum dimensions of the elements is

$$Z = \frac{S_{12}^2}{K'r_{min}}$$

or

$$Z = \frac{S_{12}^2}{[(\varkappa_1\rho_1)^{1/2} + (\varkappa_2\rho_2)^{1/2}]^2} \text{ per degree.} \tag{8.53}$$

Incorporating Eq. (8.53) for the term $K'r/S_{12}^2$,

$$\eta = \frac{\Delta T}{T_1} \frac{m/m + 1}{1 + (m + 1/ZT_1) - [\Delta T/2(m + 1)T_1]}. \tag{8.54}$$

Equation (8.54) gives the efficiency of the thermoelectric generator in terms of the properties of the thermoelectric material Z and is independent of the dimensions of the material. For maximum power transfer from the elements to the load, the resistance of the thermoelements r should be equated to the load resistance R_L. For this case one half of the voltage appears across the load resistance.

In such a case $r/R_L = 1$ and $m = 1$, is Eq. (8.54) may be rewritten

$$\eta = \frac{\Delta T}{T_1} \frac{1}{2} \left(1 + \frac{2}{ZT_1} - \frac{\Delta T}{4T_1}\right)^{-1}. \tag{8.55}$$

Simplifying,

$$\eta = \frac{\Delta T}{2T_1 + (4/Z) - (\Delta T/2)}. \tag{8.56}$$

Equation (8.56) shows that the efficiency of the thermoelectric generator depends only upon the temperature T_1 and T_0 and the figure of merit. For optimum operation of the generator, the figure of merit should be high. Looking at the definition of figure of merit, the thermoelement should have a relatively high Seebeck coefficient, low electrical resistivity, and low thermal conductivity.

In deriving Eq. (8.56) we assumed operation of the thermoelectric generator as a battery and derived the efficiency for maximum power output conditions. This may not be the maximum possible efficiency. As such, the maximum efficiency may be found if the value of m corresponding to maximum efficiency is known. This can be done by calculating $d\eta/dm$ and equating it zero or

$$\frac{d\eta}{dm} = \frac{d}{dm}\left[\frac{\Delta T}{T_1} \frac{m/m + 1}{1 + (m + 1/T_1Z) - (\Delta T/T_1)\frac{1}{2}(1/m + 1)}\right] = 0$$

or

$$\frac{d}{dm}\left[\frac{\Delta T}{T_1} \frac{m}{(1 + m) + [(m + 1)^2/ZT_1] - (\Delta T/2T_1)}\right] = 0.$$

THERMOELECTRIC ENERGY CONVERSION

Because $\Delta T/T_1 \neq 0$,

$$(1 + m) + \frac{(m+1)^2}{ZT_1} - \frac{\Delta T}{2T_1} - m\left[1 + \frac{2(m+1)}{ZT_1}\right] = 0$$

or

$$1 + \frac{m^2 + 1 + 2m}{ZT_1} - \frac{2m^2 + 2m}{ZT_1} - \frac{\Delta T}{2T_1} = 0,$$

$$1 - \frac{m^2}{ZT_1} + \frac{1}{ZT_1} - \frac{\Delta T}{2T_1} = 0,$$

or

$$\frac{m^2}{ZT_1} = 1 + \frac{1}{ZT_1} - \frac{\Delta T}{2T_1},$$

$$m^2 = ZT_1 + 1 - \frac{\Delta T Z}{2}$$

or

$$m^2 = 1 + \frac{2ZT_1 - \Delta T Z}{2},$$

or

$$m = \left(1 + \frac{T_1 + T_0}{2} Z\right)^{1/2}. \tag{8.57}$$

The optimum ratio of R_L/r for maximum efficiency is given by Eq. (8.57). Let us call this value $R_L/r = M$. Therefore,

$$M = \left(1 + Z\frac{T_1 + T_0}{2}\right)^{1/2}. \tag{8.57a}$$

Replacing m by M in Eq. (8.54),

$$\eta = \frac{\Delta T}{T_1} \frac{M/1 + M}{1 + (M + 1/ZT_1) - \frac{1}{2}(\Delta T/T_1)(1/M + 1)}.$$

Simplifying,

$$\eta = \frac{\Delta T}{T_1} \frac{M}{(1 + M) + [(M+1)^2/ZT_1] - (\Delta T/2T_1)}.$$

Multiplying numerator and denominator of the right side of the above equation by $M - 1$,

$$\eta = \frac{\Delta T}{T_1}(M - 1)\frac{M}{M^2 - 1 + [(M^2 - 1)(M + 1)/T_1 Z] - (\Delta T/2T_1)(M - 1)}.$$

Simplifying further and using Eq. (8.57a),

$$\eta = \frac{\Delta T}{T_1}\frac{M - 1}{M + (T_0/T_1)}. \tag{8.58}$$

The first term on the right side represents the thermodynamic efficiency and η_c is called the *Carnot efficiency of a reversible engine.* The second term describes the reduction in efficiency due to irreversible losses. For large values of efficiency, M should be large in comparison with unity. If $M \gg 1$, the efficiency of the generator approaches the Carnot efficiency. A large value of M means a large value of $Z(T_1 + T_0/2)$ should be attainable for the given material. Once more we conclude that a large value of Z results in a large value of the efficiency of power conversion of the generator. From Eq. (8.53),

$$Z = \frac{S_{12}^2}{[(\varkappa_1\rho_1)^{1/2} + (\varkappa_2\rho_2)^{1/2}]^2} \text{ per degree.} \qquad (8.53)$$

We conclude that for a high value of Z, the thermoelectric power of the materials should be high, the electrical conductivity high, and the thermal conductivity low. For a single material one could easily define a figure of merit

$$Z_1 = \frac{S_1^2}{\rho_1 \varkappa_1}, \qquad (8.59)$$

where the subscript 1 refers to a single material.

If ρ_1, $\varkappa_1$, and S_1 are independent of each other, it appears that in most desirable materials for thermoelectric power generation the Seebeck coefficient should be as large as possible. From solutions of the Boltzmann transport equation we shall show that the Seebeck coefficient is dependent upon mobility and concentration of charge carriers. It was shown in Chapters 3 and 4 that thermal conductivity and electrical conductivity also depend upon concentration and mobility of charge carriers. It is, therefore, necessary to deduce an optimum value of the Seebeck coefficient, corresponding to a maximum value of the figure of merit.

Metals have low figures of merit because of extremely low values of Seebeck coefficients. Semiconductors have large values of Seebeck coefficients. Semiconductors have another advantage in that elements A_1 and A_2 of the thermoelectric material could be the same material, one with an excess of electrons as charge carriers and the other with an excess of holes as charge carriers. The Seebeck coefficient S_{12} in such cases is

$$S_{12} = S_1 + S_2.$$

To date the best material for the thermoelectric power generation is PbTe with $Z = 0.5 \times 10^{-3}°\text{K}^{-1}$.

8.3 Galvanomagnetic coefficients

Galvanomagnetic effects occur when a current-carrying conductor is placed in a magnetic field. In the absence of the magnetic field, the electrical properties of a homogeneous isotropic material are described by the electric conductivity σ. According to Ohm's law,

$$J_x = \sigma \mathcal{E}_x,$$

where J_x is the current density due to an applied field $\mathcal{E}_x$. While discussing the free-electron theory of metals in Sec. 4.4, we postulated that electrons moving in a solid as a result of the applied electric field must be colliding with the positive ions or lattice. The average velocity of the electrons in the steady state due to an applied field was termed the *drift velocity*:

$$J_x = ne\langle v_x \rangle_\infty,$$

where $\langle v_x \rangle_\infty$ in Sec. 4.4 was defined as the drift velocity. It is simple to denote the drift velocity in the x axis as v_{d_x}. Hence

$$J_x = ne v_{d_x},$$

where, as discussed in Sec. 4.4,

$$v_{d_x} = \frac{e\tau}{m} \mathcal{E}_x$$

and τ is the average time between two consecutive collisions of an electron. The mobility of electrons was defined as the electron drift velocity per unit electric field. This definition of mobility requires drifting of electrons due to an applied field, so it may be appropriate to call this mobility the drift mobility μ_d:

$$\mu_d = \frac{v_{d_x}}{\mathcal{E}_x} = \frac{e\tau}{m}$$

Hence $J_x = ne\mu_d \mathcal{E}_x$ and

$$\sigma = ne\mu_d \qquad (8.60a)$$

or

$$\sigma = \frac{ne^2\tau}{m}. \qquad (8.60b)$$

In the derivation of Eq. (8.60) it was assumed that the time between two consecutive collisions on the average is the same for all the free electrons n in a unit volume. This is hardly true; therefore, a more accurate expression for σ should consider the fact that different electrons have different collision times. Let n_i electrons have a collision time τ_i on the average. The product of all number of electrons n_i and their corresponding collision times on the average τ_i in a unit volume is $\sum_i n_i \tau_i$.

By definition, $\sum_i n_i \tau_i / n$ is the average collision time $\langle \tau \rangle$ of the assembly of n electrons. It is the arithmetic average of the different collision times:

$$\langle \tau \rangle = \frac{\sum_i n_i \tau_i}{n}.$$

Hence $\sigma = (e^2/m)\sum_i n_i\tau_i$ may be rewritten

$$\sigma = \frac{ne^2}{m}\langle\tau\rangle. \tag{8.61}$$

But $\sigma = ne\mu_d$. Therefore,

$$\mu_d = \frac{e}{m}\langle\tau\rangle. \tag{8.62}$$

Equation (8.62) is a more exact expression for the drift mobility of the charge carriers.

Hall effect. If a current-carrying conductor is placed in a magnetic field, a transverse effect, discovered by Hall in 1879, is noted. This effect is called the *Hall effect.* Hall found that when a magnetic field is applied at right angles to the direction of the electric current, an electric field is set up which is perpendicular to both the direction of electric current and the applied magnetic field. It has already been discussed that the flow of electric current in a conductor is the motion of a stream of electrons with drift velocity. It is known from the Lorentz force equation that a charged particle of charge e moving with a velocity v will experience a force **F** in a magnetic field **B** given by

$$\mathbf{F} = e(\mathbf{v} \times \mathbf{B}) \tag{8.63}$$

where **B** is the vector magnetic flux density.

Let us consider a rectangular piece of a conductor which is subjected to an electric field in the x direction $\mathcal{E}_x$. As a result of the applied field there is a conventional current flow in the positive direction of x given by $\mathbf{J}_x$ (Fig. 8.5). The electrons move in the negative direction of x with a drift velocity $-v_{d_x}$. Let us further assume that all the n electrons per unit volume have the same drift velocity. The charge on the electron is $-e$, where $e = 1.6 \times 10^{-19}$ C. Hence, if the magnetic field is in the z direction with magnitude B_z,

$$\mathbf{B} = B_z \mathbf{a}_z,$$

$$\mathbf{v} = -v_{d_x}\mathbf{a}_x.$$

Therefore,

$$\mathbf{F} = -e(-v_{d_x}\mathbf{a}_x \times B_z\mathbf{a}_z) = -ev_{d_x}B_z\mathbf{a}_y, \tag{8.64}$$

where $\mathbf{a}_x$, $\mathbf{a}_y$, and $\mathbf{a}_z$ are unit vectors in the x, y, and z directions, respectively.

The force experienced by the electrons is in the negative y direction. The electrons are deflected to the left-hand surface in Fig. 8.5 and make the surface negatively charged with respect to the right-hand surface. This charge will create an electric field that will counteract the Lorentz force by exerting a force in the opposite direction. Let $\mathcal{E}_y$ be the electric field set up due to charge accumulation. In the steady state or equilibrium condition,

$$-e\mathcal{E}_y = -(-ev_{d_x}B_z).$$

The minus sign outside the parentheses is used because the two forces are in opposite directions. This becomes

$$\mathcal{E}_y = -v_{d_x}B_z\mathbf{a}_y.$$

$\mathcal{E}_y$ is in the negative y direction. Because $J_x = (-e)nv_{d_x}$,

$$\mathcal{E}_y = \frac{J_x B_z}{ne}.$$

The Hall coefficient R_H is defined as the Hall electric field created per unit electric current density per unit magnetic flux density:

$$R_H = \frac{\mathcal{E}_y}{J_x B_z} = \frac{1}{ne}. \tag{8.65}$$

Hall angle and Hall mobility. It is known from electron ballistics that the electrons in Fig. 8.5 will take a circular path in the xy plane as a result of the magnetic flux density applied in the z direction. The cyclotron frequency, i.e., the angular velocity ω, in such a case is

$$\omega = \frac{eB_z}{m},$$

where m is the mass of charge e. The electron is deflected through a certain

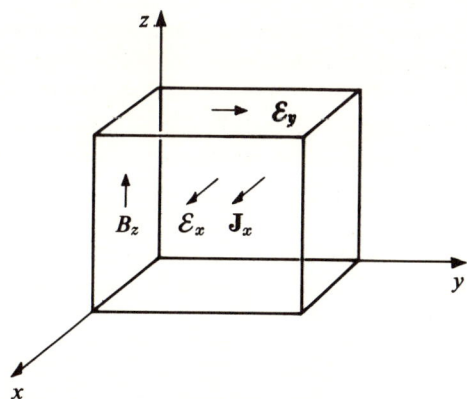

Fig. 8.5
Hall effect

angle θ before it collides with a lattice. The angle θ depends upon the collision time τ of the electrons. θ is called the *Hall angle*:

$$\theta = \omega\tau = \frac{eB_z}{m}\tau. \tag{8.66}$$

The Hall angle is also given by $\mathcal{E}_y/\mathcal{E}_x$. This can be proved as follows:

$$\mathcal{E}_y = \frac{J_x B_z}{ne} = \frac{\sigma \mathcal{E}_x B_z}{ne} = \frac{ne^2\tau}{m}\frac{1}{ne}\mathcal{E}_x B_z = \mathcal{E}_x \frac{eB_z\tau}{m}$$

or

$$\frac{\mathcal{E}_y}{\mathcal{E}_x} = \frac{eB_z}{m}\tau = \theta.$$

The Hall angle depends upon τ, the collision time, and is a property of the material. From Eq. (8.66),

$$\frac{\theta}{B_z} = \frac{e\tau}{m}. \tag{8.67}$$

Equation (8.67) has the dimensions of mobility and is called *Hall mobility* μ_H. Hall mobility is defined as the Hall angle per unit magnetic field:

$$\mu_H = \frac{\theta}{B} = \frac{e\tau}{m}. \tag{8.68}$$

It may easily be seen that the Hall mobility may also be expressed as the product of Hall coefficient and electrical conductivity:

$$\mu_H = R_H \sigma. \tag{8.69}$$

The above treatment, leading to Hall coefficient, Hall angle, and Hall mobility, has been carried out under the assumption that all the electrons have the same drift velocity and hence the same collision time. This may not be the case, however. It is this fact that leads to a value of Hall mobility different from drift mobility.

Let us now consider the case when all the electrons moving as a result of the applied electric field do not all have the same drift velocity and the same collision time τ on the average. Let n_i electrons have velocity component v_{x_i} in the x direction and let the collision time of these electrons be τ_i.

The magnitude of the Lorentz force $= B_z e v_{x_i}$. This force causes the electron to move in the y axis. Let the associated y component of the velocity be v_{y_i}. For equilibrium

$$B_z e v_{x_i} = \frac{m v_{yi}}{\tau_i}$$

or

$$v_{y_i} = \frac{eB_z}{m}\tau_i v_{x_i}.$$

GALVANOMAGNETIC COEFFICIENTS

Therefore, multiplying both sides by $n_i e$,

$$n_i e v_{y_i} = \frac{eB_z}{m} v_{x_i} n_i e \tau_i. \qquad (8.70)$$

The left side of Eq. (8.70) is the current density in the y axis due to the Hall effect. If $\mathcal{E}_y$ is the Hall field and σ_i is the conductivity of this group of electrons,

$$n_i e v_{y_i} = \sigma_i \mathcal{E}_y. \qquad (8.71)$$

Further, v_{x_i} is due to the applied electric field $\mathcal{E}_x$ and

$$\frac{m v_{x_i}}{\tau_i} = e \mathcal{E}_x.$$

Therefore,

$$v_{x_i} = \frac{e \tau_i}{m} \mathcal{E}_x. \qquad (8.72)$$

Substituting the value of $n_i e v_{y_i}$ from Eq. (8.71) and v_{x_i} from Eq. (8.72) in Eq. (8.70),

$$\sigma_i \mathcal{E}_y = \frac{eB_z}{m} \frac{e^2}{m} n_i \tau_i^2 \mathcal{E}_x.$$

Summing both sides would give us the contribution of all the i group of electrons,

$$\sum_i \sigma_i \mathcal{E}_y = \frac{eB_z}{m} \frac{e^2}{m} \mathcal{E}_x \sum_i n_i \tau_i^2.$$

Defining $\langle \tau^2 \rangle = \sum_i n_i \tau_i^2 / n$, $\langle \tau^2 \rangle$ is the average of the square of the collision time, and using $\sum_i \sigma_i = \sigma$,

$$\sigma \mathcal{E}_y = \frac{eB_z}{m} \frac{e^2}{m} \mathcal{E}_x n \langle \tau^2 \rangle$$

or

$$\mathcal{E}_y = \frac{eB_z}{m} \frac{ne^2}{m} \frac{\mathcal{E}_x}{\sigma} \langle \tau^2 \rangle.$$

Also $\mathcal{E}_x = J_x / \sigma$. Hence

$$\mathcal{E}_y = \frac{eB_z}{m} \frac{ne^2}{m} J_x \frac{\langle \tau^2 \rangle}{\sigma^2}. \qquad (8.73)$$

It was shown earlier that

$$\sigma = \frac{ne^2}{m} \langle \tau \rangle. \qquad (8.61)$$

Using this equation in Eq. (8.73),

$$\mathcal{E}_y = \frac{eB_z}{m} J_x \frac{ne^2}{m} \frac{m^2}{n^2 e^4} \frac{\langle \tau^2 \rangle}{|\langle \tau \rangle|^2}.$$

Simplifying,

$$\mathcal{E}_y = \frac{1}{ne} B_z J_x \frac{\langle \tau^2 \rangle}{|\langle \tau \rangle|^2}. \tag{8.74}$$

The Hall coefficient by definition is

$$R_H = \frac{\mathcal{E}_y}{B_z J_x} = \frac{1}{ne} \frac{\langle \tau^2 \rangle}{|\langle \tau \rangle|^2}. \tag{8.75}$$

Comparing Eqs. (8.75) and (8.65) we find that the relationship $R_H = 1/ne$ should be modified by the term $\langle \tau^2 \rangle / |\langle \tau \rangle|^2$ to account for the collision time being different for different electrons. It can be worked out with the help of the Boltzmann transport equation that for lattice scattering due to the vibration of atoms,

$$\frac{\langle \tau^2 \rangle}{|\langle \tau \rangle|^2} = \frac{3\pi}{8}. \quad \text{(See Prob. 1, page 352.)}$$

For other types of scattering $\langle \tau^2 \rangle / |\langle \tau \rangle|^2$ has different values.

The Hall mobility

$$\mu_H = R_H \sigma. \tag{8.69}$$

Using the values of R_H and σ from Eqs. (8.75) and (8.61),

$$\mu_H = \frac{e}{m} \frac{\langle \tau^2 \rangle}{\langle \tau \rangle}. \tag{8.76}$$

The ratio of Hall mobility to drift mobility is [Eqs. (8.76) and (8.62)],

$$\frac{\mu_H}{\mu_d} = \frac{\langle \tau^2 \rangle}{|\langle \tau \rangle|^2}. \tag{8.77}$$

It may sometimes be useful to write the Hall coefficient

$$R_H = \frac{\mu_H}{\mu_d} \frac{1}{ne}. \tag{8.78}$$

Hall effect in semiconductors. In a semiconductor both types of charge carriers (electrons and holes) contribute to electrical conduction. The electrical conductivity of a semiconductor was discussed earlier and is given by

$$\sigma = pe\mu_h + ne\mu_e, \tag{4.52}$$

where μ_h and μ_e are the drift mobility of holes and electrons, respectively. The Hall effects must also be modified. Referring back to Fig. 8.5, let a rectangular piece of semiconductor be subjected to an electric field $\mathcal{E}$

$= \mathcal{E}_x \mathbf{a}_x$. Both electrons and holes will drift. The drift velocities of holes and electrons are

$$\mathbf{v}_e = -v_{d x_e} \mathbf{a}_x,$$
$$\mathbf{v}_h = v_{d x_h} \mathbf{a}_x.$$

where

$$v_{d x_e} = \frac{\tau_e e \mathcal{E}_x}{m},$$

$$v_{d x_h} = \frac{\tau_h e \mathcal{E}_x}{m}.$$

τ_e and τ_h are the collision times of electrons and holes, respectively. By definition, the mobility (drift) of electrons and holes is

$$\mu_{d_e} = \frac{e \tau_e}{m},$$

$$u_{d_h} = \frac{e \tau_h}{m}.$$

Both electrons and holes experience Lorentz forces in a magnetic field,

$$\mathbf{B} = B_z \mathbf{a}_z,$$
$$\mathbf{F}_h = e(\mathbf{v}_h \times \mathbf{B}) = e v_{d x_h} B_z \mathbf{a}_x \times \mathbf{a}_z = -e v_{d x_h} B_z \mathbf{a}_y,$$
$$\mathbf{F}_e = -e(\mathbf{v}_e \times \mathbf{B}) = e v_{d x_e} B_z \mathbf{a}_x \times \mathbf{a}_z = -e v_{d x_e} B_z \mathbf{a}_y.$$

The force experienced by electrons and holes is in the same direction and both of them are deflected to the same surface. The net charge accumulation is on the left surface of Fig. 8.5. The resultant charge after recombination of electrons and holes at the surface causes the electric field in the y direction. In equilibrium the current due to the Hall field is equal and opposite to the direction due to the charge carriers deflected. The current due to the Hall field $= \sigma \mathcal{E}_y$.

The deflection current is due to both holes and electrons in the same direction and hence equals $(e p v_{y_h} - e n v_{y_e})$, where v_{y_h} is the y component of the velocity of holes due to deflection by the magnetic field. Similarly, v_{y_e} is the y component of the velocity of electrons due to deflection by the magnetic field.

In equilibrium,

$$\sigma \mathcal{E}_y = e p v_{y_h} - e n v_{y_e}. \tag{8.79}$$

v_{y_h} and v_{y_e} are related to $v_{d x_h}$ and $v_{d x_e}$ by

$$e v_{d x_h} B_z = \frac{m v_{y_h}}{\tau_h}, \tag{8.80}$$

$$e v_{d x_e} B_z = \frac{m v_{y_e}}{\tau_e}. \tag{8.81}$$

Substituting the values of v_{y_h} and v_{y_e} from Eqs. (8.80) and (8.81) in Eq. (8.79) and using the definitions of mobilities μ_{d_e} and μ_{d_h},

$$\sigma \mathcal{E}_y = e p \mu_{d_h} v_{d_{zh}} B_z - e n \mu_{d_e} v_{d_{ze}} B_z. \tag{8.82}$$

Replacing $v_{d_{zh}}$ by $\mu_{d_h} \mathcal{E}_x$ and $v_{d_{ze}}$ by $\mu_{d_e} \mathcal{E}_x$, Eq. (8.62) may be rewritten

$$\sigma \mathcal{E}_y = (e p \mu_{d_h}^2 - e n \mu_{d_e}^2) \mathcal{E}_x B_z.$$

But

$$\mathcal{E}_x = \frac{J_x}{\sigma} \quad \text{Ohm's law;}$$

hence

$$\mathcal{E}_y = \frac{e(p \mu_{d_h}^2 - n \mu_{d_e}^2)}{\sigma^2} J_x B_z$$

or

$$\mathcal{E}_y = \frac{p \mu_{d_h}^2 - n \mu_{d_e}^2}{e(p \mu_{d_h} + n \mu_{d_e})^2} J_x B_z. \tag{8.83}$$

Using the definition of the Hall coefficient,

$$R_H = \frac{\mathcal{E}_y}{J_x B_z} = \frac{p \mu_{d_h}^2 - n \mu_{d_e}^2}{e(p \mu_{d_h} + n \mu_{d_e})^2}. \tag{8.84}$$

Equation (8.84) has been derived with the assumption that all the electrons have the same collision time τ_e and all the holes the same collision time τ_h. However, if different groups of charge carriers have different collision times and we assume lattice scattering due to atomic vibration, Eq. (8.84) has to be multiplied by $3\pi/8$ on the right side and

$$R_H = \frac{3\pi}{8} \frac{p \mu_{d_h}^2 - n \mu_{d_e}^2}{e(p \mu_{d_h} + n \mu_{d_e})^2}. \tag{8.85}$$

In Eq. (8.85) if the charge carriers are predominantly of one type, say electrons, $n \gg p$ and also $\mu_{d_h} > \mu_{d_e}$:

$$R_H = -\frac{3\pi}{8} \frac{1}{ne},$$

which has the same magnitude as Eq. (8.65). Expressing the ratio of the drift mobilities $\mu_{d_e}/\mu_{d_h} = b$, Eq. (8.85) may be written

$$R_H = \frac{3\pi}{8} \frac{p - nb^2}{(p + nb)^2 e}. \tag{8.86}$$

The numerator in Eq. (8.86) tells us that the Hall voltage and Hall coefficient is generally smaller for intrinsic specimens than for extrinsic semiconductors. The sign of the Hall constant depends upon the relative mobilities of holes and electrons. The sign of the Hall voltage may change with temperature. This may occur in P-type material at the onset of intrinsic conductivity if b is greater than 1.

The Hall coefficient or constant $R_H = 0$ if

$$p = nb^2. \qquad (8.87)$$

The Hall effect is used to determine the concentration of charge carriers.

Magnetoresistance. The change of resistance in a magnetic field is called the *magnetoresistance effect.* There is an increase in resistance of a conductor when a magnetic field is applied. While discussing the Hall effect we saw that charge carriers are deflected as a result of the application of a magnetic field at right angles to the flow of charge carriers. If all the charge carriers have the same drift velocity, the Hall electric field exerts an equal and opposite force on the charge carriers, and in an equilibrium condition no more charge carriers are deflected. There will be no component of motion of electrons or, in general, charge carriers in the direction of the Hall field, and hence there is no change in conductivity or resistivity of the material expected in the presence of the magnetic field. However, if there is a distribution of velocities, only those charge carriers whose velocity is precisely the value that satisfies the condition of equal and opposite force by the Hall field will be undeflected. Deflection of those charge carriers which have larger or smaller values of velocity than this precise value will occur. The deflection reduces the component of motion in the direction of the applied electric field, and therefore the resistance of the material will be higher.

If ρ_0 is the resistivity with no applied magnetic field and ρ is the resistivity in the presence of a magnetic field, the magnetoresistance coefficient M is defined as the change in resistivity per unit of no field resistivity:

$$M = \frac{\rho - \rho_0}{\rho_0} = \frac{\Delta \rho}{\rho_0}. \qquad (8.88)$$

Because M arises due to the fact that all the charge carriers do not have the same drift velocity, we shall compute M by considering a group of charge carriers n_i having drift velocities v_{x_i}. The component of velocity in the y direction v_{y_i} is given by

$$B_z e v_{x_i} = \frac{m v_{y_i}}{\tau_i},$$

$$v_{y_i} = \frac{eB_z}{m} \tau_i v_{x_i}.$$

eB_z/m is the cyclotron frequency ω and hence

$$v_{y_i} = \omega \tau_i v_{x_i}.$$

Multiplying both sides of the above equation by $n_i e$,

$$n_i e v_{y_i} = \omega \tau_i n_i e v_{x_i}$$

or

$$J_{y_i} = \omega \tau_i J_{x_i}.$$

The y component of the current due to these n_i charge carriers is related to the x component by $\omega \tau_i$. The total current density due to n_i electrons is

$$J_i = (J^2{}_{x_i} + J^2{}_{y_i}) = J_{x_i}(1 + \tau_i^2 \omega^2)^{1/2}.$$

If $\omega \tau_i \ll 1$, assuming the Hall angle is small,

$$(1 + \tau_i^2 \omega^2)^{1/2} \simeq 1 + \frac{\tau_i^2 \omega^2}{2}$$

or

$$J_i \simeq J_{x_i}\left(1 + \frac{\tau_i^2 \omega^2}{2}\right).$$

The current density without the applied magnetic field is J_{x_i} and the current density with the applied field is J_i. $M = (\rho - \rho_0)/\rho_0$ may be rewritten for the v_i carriers as

$$M_i \simeq \frac{J_{x_i} - J_i}{J_{x_i}} = -\frac{J_{x_i} \omega^2 \tau_i^2}{2 J_{x_i}}. \tag{8.89}$$

Because

$$J_{x_i} = \sigma_i \mathcal{E}_x = \frac{n_i e^2 \tau_i}{m} \mathcal{E}_x,$$

$$M_i \simeq -\frac{1}{2} \frac{(n_i e^2 \tau_i/m) \omega^2 \tau_i^2 \mathcal{E}_x}{(n_i e^2 \tau_i/m) \mathcal{E}_x},$$

where $\mathcal{E}_x$ is the applied electric field and $\omega^2 = e^2 B_z^2/m^2$. Therefore,

$$M_i \simeq -\frac{1}{2} \frac{(n_i e^2/m) \tau_i^3 (e^2 B_z^2/m^2)}{n_i \tau_i (e^2/m)}.$$

The total magnetoresistance coefficient is

$$M \simeq \left(-\frac{1}{2} \frac{\sum_i n_i \tau_i^3}{\sum_i n_i \tau_i}\right) \frac{e^2 B_z^2}{m^2}.$$

Defining a new average

$$\langle \tau^3 \rangle = \frac{\sum_i n_i \tau_i^3}{n},$$

$$M \simeq -\frac{1}{2} \frac{e^2 B_z^2}{m^2} \frac{\langle \tau^3 \rangle}{\langle \tau \rangle}. \tag{8.90}$$

The magnetoresistance mobility μ_m is defined as the square root of magnetoresistance coefficient per unit magnetic field:

$$\mu_m = \frac{M^{1/2}}{B}$$

or

$$\mu_m \simeq \frac{e}{m} \left|\frac{\langle \tau^3 \rangle}{\langle \tau \rangle}\right|^{1/2}. \qquad (8.91)$$

It may be easy to check that drift mobility, Hall mobility, and magnetoresistance mobilities have the same direction. The difference in the three mobilities is the way collision time has been averaged:

$$\frac{\mu_m}{\mu_d} \simeq \left|\frac{\langle \tau^3 \rangle}{\langle \tau \rangle}\right|^{1/2} \frac{1}{\langle \tau \rangle} = \frac{|\langle \tau^3 \rangle|^{1/2}}{|\langle \tau \rangle|^{3/2}}. \qquad (8.92)$$

8.4 Electrical and heat current densities; thermoelectric power

In this section we shall develop expressions for the electrical current density, heat current density, and thermoelectric power using the free-electron theory. The free-electron theory, which was introduced earlier when discussing metals and semiconductors, assumes the electrons to be free particles, as molecules are in a gas. The theory is made more realistic in the case of solids by taking Fermi-Dirac statistics in place of Maxwell-Boltzmann statistics, the latter being used in gases. Further, the Pauli exclusion principle and the dual nature of the charge carrier is taken into account in considering the density of states. The fact that charge carriers move in a periodic potential in a solid is taken into account by replacing the mass of the free charge carriers by the effective mass m^*. It is assumed that the solids are isotropic and have one type of charge carrier.

Relationship between concentration of charge carriers and Fermi level. In a free carrier gas the number of states between energy E and $E + dE$ was defined as $g(E)\, dE$ and is given by Eq. (4.58), derived earlier:

$$g(E)\, dE = (2m^*)^{3/2} \frac{4\pi}{h^3} E^{1/2}\, dE, \qquad (4.58)$$

where m^* is the effective mass of the carrier. The probability of occupation of a state of energy E according to Fermi-Dirac statistics is given by Eq. (2.33):

$$p(E) = \left[\exp\left(\frac{E - E_f}{kT}\right) + 1\right]^{-1}, \qquad (2.33)$$

where E_f is the Fermi energy level.

At temperatures which are not very high it is usually found that $E - E_f \geq 4kT$, or assuming nondegenerate conditions we have

$$p(E) = \exp\left(\frac{E_f - E}{kT}\right).$$

Hence the number of charge carriers dn between energy level E and $E + dE$ is given by the product of the density of available states [Eq. (4.58)] and the probability of occupation of a state [Eq. (2.33)]:

$$dn = g(E)\,dE\,p(E)$$

or

$$dn = (2m^*)^{3/2}\frac{4\pi}{h^3}E^{1/2}e^{E_f/kT}e^{-E/kT}\,dE. \tag{8.93}$$

The total number of charge carriers is obtained by integrating the right side of Eq. (8.93) over all possible values of energy from $E = 0$ to $E \to \infty$. Hence

$$n = \int dn = \int_0^\infty (2m^*)^{3/2}\frac{4\pi}{h^3}e^{E_f/kT}E^{1/2}e^{-E/kT}\,dE,$$

$$n = \frac{4\pi(2m_e^*)^{3/2}}{h^3}e^{E_f/kT}\int_0^\infty e^{-E/kT}E^{1/2}\,dE,$$

or

$$= \frac{2(2\pi m^*kT)^{3/2}}{h^3}e^{E_f/kT}.$$

Hence the Fermi energy level is

$$E_f = kT\ln\frac{nh^3}{2(2\pi m^*kT)^{3/2}}. \tag{8.94}$$

The charge-carrier concentration can be determined by a measurement of the Hall coefficient [Eq. (8.65)], so Eq. (8.94) may be used to determine the Fermi energy level.

Electrical conductivity. We have discussed earlier the fact that if the solid is subjected to an electric field, the charge carriers have a drift velocity due to the applied electric field. The energy of the electron changes, the density of states between energy E and $E + dE$ remains the same, but $p(E)$, the probability of occupation of a state of energy E, changes to a new value $p'(E)$. The distribution of charge carriers is therefore obtained by $p'(E)$. If the applied electric field is removed, $p'(E)$ reverts back to $p(E)$. The rate at which the field distribution reverts back is

$$\frac{dp}{dt} = \frac{p'(E) - p(E)}{\tau},$$

CURRENT DENSITIES; THERMOELECTRIC POWER

where it is assumed that $p'(E) - p(E)$ is small and τ is the relaxation-time characteristic of the system, which is a function of the scattering mechanism bringing the system back to equilibrium.

If a charge carrier (an electron) moves for a time dt without scattering, its drift velocity dv_d is given by

$$dv_d = -\frac{e\mathcal{E}}{m^*}dt,$$

where $\mathcal{E}$ is the applied field and $-e$ the charge of the electron. During its drift the electron is scattered by interactions with imperfections or lattice vibrations. The scattering processes tend to restore equilibrium.

In equilibrium or steady state,

$$\left.\frac{dp}{dt}\right|_{\text{drift}} + \left.\frac{dp}{dt}\right|_{\text{scattering}} = 0. \tag{8.95}$$

The first term of Eq. (8.95) represents changes in $p(E)$ as a result of change in drift velocity and may be written

$$\left(\frac{dp}{dt}\right)_{\text{drift}} = \frac{dp}{dv_d}\frac{dv_d}{dt}.$$

Because

$$\frac{dv_d}{dt} = -\frac{e\mathcal{E}}{m^*},$$

$$\left(\frac{dp}{dt}\right)_{\text{drift}} = -\frac{e\mathcal{E}}{m^*}\frac{dp}{dv_d}.$$

The second term of Eq. (8.95) represents changes in $p(E)$ as a result of scattering mechanisms:

$$\left(\frac{dp}{dt}\right)_{\text{scattering}} = \frac{p'(E) - p(E)}{\tau}.$$

Hence a solid subjected to an electric field satisfies the condition

$$\left.\frac{dp}{dt}\right|_{\text{drift}} + \left.\frac{dp}{dt}\right|_{\text{scattering}} = 0 \quad \text{in equilibrium,}$$

$$\frac{dp}{dv_d}\frac{dv_d}{dt} + \frac{p' - p}{\tau} = 0,$$

$$-\frac{e\mathcal{E}}{m^*}\frac{dp(E)}{dv_d} + \frac{p'(E) - p(E)}{\tau} = 0,$$

or

$$p'(E) = p(E) + \frac{e\tau\mathcal{E}}{m^*}\frac{dp(E)}{dv_d}, \tag{8.96}$$

where $p'(E)$ is the steady-state probability function in the presence of the applied field. It can be easily seen that when $\mathcal{E} = 0$,

$$p'(E) = p(E).$$

Now if an electric field $\mathcal{E}_x$ is applied in the x direction, the current density contribution dJ_x due to carriers dn is

$$dJ_x = -ev_x \, dn,$$

where v_x is the x component of the velocity of an electron of energy E.
$dn = p'(E)g(E) \, dE$, because $p'(E)$ is different from $p(E)$ in the presence of an applied field. dn is the number of electrons between energy E and $E + dE$. Hence

$$dJ_x = -ev_x p'(E) g(E) \, dE$$

and the total current density is obtained by integrating

$$J_x = \int_0^\infty - ev_x p'(E) g(E) \, dE.$$

From Eq. (8.96) $p'(E)$ for an applied field in the x direction is

$$p'(E) = p(E) + \frac{e\tau\mathcal{E}_x}{m^*} \frac{dp(E)}{dv_x},$$

and hence

$$J_x = -e \int_0^\infty v_x \left[p(E) + \frac{e\tau\mathcal{E}_x}{m^*} \frac{dp(E)}{dv_x} \right] g(E) \, dE. \tag{8.97}$$

Rewriting Eq. (8.97),

$$J_x = -e \int_0^\infty v_x p(E) g(E) \, dE - e \int_0^\infty v_x \frac{e\tau\mathcal{E}_x}{m^*} \frac{dp(E)}{dv_x} g(E) \, dE. \tag{8.97}$$

The first term on the right side is the current density in the x direction in the absence of an applied electric field and must vanish, because the current density is zero in such a case. Hence

$$J_x = -e \int_0^\infty v_x \frac{e\tau\mathcal{E}_x}{m^*} g(E) \, dE \frac{dp(E)}{dv_x} \tag{8.98}$$

or

$$J_x = \mathcal{E}_x \left[-\int_0^\infty \frac{e^2 \tau}{m^*} v_x \frac{dp}{dv_x} g(E) \, dE \right].$$

The term in parentheses is the electrical conductivity σ:

$$\sigma = -\int_0^\infty \frac{e^2 \tau}{m^*} v_x \frac{dp}{dv_x} g(E) \, dE. \tag{8.99}$$

CURRENT DENSITIES; THERMOELECTRIC POWER

To evaluate dp/dv_x we write the equation of energy of a charge carrier. The charge carrier is free and its total energy is the kinetic energy

$$E = \tfrac{1}{2}m^*v^2 = \tfrac{1}{2}m^*(v_x^2 + v_y^2 + v_z^2), \tag{8.100}$$

where v_x, v_y, and v_z are the components of velocity in the x, y, and z directions. In the absence of an applied electric field there are thermal velocities. However, when an electric field is applied in the x direction, v_x gets an additional drift component. The drift component is very small compared to the thermal velocity component and one can on the average assume that

$$v_x^2 = v_y^2 = v_z^2$$

and therefore on the average

$$\tfrac{1}{2}m^*v_x^2 = E/3. \tag{8.101}$$

From Eq. (8.100) the applied field only affects the velocity component in the x direction. The change in energy dE during a change in v_x is given by differentiating the expression only with respect to v_x:

$$dE = m^* v_x \, dv_x. \tag{8.102}$$

According to Eq. (8.93),

$$p(E) = \exp\left(\frac{E_f - E}{kT}\right).$$

Differentiating with respect to v_x and assuming a constant temperature,

$$\frac{dp(E)}{dv_x} = e^{E_f/kT}\left(-\frac{1}{kT} e^{-E/kT}\right)\frac{dE}{dv_x}.$$

Using dE/dv_x from Eq. (8.102),

$$\frac{dp(E)}{dv_x} = m^* v_x \left(-\frac{1}{kT}\right) e^{-E/kT} e^{E_f/kT},$$

and

$$v_x \frac{dp(E)}{dv_x} = m^* v_x^2 \left(-\frac{1}{kT}\right) \exp\left(\frac{E_f - E}{kT}\right).$$

Using Eq. (8.101) and writing $2E/3$ for $m^* v_x^2$,

$$v_x \frac{dp(E)}{dv_x} = -\frac{1}{kT}\frac{2E}{3} \exp\left(\frac{E_f - E}{kT}\right), \tag{8.102a}$$

and hence

$$\sigma = \int_0^\infty \frac{e^2 \tau}{m^*} \frac{1}{kT} \frac{2E}{3} \exp\left(\frac{E_f - E}{kT}\right) g(E) \, dE. \tag{8.103}$$

Substituting $g(E) \, dE$ in Eq. (8.103),

$$\sigma = \frac{2}{3} \frac{4\pi(2m^*)^{3/2}}{h^3} \frac{e^2}{m^*} \frac{1}{kT} \int_0^\infty \tau E^{3/2} \, dE \, \exp\left(\frac{E_f - E}{kT}\right). \tag{8.104}$$

τ, the collision time, is a characteristic of the system, and depending upon the time of scattering mechanism involved, varies in some fashion with the energy E. Let us assume that

$$\tau = aE^p,$$

where a and p are constants for a given system. Hence

$$\sigma = \frac{8\pi}{3} \frac{(2m^*)^{3/2}}{h^3} \frac{e^2}{m^*} \frac{e^{E_f/kT}}{kT} a \int_0^\infty E^{p+\frac{3}{2}} e^{-E/kT} dE.$$

Using Eq. (8.94),

$$\sigma = \frac{4\pi}{3} \frac{e^2}{m^*} \frac{n}{(\pi kT)^{3/2}} \frac{1}{kT} a \int_0^\infty E^{p+\frac{3}{2}} e^{-E/kT} dE.$$

Let us normalize by substituting

$$\eta = E/kT.$$

Differentiating, $d\eta(kT) = dE$. Therefore,

$$\sigma = \frac{4}{3} \frac{ne^2}{\pi^{1/2} m^*} \frac{a}{(kT)^{3/2}} \frac{(kT)^{p+\frac{3}{2}}}{kT} \int_0^\infty \eta^{p+\frac{3}{2}} e^{-\eta} d\eta,$$

and simplifying,

$$\sigma = \frac{4ne^2}{3\pi^{1/2} m^*} a(kT)^p \int_0^\infty \eta^{p+\frac{3}{2}} e^{-\eta} d\eta. \qquad (8.105)$$

We see that the integral is a gamma function,

$$\int_0^\infty \eta^{p+3/2} e^{-\eta} d\eta = \overline{|p+\tfrac{5}{2}}.$$

The conductivity

$$\sigma = \frac{4}{3} \frac{ne^2}{\pi^{1/2} m^*} a(kT)^p \overline{|p+\tfrac{5}{2}}. \qquad (8.106)$$

From Eq. (8.106) we can conclude that the value of electrical conductivity σ depends upon constants a and p, which depend upon the collision time of the system τ. If v is the velocity of the electron, the mean free path λ may be defined as

$$\lambda = v\tau.$$

Consider now a special case in which the mean free path λ may be independent of energy: $\lambda = \tau v$ but $E = \tfrac{1}{2} m^* v^2$. Therefore,

$$\lambda = \tau \left(\frac{2E}{m^*}\right)^{1/2},$$

$$\tau = \lambda \left(\frac{m^*}{2E}\right)^{1/2} = aE^p. \qquad (8.107)$$

Therefore the constants a and p in this special case are

$$a = \lambda \left(\frac{m^*}{2}\right)^{1/2},$$

$$p = -\tfrac{1}{2}.$$

For $p = -\tfrac{1}{2}$, $\overline{|p + 5/2|} = \overline{|2|} = 1$. Hence

$$\sigma = \frac{4}{3} \frac{ne^2\lambda}{(2\pi m^* kT)^{1/2}}. \tag{8.108}$$

In the theory of metals and semiconductors we derived the expression

$$\sigma = ne\mu_d,$$

where μ_d is the drift mobility of charge carriers. Therefore,

$$\mu_d = \frac{4}{3} \frac{e\lambda}{(2\pi m^* kT)^{1/2}}. \tag{8.109}$$

All other quantities except λ and T in Eq. (8.109) are constant. For a given temperature, a knowledge of the mean free path leads us to an estimation of the drift mobility of the charge carriers in the solid. The same procedure, i.e., change in the probability function as a result of the applied field, leads us to the expressions for thermal conductivity and thermoelectric power if the applied field is in a temperature gradient.

Thermoelectric power and thermal conductivity. Both these quantities are properties of a solid subjected to a temperature gradient. It is known from earlier discussion that a temperature gradient results in the establishment of a thermoelectric voltage across the specimen. Thermoelectric power or the Seebeck coefficient was defined as the thermoelectric electric field per unit temperature gradient: $S = \mathcal{E}/\mathrm{grad}\, T$ if the electric current density J is zero. Considering a one-dimensional temperature gradient,

$$\mathrm{grad}\, T = \frac{dT}{dx},$$

the Seebeck voltage is also in the x direction and

$$S = + \left(\frac{\mathcal{E}_x}{dT/dx}\right)_{J_x = 0},$$

where J_x is the electrical current density along the x axis.

The Seebeck electric field is present in the presence of a temperature gradient, so

$$\left.\frac{dp}{dt}\right|_{\text{drift}} = \left.\frac{dp}{dt}\right|_{\text{Seebeck elec. field}} + \left.\frac{dp}{dt}\right|_{\text{temp. grad.}}$$

$$\left.\frac{dp}{dt}\right|_{\text{elec. field}} = -\frac{e\mathcal{E}_x}{m^*}\frac{dp}{dv_x},$$

where $\mathcal{E}_x$ is the Seebeck electric field,

$$\left.\frac{dp}{dt}\right|_{\text{temp. grad.}} = \frac{dp}{dx}\frac{dx}{dt} = \frac{dp}{dx}v_x.$$

p is a function of x, because there is a thermal gradient along the x direction. Hence

$$\left.\frac{dp}{dt}\right|_{\text{drift}} = -\frac{e\mathcal{E}_x}{m^*}\frac{dp}{dv_x} + \frac{dp}{dx}v_x.$$

Again

$$\left.\frac{dp}{dt}\right|_{\text{scattering}} = \frac{p' - p}{\tau}.$$

In equilibrium or steady state,

$$\left.\frac{dp}{dt}\right|_{\text{drift}} + \left.\frac{dp}{dt}\right|_{\text{scattering}} = 0.$$

Hence

$$\frac{p' - p}{\tau} + \left(-\frac{e\mathcal{E}_x}{m^*}\frac{dp}{dv_x} + \frac{dp}{dx}v_x\right) = 0$$

or

$$p'(E) = p(E) + \tau\left(\frac{e\mathcal{E}_x}{m^*}\frac{dp}{dv_x} - v_x\frac{dp}{dx}\right). \tag{8.110}$$

The electric current density dJ_x due to electrons between the energy E and $E + dE$ is

$$dJ_x = -ev_x\,dn,$$

where

$$dn = p'(E)\,g(E)\,dE.$$

The total electric current density is obtained by integrating over all possible values of energy from $E = 0$ to $E \to \infty$. Therefore,

$$J_x = -e\int_0^\infty v_x p'(E)g(E)\,dE. \tag{8.111}$$

Substituting the value of $p'(E)$ from Eq. (8.110) in Eq. (8.111),

$$J_x = -e\int_0^\infty v_x\left[p(E) + \tau\left(\frac{e\mathcal{E}_x}{m^*}\frac{dp}{dv_x} - v_x\frac{dp}{dx}\right)\right]g(E)\,dE.$$

CURRENT DENSITIES; THERMOELECTRIC POWER

The part of the integral

$$-e \int_0^\infty v_x p(E) g(E)\, dE$$

is the current density without any electric field and must be zero. Hence

$$J_x = -e \int_0^\infty v_x \tau \left(\frac{e\mathcal{E}_x}{m^*} \frac{dp}{dv_x} - v_x \frac{dp}{dx} \right) g(E)\, dE. \tag{8.112}$$

In the presence of a temperature gradient, the Seebeck coefficient S is defined when $J_x = 0$. To derive an expression for S the value of J_x is set equal to zero, or

$$0 = -e \int_0^\infty v_x \tau \left(e\frac{\mathcal{E}_x}{m^*} \frac{dp}{dv_x} - v_x \frac{dp}{dx} \right) g(E)\, dE. \tag{8.113}$$

Equation (8.113) when solved for $\mathcal{E}_x/(dT/dx)$ would give the Seebeck coefficient.

It was shown earlier that for a nondegenerate material

$$v_x \frac{dp}{dv_x} = -\frac{2}{3kT} E \exp\left(\frac{E_f - E}{kt}\right). \tag{8.102a}$$

It now remains to express dp/dx in a more convenient form incorporating energy E and the temperature gradient:

$$\frac{dp}{dx} = \frac{dp}{dE} \frac{dE}{dT} \frac{dT}{dx}. \tag{8.114}$$

It now remains to calculate dE/dT in Eq. (8.114).

Now

$$p(E) = \exp\left(\frac{E_f - E}{kT}\right),$$

which shows that p is a function of both E and T. Hence

$$\frac{dE}{dT} = \frac{\partial p/\partial T}{\partial p/\partial E} = \frac{p[(E/kT^2) - (E_f/kT^2)]}{p(-1/kT)},$$

$$\frac{dE}{dT} = \frac{E_f}{T} - \frac{E}{T},$$

or

$$-\frac{dE}{dT} = \frac{E}{T} + T \frac{d}{dT}\frac{(E_f)}{T}.$$

Therefore,

$$\frac{dp}{dx} = -\frac{dT}{dx} \frac{dp}{dE} \left[T \frac{d}{dT}\left(\frac{E_f}{T}\right) + \frac{E}{T} \right]. \tag{8.115}$$

Using Eqs. (8.102a) and (8.115), Eq. (8.113) may be rewritten

$$0 = -e \int_0^\infty \frac{\tau e \mathcal{E}_x}{m^*}\left(-\frac{2}{3kT}\right) E \exp\left(\frac{E_f - E}{kT}\right) g(E)\, dE$$

$$+ e \int_0^\infty v_x^2 \tau \frac{dT}{dx} \frac{dp}{dE}\left[T \frac{d}{dT}\left(\frac{E_f}{T}\right) + \frac{E}{T}\right] g(E)\, dE \quad (8.116)$$

or

$$\int_0^\infty \frac{\tau e \mathcal{E}_x}{m^*}\left(-\frac{2}{3kT}\right) E \exp\left(\frac{E_f - E}{kT}\right) g(E)\, dE$$

$$= -\int_0^\infty v_x^2 \tau \frac{dT}{dx} \frac{dp}{dE}\left[T \frac{d}{dT}\left(\frac{E_f}{T}\right) + \frac{E}{T}\right] g(E)\, dE. \quad (8.117)$$

Both the left and right sides of Eq. (8.117) contain τ within the integral sign. τ is some function of the energy depending upon the type of scattering mechanism. To evaluate the right-side integral of the equation it is helpful to define the integral having a value K_i by

$$K_i = \int_0^\infty \tau g(E) E^i \frac{dp}{dE}\, dE \quad (8.118)$$

for the values $i = 1, 2, 3, \ldots$.

$$K_1 = \int_0^\infty \tau g(E) \frac{dp}{dE} E\, dE,$$

$$K_2 = \int_0^\infty \tau g(E) \frac{dp}{dE} E^2\, dE,$$

$$K_3 = \int_0^\infty \tau g(E) \frac{dp}{dE} E^3\, dE,$$

and so on. Assume again that τ is a function of energy given by

$$\tau = aE^p,$$

and, therefore,

$$K_i = a \int_0^\infty E^{p+i} \frac{dp}{dE} g(E)\, dE. \quad (8.119)$$

Before making use of the definition of K_i in solving Eq. (8.117), it is useful to evaluate the integral
Because

$$p(E) = \exp\left(\frac{E_f - E}{kT}\right),$$

$$\frac{dp}{dE} = -\frac{1}{kT} \exp\left(\frac{E_f - E}{kT}\right).$$

Also,
$$g(E)\,dE = \frac{4\pi}{h^3}(2m^*)^{3/2}E^{1/2}\,dE,$$

and therefore
$$K_i = -\frac{4\pi}{h^3}\frac{(2m^*)^{3/2}}{kT}e^{E_f/kT}a\int_0^\infty E^{p+i+\frac{1}{2}}e^{-E/kT}\,dE.$$

Normalizing and substituting,
$$\frac{E}{kT} = \eta,$$

$$K_i = -\frac{4\pi}{h^3}\frac{(2m^*)^{3/2}}{kT}e^{E_f/kT}(kT)^{p+i+\frac{3}{2}}a\int_0^\infty \eta^{p+i+\frac{1}{2}}e^{-\eta}\,d\eta.$$

Substituting the value of E_f in terms of n from Eq. (8.94) and noting that
$$\int_0^\infty \eta^{p+i+\frac{1}{2}}e^{-\eta}\,d\eta = \overline{|p+i+\tfrac{3}{2}},$$

$$K_i = -\frac{2}{\pi^{1/2}}an(kT)^{p+i-1}\overline{|p+i+\tfrac{3}{2}}.$$

Therefore,
$$K_1 = -\frac{2}{\pi^{1/2}}an(kT)^p\overline{|p+\tfrac{5}{2}},$$

$$K_2 = -\frac{2}{\pi^{1/2}}an(kT)^{p+1}\overline{|p+\tfrac{7}{2}},$$

$$K_3 = -\frac{2}{\pi^{1/2}}an(kT)^{p+2}\overline{|p+\tfrac{9}{2}}.$$

Now we proceed to solve the right side of Eq. (8.117),
$$-\left\{\int_0^\infty v_x^2\,\tau\,\frac{dT}{dx}\frac{dp}{dE}\left[T\frac{d}{dT}\left(\frac{E_f}{T}\right) + \frac{E}{T}\right]g(E)\,dE\right\}$$

$$= -\left[\frac{dT}{dx}T\frac{d}{dT}\left(\frac{E_f}{T}\right)\int_0^\infty v_x^2\tau\frac{dp}{dE}g(E)\,dE + \frac{1}{T}\frac{dT}{dx}\int_0^\infty v_x^2\tau E\frac{dp}{dE}g(E)\,dE\right],$$

Because $\tfrac{1}{2}m^*v_x^2 \simeq E/3$, the right side of Eq. (8.117) equals
$$-\left[\frac{2}{3}\frac{1}{m^*}\frac{dT}{dx}T\frac{d}{dT}\left(\frac{E_f}{T}\right)\int_0^\infty E\tau\frac{dp}{dE}g(E)\,dE + \frac{2}{3m^*}\frac{1}{T}\frac{dT}{dx}\int_0^\infty \tau E^2\frac{dp}{dE}g(E)\,dE\right],$$

or the right side of Eq. (8.117) equals
$$\frac{2}{3m^*}\frac{dT}{dx}\left[T\frac{d}{dT}\left(\frac{E_f}{T}\right)K_1 + \frac{K_2}{T}\right].$$

The left side of Eq. (8.117) equals

$$\frac{2e}{3m^*} \mathcal{E}_x \int_0^\infty -\frac{1}{kT} \exp\left(\frac{E_f - E}{kT}\right) \tau E g(E) \, dE,$$

or the left side of Eq. (8.117) equals

$$\frac{2e}{3m^*} \mathcal{E}_x \int_0^\infty \tau \frac{dp}{dE} E g(E) \, dE = -\frac{2e}{3m^*} \mathcal{E}_x K_1.$$

Hence

$$-\frac{2e}{3m^*} \mathcal{E}_x K_1 = \frac{2}{3m^*}\left[T \frac{d}{dT}\left(\frac{E_f}{T}\right) K_1 + \frac{K_2}{T}\right]\frac{dT}{dx}$$

or

$$\left.\frac{\mathcal{E}_x}{dT/dx}\right)_{J_x=0} = -\frac{1}{e}\left(\frac{K_2}{K_1 T} - \frac{E_f}{T}\right). \qquad (8.120a)$$

Therefore,

$$S = -\frac{k}{e}\left(\frac{1}{kT}\frac{K_2}{K_1} - \frac{E_f}{kT}\right).$$

Substituting the values of K_2 and K_1,

$$S = -\frac{k}{e}\left(\frac{\lvert p + \frac{7}{2}}{\lvert p + \frac{5}{2}} - \frac{E_f}{kT}\right). \qquad (8.120)$$

Equation (8.120) is the Seebeck coefficient of a solid. It depends upon p, which is dependent upon the type of scattering mechanism. For scattering due to vibrating atoms, the mean free path is independent of energy and $p = -\frac{1}{2}$.

Therefore,

$$S = -\frac{k}{e}\left(\frac{\lvert 3}{\lvert 2} - \frac{E_f}{kT}\right)$$

or

$$S = -\frac{k}{e}\left(2 - \frac{E_f}{kT}\right). \qquad (8.121)$$

For an ionic lattice it is found that the mean free path $\lambda \propto E$ at relatively large temperatures, and hence

$$v\tau \propto E.$$

Because $v \propto E^{1/2}$, $\tau \propto E^{1/2}$, and hence $p = \frac{1}{2}$. Equation (8.120) for ionic lattices reduces to

$$S = -\frac{k}{e}\left(3 - \frac{E_f}{kT}\right). \qquad (8.122)$$

CURRENT DENSITIES; THERMOELECTRIC POWER

The minus signs for the values of S in Eqs. (8.121) and (8.122) are due to the assumption that the charge carriers are electrons. For holes the Seebeck coefficient is positive. The Fermi level increases with carrier concentration,

$$E_f = kT \ln\left[\frac{nh^3}{2(2\pi m^* kT)^{3/2}}\right], \qquad (8.94)$$

so the Seebeck coefficient decreases with increasing carrier concentration. It may be more appropriate to write

$$S = -\frac{k}{e}\left\{\frac{\left|p+\frac{7}{2}\right|}{\left|p+\frac{5}{2}\right|} - \ln\left[\frac{nh^3}{2(2\pi m^* kT)^{3/2}}\right]\right\}. \qquad (8.123)$$

The carrier concentration is very large in metals and hence the Seebeck coefficient is very low.

Thermal conductivity. The thermal current density J_Q has been defined as the amount of heat in joules flowing across a square meter of a surface per second. If dn is the number of charge carriers with energy between E and $E + dE$ or approximately of energy E and velocity v_x, the thermal current density due to charge carriers is

$$dJ_Q = v_x E \, dn.$$

It should be mentioned that the thermal current density is caused by a temperature gradient.

In the presence of a temperature gradient,

$$dn = p'(E) g(E) \, dE,$$

where

$$p'(E) = p(E) + \tau\left(\frac{e\mathcal{E}_x}{m^*}\frac{dp}{dv_x} - v_x\frac{dp}{dx}\right).$$

Hence

$$dJ_Q = v_x E\left\{p(E) + \tau\left[\frac{e\mathcal{E}_x}{m^*}\frac{dp}{dv_x} - v_x\frac{dp}{dx}\right]\right\} g(E) \, dE.$$

The total thermal current density can be obtained by integrating the right side of the above equation for all values of energy from $E = 0$ to $E \to \infty$:

$$J_Q = \int_0^\infty v_x E\left\{p(E) + \tau\left[\frac{e\mathcal{E}_x}{m^*}\frac{dp}{dv_x} - v_x\frac{dp}{dx}\right]\right\} g(E) \, dE. \qquad (8.124)$$

$\int_0^\infty v_x E p(E) g(E) \, dE$ is the thermal current density in the absence of a temperature gradient and hence must be zero. Therefore,

$$J_Q = \int_0^\infty v_x E_\tau \left(\frac{e\mathcal{E}_x}{m^*} \frac{dp}{dv_x} - v_x \frac{dp}{dx} \right) g(E) \, dE, \qquad (8.125)$$

$$J_Q = \int_0^\infty \frac{e\mathcal{E}_x}{m^*} v_x E_\tau \left(\frac{dp}{dv_x} \right) g(E) \, dE - \int_0^\infty v_x^2 E_\tau \frac{dp}{dx} g(E) \, dE,$$

or

$$J_Q = \frac{e\mathcal{E}_x}{m^*} \int_0^\infty v_x \frac{dp}{dv_x} \tau E g(E) \, dE - \int_0^\infty v_x^2 E_\tau \frac{dp}{dx} g(E) \, dE.$$

It was shown earlier that

$$v_x \frac{dp}{dv_x} = -\frac{2E}{3kT} \exp\left(\frac{E_f - E}{kT} \right) = -\frac{2E}{3} \frac{dp}{dE}$$

and $v_x^2 = 2E/3m^*$. Also

$$\frac{dp}{dx} = -\frac{dT}{dx} \frac{dp}{dE} \left[T \frac{d}{dT}\left(\frac{E_f}{T} \right) + \frac{E}{T} \right]$$

and

$$J_Q = \frac{e\mathcal{E}_x}{m^*} \int_0^\infty \left(-\frac{2}{3} \right) E^2 \tau \frac{dp}{dE} g(E) \, dE$$

$$+ \frac{2}{3m^*} \int_0^\infty \frac{dT}{dx} E^2 \tau \frac{dp}{dE} \left[T \frac{d}{dT}\left(\frac{E_f}{T} \right) + \frac{E}{T} \right] g(E) \, dE,$$

$$J_Q = -\frac{2e\mathcal{E}_x}{3m^*} \int_0^\infty E^2 \tau \frac{dp}{dE} g(E) \, dE + \frac{2}{3m^*} \frac{dT}{dx} \frac{1}{T} \int_0^\infty E^3 \tau \frac{dp}{dE} g(E) \, dE$$

$$- \frac{2}{3m^*} \frac{E_f}{T} \frac{dT}{dx} \int_0^\infty E^2 \tau \frac{dp}{dE} g(E) \, dE,$$

or

$$J_Q = \frac{2}{3m^*} \left(-e\mathcal{E}_x K_2 + \frac{dT}{dx} \frac{K_3}{T} - \frac{E_f}{T} K_2 \frac{dT}{dx} \right). \qquad (8.126)$$

A temperature causes a Seebeck voltage, and hence substituting $\mathcal{E}_x$ from Eq. (8.120a) in Eq. (8.126),

$$J_Q = -\frac{2}{3m^*} \left[\left(\frac{K_2}{K_1 T} - \frac{E_f}{T} \right) \frac{dT}{dx} K_2 + \frac{dT}{dx} \frac{K_3}{T} - \frac{E_f}{T} K_2 \frac{dT}{dx} \right]$$

Simplifying,

$$J_Q = -\frac{2}{3m^* T} \left(\frac{K_2^2}{K_1} - K_3 \right) \frac{dT}{dx}. \qquad (8.127)$$

BOLTZMANN'S TRANSPORT EQUATION

The electronic contribution to thermal conductivity κ_{elect} is defined by Fourier's law,

$$\kappa_{elect} = -\frac{J_Q}{dT/dx} = \frac{2}{3m^*T}\left(\frac{K_2^2}{K_1} - K_3\right).$$

Substituting the values of K_1, K_2, and K_3,

$$\kappa_{elect} = \frac{2}{3m^*T}\left(-\frac{2}{\pi^{1/2}}an\right)\left[(kT)^{p+2}\frac{\overline{|p+\tfrac{7}{2}|}\,\overline{|p+\tfrac{7}{2}|}}{\overline{|p+\tfrac{5}{2}|}} - (kT)^{p+2}\overline{|p+\tfrac{9}{2}|}\right],$$

$$\kappa_{elect} = -\frac{2}{3m^*T}\left(-\frac{2}{\pi^{1/2}}an\right)(kT)^{p+2}[(p+\tfrac{5}{2})\overline{|p+\tfrac{7}{2}|} - (p+\tfrac{7}{2})\overline{|p+\tfrac{7}{2}|}],$$

or

$$\kappa_{elect} = \frac{4an}{3\pi^{1/2}m^*}\frac{1}{T}(kT)^{p+2}\overline{|p+\tfrac{7}{2}|}.$$

In case we assume scattering due to lattice vibrations $p = -\tfrac{1}{2}$ and $a = (m^*/2)^{1/2}\lambda$,

$$\kappa_{elect} = \frac{8an}{3\pi^{1/2}m^*T}(kT)^{3/2} = \frac{8n}{3\pi^{1/2}m^*T}\left(\frac{m^*}{2}\right)^{1/2}(kT)^{3/2}\lambda$$

or

$$\kappa_{elect} = \frac{8nk^2T\lambda}{3(2\pi m^*kT)^{1/2}}. \tag{8.128}$$

Using the definition of drift mobility,

$$\mu_d = \frac{4}{3}\frac{e\lambda}{2(2\pi m^*kT)^{1/2}},$$

$$\kappa_{elect} = 2\frac{k^2T}{e}n\mu_d = 2\frac{k^2}{e^2}Tne\mu_d,$$

or

$$\kappa_{elect} = \sigma\left(\frac{2k^2}{e^2}T\right).$$

The Wiedemann-Franz ratio κ_{elect}/σ equals $2(k^2/e^2)T$. The Wiedemann-Franz law was discussed in Sec. 4.3. This value differs from those given by Eqs. (4.28) and (4.38) by a constant.

8.5 Boltzmann's transport equation

The flow of heat or electricity disturbs the distribution of charge carriers in a solid. The change occurs in $p(E)$. If $p_1(E)$ is the new probability function in the presence of external effects such as a temperature gradient or an electrical gradient due to scattering, the function $p_1(E)$ changes to $p(E)$ in collision time τ. Hence

$$\left.\frac{dp}{dt}\right|_{\text{scattering}} = \frac{p_1(E) - p(E)}{\tau}.$$

In the presence of an external gradient a steady state is reached if

$$\left.\frac{dp}{dt}\right|_{\text{drift}} + \left.\frac{dp}{dt}\right|_{\text{scattering}} = 0.$$

In one dimension, if both electric field and temperature are present in the x axis, it has been discussed that

$$\left.\frac{dp}{dt}\right|_{\text{drift}} = -\frac{e\mathcal{E}_x}{m^*}\frac{dp}{dv_x} + v_x\frac{dp}{dx}.$$

Extending to three dimensions,

$$\left.\frac{dp}{dt}\right|_{\text{drift}} = -\left(\frac{e\mathcal{E}}{m^*}\right)\cdot\nabla_v p + \mathbf{v}\cdot\nabla_r p,$$

where a variable $\mathbf{r}$ has been introduced for a one-dimensional variable, $\mathbf{v}$ is the velocity of the charge carrier represented vectorially, and $\mathcal{E}$ is the electric field:

$$\mathcal{E} = \mathcal{E}_x\mathbf{a}_x + \mathcal{E}_y\mathbf{a}_y + \mathcal{E}_z\mathbf{a}_z,$$

$$\mathbf{v} = v_x\mathbf{a}_x + v_y\mathbf{a}_y + v_z\mathbf{a}_z,$$

$$\mathbf{r} = x\mathbf{a}_x + y\mathbf{a}_y + z\mathbf{a}_z.$$

$\mathbf{a}_x$, $\mathbf{a}_y$, and $\mathbf{a}_z$ are unit vectors in the x, y, and z directions. Hence in equilibrium or steady state,

$$-\frac{e\mathcal{E}}{m^*}\cdot\nabla_v p + \mathbf{v}\cdot\nabla_r p + \frac{p' - p}{\tau} = 0$$

or

$$p' = p + \tau\left(\frac{e\mathcal{E}}{m^*}\cdot\nabla_v p - \mathbf{v}\cdot\nabla_r p\right). \tag{8.129}$$

In case a magnetic field is present, the force on an electron can be written

$$\mathbf{F} = -e(\mathcal{E} + \mathbf{v}\times\mathbf{B}),$$

where $\mathbf{B}$ is the vector magnetic flux density. Hence in the presence of a magnetic field

$$\left.\frac{dp}{dt}\right|_{\text{drift}} = -\frac{e}{m^*}(\mathcal{E} + \mathbf{v}\times\mathbf{B})\cdot\nabla_v p + \mathbf{v}\cdot\nabla_r p,$$

and the new probability function is given by

$$p' = p + \left[\frac{e}{m^*}(\mathcal{E} + \mathbf{v} \times \mathbf{B})\cdot\nabla_v p - \mathbf{v}\cdot\nabla_r p\right]\tau \qquad (8.130)$$

This is the Boltzmann transportation equation for electrons in magnetic and electric fields. If the electric field, magnetic field, and temperature field are simultaneously present, Eq. (8.130) must be solved to obtain the new probability function. It is not within the scope of this book to solve the general Boltzmann equation. The reader is referred to the Bibliography for solutions of the equation.

8.6 Thermomagnetic coefficients; Nernst-Ettinghausen generator

While discussing the galvanomagnetic effects we considered the solid to be at one temperature. Now if a temperature gradient across the solid exists, a thermal current flows. Galvanomagnetic and thermomagnetic effects occur when a conductor carrying an electric or thermal current is placed in a magnetic field. Like galvanomagnetic effects, thermomagnetic effects are also transverse effects. They are the interaction of mutually perpendicular magnetic fields and thermal and potential gradients.

Nernst effect. Consider a rectangular plate of the specimen of thickness t along the z axis and length l along the x axis. Let w be the width of the specimen along the y direction. Let us suppose that a temperature gradient is set up along the x direction and a magnetic field of flux density B_z applied in the z direction. The Nernst effect is the establishment of a voltage between sides A and B. If electrons are the charge carriers, the direction of potential gradient is the y direction. In equilibrium an equal and opposite force is exerted on the electrons and if all the electrons had the same velocity no more electrons would be deflected. The Nernst coefficient Q is defined by

$$\mathcal{E}_y = -QB_z\frac{dT}{dx}, \qquad (8.131)$$

where B_z is the flux density and dT/dx is the temperature gradient. It can be worked out from the Boltzmann transport equation that for a nondegenerate semiconductor with scattering due to vibrating atoms ($p = -\frac{1}{2}$) and one type of charge carrier,

$$Q = -\frac{3\pi}{16}\frac{k}{e}\mu_d, \qquad (8.132)$$

where k is the Boltzmann constant, e the electron charge, and μ_d the drift mobility.

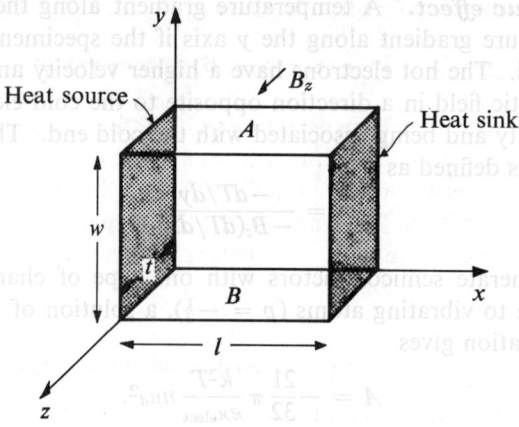

Fig. 8.6
Ettinghausen effect

Ettinghausen effect. In the specimen of Fig. 8.6, if a conventional current density J_x exists in the x axis and the specimen is placed in a magnetic field, a temperature difference will be noted between surfaces A and B. The Ettinghausen coefficient P is defined by

$$\frac{dT}{dy} = -PJ_xB_z, \qquad (8.132)$$

where dT/dy is the temperature gradient in the y direction.

The velocity distribution of electronic current explains the Ettinghausen effect. Those electrons which have just the right velocity to match the equal and opposite force of the Hall field are not deflected. The faster electrons are deflected in one direction and the slower in the opposite direction, causing a temperature difference between faces A and B.

For nondegenerate semiconductors with one type of charge carrier and scattering due to vibrating atoms, a solution of the Boltzmann transport equation gives

$$P = -\frac{3\pi}{16}\left(\frac{k}{e}\right)\frac{T}{\varkappa_{\text{elect}}}\mu_d, \qquad (8.133)$$

where T is the temperature of the specimen and $\varkappa_{\text{elect}}$ is the electronic thermal conductivity. From Eqs. (8.132) and (8.133) we find that Q, the Nernst coefficient, is related to the Ettinghausen effect,

$$Q = \frac{P\varkappa_{\text{elect}}}{T}. \qquad (8.134)$$

NERNST-ETTINGHAUSEN GENERATOR

Righi-Leduc effect. A temperature gradient along the x axis results in a temperature gradient along the y axis if the specimen is placed in a magnetic field. The hot electrons have a higher velocity and are deflected by the magnetic field in a direction opposite to the cold electrons, having a lower velocity and being associated with the cold end. The Righi-Leduc coefficient A is defined as

$$A = \frac{-dT/dy}{-B_z(dT/dx)}. \tag{8.135}$$

For nondegenerate semiconductors with one type of charge carrier and scattering due to vibrating atoms ($p = -\frac{1}{2}$), a solution of the Boltzmann transport equation gives

$$A = -\frac{21}{32}\pi \frac{k^2 T}{e\chi_{\text{elect}}} n\mu_d{}^2. \tag{8.136}$$

A is related to P by

$$\frac{A}{P} = \tfrac{7}{2} k n \mu_d.$$

Because $ne\mu_d = \sigma$,

$$\frac{A}{P} = \frac{7}{2} \frac{k}{e} \sigma. \tag{8.137}$$

It may again be stressed that we assume an isotropic solid.

Nernst-Ettinghausen generator. Referring to Fig. 8.6, we consider the case of a temperature gradient existing along the x axis when a magnetic field is applied along the z direction.

The voltage developed between surfaces A and B is $W\mathcal{E}_y$, where

$$\mathcal{E}_y = QB_z \frac{\Delta T}{l}.$$

ΔT is the temperature difference between the heat source and heat sink. Therefore,

$$V_{AB} = WQB_z \frac{\Delta T}{l}.$$

Suppose an external resistance R_L is connected across A and B. A current will start flowing through the load. If R_s is the resistance of the specimen,

$$I_L = \frac{V_{AB}}{R_L + R_S} = \frac{WQB_z \Delta T}{l(R_L + R_S)}$$

or

$$I_L = \frac{WQB_z \Delta T}{R_S l[(R_L/R_S) + 1]}.$$

Now if $R_L/R_S = m$, the load resistance is m times the specimen resistance,

$$I_L = \frac{WQB_z \Delta T}{R_S l(m+1)}.$$

Further, if σ is the electrical conductivity of the specimen,

$$R_S = \frac{1}{\sigma}\frac{W}{lt}.$$

Therefore,

$$I_L = \frac{QB_z \Delta T \sigma t}{m+1}. \qquad (8.138)$$

The power output of the generator will be

$$P_{\text{output}} = I_L^2 R_L,$$

because the device is converting the heat supplied at the heat source into electrical energy; the action is called *generation*.

Hence

$$P_{\text{output}} = \left(\frac{QB_z \Delta T \sigma t}{m+1}\right)^2 R_L$$

or

$$P_{\text{output}} = \frac{m}{(m+1)^2} Q^2 B_z^2 (\Delta T)^2 \sigma^2 t^2 R_S,$$

or using $R_S = (1/\sigma)(W/lt)$,

$$P_{\text{output}} = \frac{m}{(m+1)^2} \frac{Q^2 B_z^2 (\Delta T)^2 W t \sigma}{l}. \qquad (8.139)$$

The input power which must be supplied to the hot junction is made up of three terms:

1. The amount of heat conducted from the hot junction to the cold junction per second $= \varkappa \Delta T(Wt/l) = P_C$.
2. The Joule heating caused by the flow of current I_L through the specimen. This is a negative term, because this heat is not to be supplied by the source P_J.
3. The heat necessary to replace the Ettinghausen cooling due to passage of current I. This effect tries to cool the hot junction and heat the cold junction. The temperature difference between the heat source and sink due to the Ettinghausen effect is

$$\Delta T_E = \frac{dT}{dx} l = -PB_z J_L l,$$

where $J_L = I_L/lt$. Therefore,

$$\Delta T_E = -\frac{PB_z I_L}{t}.$$

This temperature difference enhances the flow of heat. This energy is extracted from the electric current and must therefore be supplied by the heat source. Assuming a small temperature difference between the heat source and heat sink, the heat supplied to overcome Ettinghausen cooling is

$$Q_E = \frac{\varkappa W t}{l} \Delta T_E,$$

$$Q_E = \frac{\varkappa W t}{l} \frac{P B_z I_L}{t}.$$

But $P = QT/\varkappa_{elec}$ or simply $QT/Q_E\varkappa$. Therefore, heat supplied per second to overcome Ettinghausen cooling $Q_E = (\varkappa W t/l)(QT/\varkappa) B_z(I_L/t)$ or $P_E = (W/l) QT B_z I_L$.

The hot-junction temperature is not very different from the cold junction (assumed), so $T \simeq T_H$. Hence

$$P_E = \frac{W}{l} Q T_H B_z I_L. \tag{8.140}$$

Substituting the value of I_L from Eq. (8.138) in Eq. (8.140),

$$P_E = Q^2 B_z^2 T_H \frac{\Delta T \sigma t}{l(m+1)} \frac{W}{}.$$

The heat conducted is

$$P_C = \varkappa \Delta T \frac{W t}{l}.$$

The Joule heating per second, if assumed to contribute one half to each end, is

$$P_J = -\tfrac{1}{2} I_L^2 R = -\frac{1}{2} \frac{Q^2 B_z^2 (\Delta T)^2 \sigma^2 t^2}{(m+1)^2} \frac{1}{\sigma} \frac{W}{lt}$$

or

$$P_J = -\frac{1}{2} \frac{Q^2 B_z^2 \sigma (\Delta T)^2}{(m+1)^2} \frac{W t}{l}.$$

Hence the total heat input per second is

$$P_{input} = P_E + P_C + P_J.$$

Therefore,

$$P_{input} = \frac{W}{lt} \left[\varkappa \Delta T + \frac{Q^2 B_z^2 \sigma T_H \Delta T}{m+1} - \frac{1}{2} \frac{Q^2 B_z^2 \sigma (\Delta T)^2}{(m+1)^2} \right]$$

The efficiency of the thermomagnetic Nernst generator is

$$\eta = \frac{P_{output}}{P_{input}} = \frac{[m/(m+1)^2]\sigma Q^2 B_z^2 (\Delta T)^2}{\varkappa \Delta T + [Q^2 B_z^2 \sigma T_H \Delta T/(m+1)] - \tfrac{1}{2}[Q^2 B_z^2 \sigma (\Delta T)^2/(m+1)^2]}.$$

Calling $C = Q^2 B_z^2 \sigma / \kappa$,

$$\eta = \frac{mC \Delta T}{(m+1)^2 + (m+1)CT_H - (C/2)\Delta T}.$$

The condition for maximum efficiency, $d\eta/dm = 0$, gives

$$m^2 = 1 + C\left(\frac{T_H + T_C}{2}\right),$$

where $(T_H + T_C)/2$ is the mean temperature.

C is called the *thermomagnetic figure of merit* and is of the order of 10^{-5} for normal magnetic fields of approximately 1 Wb/m². Intrinsic semiconductors show great promise but only at high magnetic fields. Intensive research for suitable materials with high C is called for. A high figure of merit C results in a high value of efficiency.

Problem 1

a. If the average relaxation time is defined by

$$\langle \tau \rangle = -\frac{2}{3} \frac{\int_0^\infty \tau E g(E)(\partial p/\partial E) dE}{\int_0^\infty g(E) p(E)\, dE},$$

show that for a nondegenerate case with $\tau = aE^p$, the average value $\langle \tau^n \rangle$ is given by

$$\langle \tau^n \rangle = \frac{a^n (kT)^{np} \lceil np + \tfrac{5}{2}}{\lceil \tfrac{5}{2}}.$$

b. Show that for $p = -\tfrac{1}{2}$ and $p = \tfrac{3}{2}$ the ratios of Hall mobility to drift mobility are $3\pi/8$ and $315\pi/512$, respectively.

c. Determine the ratio of magnetoresistive mobility to drift mobility for $p = -\tfrac{1}{2}$ and $p = \tfrac{3}{2}$.

d. Using the expression of electrical conductivity,

$$\sigma = \frac{ne^2}{m^*} \langle \tau \rangle,$$

justify the definition of average relaxation time given in part **a**.

Ans.   **c.** $3\sqrt{\pi}/4$; $(3/4)(10\pi/3)^{1/2}$

Problem 2

a. Show that for an isotropic material with one type of charge carrier, the carrier density n is given by

$$n = \frac{4}{\pi^{1/2}}\left(\frac{2\pi m^* kT}{h^2}\right)^{3/2} F_{1/2}(\eta),$$

where

$$F_r(\eta) = \int_0^\infty \xi^r \frac{1}{e^{\xi-\eta}+1} d\xi$$

is the Fermi integral and $\eta = E_f/kT$.

b. Plot n as a function of η with the help of tables of Fermi integrals.

Problem 3

Show that for an isotropic material with one type of charge carrier the electrical conductivity σ is given by

$$\sigma = \frac{2}{3}\frac{ne^2}{m^*}a(kT)^p\left(p+\frac{3}{2}\right)\frac{F_{p+\frac{1}{2}}}{F_{1/2}}.$$

Problem 4

Show that the relative Seebeck coefficient of two nondegenerate N-type semiconducting materials with different impurity dopings is given by $S_{12} = 86.2 \times 10^{-6} \ln(n_1/n_2)$ V/°K, where n_1 and n_2 and the electronic carrier concentrations per unit volume in the two materials.

Problem 5

a. Determine the relative Seebeck coefficients of two N-type samples of germanium at room temperature if the Fermi levels in the two samples are located at 0.5 and 0.6 eV above the conduction band.

b. Determine the relative Peltier coefficient for part a of the problem.

Ans. **a.** 344.8×10^{-6} V/deg **b.** 0.10344 V

Problem 6

With $T_0 = 300°$K and $\Delta T = 10, 20, 30, 40, 50, 100,$ and $200°$K, plot a series of curves for the optimum efficiency of a thermoelectric generator with Z as parameter. Assume values of $Z = 0.001, 0.002, 0.003, 0.004,$ and 0.005.

Problem 7

Show that for an infinite-stage thermoelectric generator, the optimum efficiency is given by

$$\eta_{opt} = 1 - \exp\left[-\int_{T_0}^{T_1} \frac{(1+ZT)^{1/2}-1}{(1+ZT)^{1/2}+1}\frac{dT}{T}\right].$$

Problem 8

With $T_0 = (n/4\pi)^{2/3} (h^2/2m^*k)$, plot the Seebeck coefficient as a function of T/T_0 with the scattering index as parameter (use the general expression for S).

Problem 9

The continuity equation for heat current in steady state is

$$\frac{d}{dx}\left[\varkappa(T)\frac{dT}{dx}\right] + J\tau(T)\frac{dT}{dx} + J^2\rho(T) = 0,$$

where J is the current density. Assuming that $\varkappa$, τ, and ρ are independent of temperature, find the temperature as a function of x in a bar of length L.

Ans.

$$T = T_1 - \frac{J\rho x}{\tau} + \left(\frac{J\rho L}{\tau} - T_1 + T_0\right)\left(\frac{1 - e^{-J\tau x/\varkappa}}{1 - e^{-J\tau L/\varkappa}}\right).$$

9

Quantum Electronics

9.1 Why quantum electronics for engineers?

The quantum mechanics of microscopic charged particles, electromagnetic fields, and their mutual interaction has been the subject of intensive investigation by the scientists of the world for the past half century. As pointed out by Sommerfeld in 1948, it is amazing that so many advances in electrical and electronic technology had taken place up to that time using simple physical models which completely ignored quantum mechanics. We have seen in the preceding chapters how often the simple concept of the electron as a charged classical particle has yielded results agreeing quite closely with experiment, and with results obtained using quantum-mechanical models. Ohm's law and electrical conduction are understandable to a certain extent in terms of a classical model. Most devices that depend on electronic motion in a partial vacuum can be understood using classical models — devices such as the vacuum triode, the klystron, and the magnetron.

We have already seen, however, that certain phenomena can be successfully understood only on the basis of quantum-mechanical models. Electronic motion in semiconductors requires a quantum-mechanical treatment, especially with regard to certain special aspects such as tunneling. There are, in addition, a certain class of phenomena which have been given a special classification, called *quantum electronics*. These phenomena generally involve interactions between individual atoms or molecules and electromagnetic fields, and no classical model affords even an approximate explanation. In many cases the quantum theory is applied only to the atomic system, and the interaction between the electromagnetic field and the system, but not to the electromagnetic field itself. It is this so-called *semiclassical treatment* that will be considered here.

Although the domain encompassed by the words "quantum electronics" is still somewhat vague, the following quotation from C. H. Townes' opening remarks at the first Conference on Quantum Electronics in 1959 serves as a fairly good guide line: "However, we are concerned here primarily with resonance spectroscopy and resonant interactions examined by approximately coherent electromagnetic waves."

9.2 Energy levels in atomic systems

It was seen in Chapter 1 that a quantum-mechanical analysis of the hydrogen atom yielded discrete energy levels. The fact that discrete levels exist cannot be understood in terms of classical models. Because quantum-electronic devices depend on the existence of discrete energy levels in atomic or molecular systems, the necessity for quantum mechanics is obvious. In this section we wish to discuss more fully what types of atomic systems are most suitable for use in quantum-electronic devices and the characteristics of their energy levels.

It has been shown that when one considers a solid made up of many interacting atomic systems in a regular lattice, the discrete atomic-energy levels get spread into bands. Because in most cases involving laser action one is interested in transitions between discrete levels, one must deal with atomic systems in the gaseous state or at least with weakly interacting atoms or molecules. The latter can be obtained by lightly doping a crystal lattice with an ionic impurity. Although the energy levels of the ionic impurities are distorted by the field of the crystal lattice, they remain discrete levels and do not become bands. The most obvious example is the case of ruby, where triply ionized chromium ions are distributed within an aluminum oxide lattice.

For the sake of the following discussion, let us just assume the existence of systems with discrete energy levels, the different levels being characterized not only by an energy quantum number but also total angular momentum and projection of angular momentum quantum numbers. We shall consider later, in more detail, the various types of systems that are of practical interest.

Transitions between energy levels. Consider an assembly of weakly coupled atomic systems, every one of which has two possible energy states,

E_1 and E_2. In thermodynamic equilibrium the relative propulations of the two energy states will be governed by Boltzmann's law:

$$\frac{N_2}{N_1} = \exp\left(-\frac{E_2 - E_1}{kT}\right), \tag{9.1}$$

where N_i is the population of the state with energy E_i.

The purpose of considering the systems as weakly interacting was to allow for some mechanism for bringing the whole system into thermodynamic equilibrium. Otherwise the individual atomic systems would maintain any energy distribution into which they were originally put and not change with time. Because the individual atomic system we are considering are distinguishable, the Boltzmann distribution is appropriate.

We shall now consider the interaction of an individual atom, ion, or molecule (referred to previously as an *atomic system*) with an electromagnetic field. This interaction will be treated on a semiclassical basis, although at times reference will be made to photons, which correspond to quanta of electromagnetic radiation which result from the quantization of the electromagnetic field. The types of processes we shall consider here involve the absorption and emission of single quanta by one of the atomic systems. Because energy must be conserved, it is obvious that a photon must be of energy $E_2 - E_1 = h\nu$ to interact with the system. In a two-level system the atom must be in its low-energy state to absorb a photon, and it must be in its high-energy state to emit a photon. Although absorption has to take place in the presence of a stimulating field, emission can take place whether an electromagnetic field is present or not. However, the probability for emission will depend on whether or not a field is present to stimulate the emission. The process of emission in the presence of a photon field is called *stimulated emission,* and in the absence of a field it is called *spontaneous emission.* The calculation of the probabilities for the processes of resonance absorption, stimulated emission, and spontaneous emission will be carried out below.

Transition probabilities for emission and absorption. A complete quantum-mechanical treatment of the absorption and emission of electromagnetic radiation by atomic system is beyond the scope of this text. Enough of a derivation will be given to acquaint the reader with some of the calculational techniques involved as well as some of the approximations that must be made.

Let us first consider the process by which a quantum of radiation is absorbed by an atomic system consisting of a single electron moving in a central force field which results from the interaction of the electron with the nucleus and all other electrons present in the atom. The Hamiltonian of the system in the absence of any perturbing electromagnetic field will be called H_0 and the time-independent Schrödinger equation is written

$$H_0 \Psi_n = E_n \Psi_n, \qquad (9.2)$$

$$\Psi_n = u_n e^{-jE_n t/\hbar}.$$

where Ψ_n are the eigenfunctions corresponding to the stationary states of the atom with energies E_n. In the presence of a perturbing external electromagnetic field the Hamiltonian can be written as the sum of H_0 plus the time-dependent part H_1 due to the externally applied field. Because H_1 depends on time, we must use the time-dependent Schrödinger equation,

$$j\hbar \frac{\partial \Psi}{\partial t} = H_{\text{tot}} \Psi = (H_0 + H_1)\Psi. \qquad (9.3)$$

We may express Ψ as a sum of the unperturbed wave function Ψ_n, with the expansion coefficients $a_n(t)$ being functions of time:

$$\Psi = \sum_n a_n(t) \Psi_n. \qquad (9.4)$$

Substituting this in the wave equation,

$$j\hbar \sum_n \left[a_n \frac{\partial \Psi_n}{\partial t} + \Psi_n \frac{\partial a_n}{\partial t} \right] = \sum_n a_n E_n \Psi_n + \sum_n a_n H_1 \Psi_n \qquad (9.5)$$

or

$$j\hbar \sum_n \left[-a_n \left(\frac{jE_n}{\hbar} \right) \Psi_n + \frac{da_n}{dt} \Psi_n \right] = \sum_n a_n E_n \Psi_n + \sum_n a_n H_1 \Psi_n. \qquad (9.6)$$

If we multiply all terms by Ψ_m^* and integrate over all space, making use of the orthonormality of the Ψ_n's,

$$j\hbar \left(-j \frac{E_m}{\hbar} a_m + \frac{da_m}{dt} \right) = a_m E_m + \sum_n a_n \int \Psi_m^* H_1 \Psi_n \, dv. \qquad (9.7)$$

Therefore, one can write

$$\frac{da_m}{dt} = -\frac{j}{\hbar} \sum_n a_n \int \Psi_m^* H_1 \Psi_n \, dv \qquad (9.8)$$

or

$$\frac{da_m}{dt} = -\frac{j}{\hbar} \sum_n a_n e^{j\omega_{mn} t} \int u_m^* H_1 u_n \, dv,$$

$$\omega_{mn} = \frac{E_m - E_n}{\hbar}. \qquad (9.9)$$

The above system of equations is equivalent to the Schrödinger equation. We now replace H_1 by λH_1 and express the a's as a power series in λ:

$$a_n(t) = a_n^{(0)}(t) + \lambda a_n^{(1)}(t) + \lambda^2 a_n^{(2)}(t) + \cdots. \tag{9.10}$$

We substitute this expression for $a_n(t)$ in Eq. (9.8) and equate coefficients of equal power of λ:

$$\frac{da_m^{(0)}}{dt} = 0, \tag{9.11}$$

$$\frac{da_m^{(l+1)}}{dt} = -\frac{j}{\hbar} \sum_n a_n^{(l)} e^{j\omega_{mn} t} \int u_m^* H_1 u_n \, dv.$$

The zero-order terms $a_m^{(0)}$ are constant and their values indicate the initial conditions of the system. We shall be interested in the results of first-order perturbation theory corresponding to single quantum processes. Furthermore, we shall assume that the interacting field is constant over the extent of the atom: the so-called *electric dipole approximation*. For the present we shall also assume monochromatic radiation of frequency ω. Because of the last two assumptions we can write the perturbing Hamiltonian as follows:

$$H_1 = eEz \cos \omega t. \tag{9.12}$$

This assumes that the electric field in the radiation interacts with a particle of charge e and that the electric field is directed in the z direction.

The expansion coefficients can now be written

$$\frac{da_m^{(1)}}{dt} = +\frac{j}{\hbar} \sum_n a_n^{(0)} e^{j\omega_{mn} t} \int u_m^* eEz \cos \omega t \, u_n \, dv. \tag{9.13}$$

If only two states are involved, the system is originally in state n, and we want to find $a_m(t)$ at a time t after the perturbation is turned on, we can first find $da_{(m)}(1)/dt$:

$$\frac{da_m^{(1)}}{dt} = +\frac{j}{\hbar} \sum_n a_n^{(0)} e^{j\omega_{mn} t} \int u_m^* eEz \cos \omega t \, u_n \, dv. \tag{9.14}$$

Then

$$a_m^{(1)} = \frac{eE a_n^{(0)}}{2\hbar} \int u_m^* z u_n \, dv \left[-\frac{e^{j(\omega - \omega_{mn})t} - 1}{\omega - \omega_{mn}} + \frac{e^{-j(\omega - \omega_{mn})t} - 1}{\omega + \omega_{mn}} \right]. \tag{9.15}$$

We are interested in terms for which $\omega \approx \omega_{mn}$. Also, $a_n^{(0)} = 1$:

$$a_m = \frac{eE}{2\hbar} \int u_m^* z u_n \, dv \left[\frac{e^{j(\omega - \omega_{mn})t} - 1}{\omega - \omega_{mn}} \right], \tag{9.16}$$

$$|a_m|^2 = \left(\frac{eE}{2\hbar}\right)^2 \left[\int u_m^* z u_n \, dv \right]^2 \left[\frac{2 - 2\cos(\omega - \omega_{mn})t}{(\omega - \omega_{mn})^2} \right]. \tag{9.17}$$

This expression is valid only when a_m is small compared to a_n, which we have assumed is equal to 1. If we assume that the driving radiation is broad-band compared to the width of the states involved, we can write

$$|a_m|^2 = \left(\frac{eE}{2\hbar}\right)^2 \left[\int u_m^* z u_n \, dv\right]^2 \int_{-\infty}^{\infty} \frac{4 \sin^2 \frac{1}{2}(\omega - \omega_{mn})t}{(\omega - \omega_{mn})^2} \, d\omega \quad (9.18)$$

$$= \left(\frac{eE}{2\hbar}\right)^2 \left[\int u_n^* z u_n \, dv\right]^2 2\pi t.$$

The transition probability per unit time is then

$$W_{nm} = \frac{\pi e^2 E^2}{2\hbar^2} \left[\int u_m^* z u_n \, dv\right]^2. \quad (9.19)$$

We would have obtained the same result if E_n had been greater than E_m, so the transition probability for stimulated emission is the same as for absorption. Expressions can be worked out in a similar manner for the case of monochromatic radiation in the presence of broad states.

If the integral $\int \Psi_2 r_\alpha \Psi_1 \, dv$ is zero for $r_\alpha = x, y,$ and z, then the transition from E_2 to E_1 is said to be forbidden in the electric dipole approximation. If higher order terms are used to describe the electromagnetic field, the corresponding integral may not be zero, in which case magnetic dipole or higher multipolarity transitions may take place. However, if Ψ_m and Ψ_n are both spherically symmetric, the transition $\Psi_m \to \Psi_n$ is said to be strictly forbidden, and transitions can take place only through multiple quantum processes which must be calculated using higher-than-first-order perturbation theory.

As we have shown, the processes of absorption and stimulated emission in the presence of electromagnetic radiation are reciprocal processes that have the same transition probability. The transition probability for spontaneous emission cannot be calculated in terms of the semiclassical theory used previously, but it can be deduced using thermodynamic principles in a manner employed by Einstein. Going back to the system considered earlier, the distribution of particles between the two energy levels is given by

$$\frac{N_2}{N_1} = \exp\left(-\frac{E_2 - E_1}{kT}\right). \quad (9.1)$$

If one writes the transition probability per unit time as the product of the intensity of radiation $I(\omega_{12})$, which is proportional to E^2 in expression (9.19), and a quantity B_{12}, called the *Einstein B coefficient*, then

$$W_{12} = I(\omega_{12}) B_{12}. \quad (9.20)$$

If one considers only the processes of stimulated emission, stimulated absorption, and spontaneous emission, then when the radiation field and the atomic system are in thermal equilibrium there must be as many atoms making the transition from E_1 to E_2 as from E_2 to E_1. Spontaneous absorption obviously is energetically impossible. The probability of emission is proportional to the number of atoms of energy E_2 times the transition probability for spontaneous plus stimulated emission. This must be equal to the probability for absorption, which is proportional to the number of atoms of energy E_1 times the transition probability for absorption. Therefore, one can write

$$I(\omega_{12})B_{12}N_1 = I(\omega_{12})B_{21}N_2 + A_{21}N_2, \qquad (9.21)$$

where A_{21} is the probability per unit time for spontaneous emission by a single atom. Making the substitution that $B_{12} = B_{21} = B$,

$$A_{21} = \frac{I(\omega_{12})B(N_1 - N_2)}{N_2} = I(\omega_{12})B\left(\frac{N_1}{N_2} - 1\right). \qquad (9.22)$$

The intensity of radiation in equilibrium with a system of atomic oscillation is given by Planck's law,

$$I(\omega) = \frac{\hbar\omega^3}{\pi^2 c^3 (e^{\hbar\omega/kT} - 1)}. \qquad (9.23)$$

Therefore, substituting Eqs. (9.1) and (9.23) in Eq. (9.22) one gets

$$A = \frac{\hbar\omega^3}{\pi^2 c^3 (e^{\hbar\omega/kT} - 1)} B(e^{\hbar\omega/kT} - 1)$$

$$= B \frac{\hbar\omega^3}{\pi^2 c^3}. \qquad (9.24)$$

Because of the dependence of spontaneous emission on the cube of the frequency, it is clear that as one goes from microwave to optical frequencies the relative role of spontaneous emission becomes more and more important. At optical frequencies spontaneous emission is the principal process by which the system relaxes to equilibrium.

Spin-lattice relaxation. Another competing transition process is the one that involves the presence of lattice vibrations or phonons. This type of transition occurs when the atomic systems are distributed in the lattice of some host crystal. If the crystal lattice is in thermal equilibrium, the presence of coupling between the active ions and the host lattice, even though it is weak, tends to bring the ionic energy distribution into thermal equilibrium. At microwave frequencies this is the dominant relaxation process, and the time constant associated with it is called the *spin-lattice relaxation time*. At optical frequencies spontaneous emission is usually the dominant relaxation process, although transitions involving phonons also occur. We shall call the transition probabilities involving phonon processes for the two-level system w_{12} and w_{21}. In terms of this interaction, neglecting all other interactions,

$$\frac{dN_1}{dt} = -w_{12}N_1 + w_{21}N_2,$$

$$\frac{dN_2}{dt} = w_{12}N_1 - w_{21}N_2. \tag{9.25}$$

When the whole system is in thermal equilibrium,

$$\frac{dN_1}{dt} = \frac{dN_2}{dt} = 0.$$

Therefore,

$$\frac{w_{12}}{w_{21}} = \frac{N_2}{N_1} = \exp\left(\frac{-E_2 + E_1}{kT}\right). \tag{9.26}$$

This result follows from the fact that the system of atomic particles in thermodynamic equilibrium is exchanging energy with the lattice and both systems must maintain their equilibrium distribution in the steady state.

Rate Equations. We have now considered the existence of energy levels in atomic systems, the mechanisms by which transitions between energy levels are made, and the thermal equilibrium population of the various energy levels. We shall now consider the energy levels of a system of atomic particles in a nonequilibrium situation. For a two-level system the rate equation can be written

$$\frac{dN_1}{dt} = -W_{12}N_1 + (W_{21} + A_{21})N_2 - w_{12}N_1 + w_{21}N_2,$$

$$\frac{dN_2}{dt} = W_{12}N_1 - (W_{21} + A_{21})N_2 + w_{12}N_1 - w_{21}N_2. \tag{9.27}$$

The conservation of the total number of particles can be stated

$$N_1 + N_2 = N_0, \tag{9.28}$$

where N_0 is the total number of active particles in the system. This condition is obviously satisfied by the two rate equations, because

$$\frac{dN_0}{dt} = \frac{dN_1}{dt} + \frac{dN_2}{dt} = 0. \tag{9.29}$$

These equations can be generalized to an n-level system. For the ith state, because $W_{ij} = W_{ji}$,

$$\frac{dN_i}{dt} = \sum_{j=1}^{n} W_{ij}(N_j - N_i) + \sum_{j>i} A_{ji}N_j - \sum_{j<i} A_{ij}N_i + \sum_{i}(w_{ji}N_j - w_{ij}N_i). \quad (9.30)$$

It is the above set of equations which must be satisfied when a given set of atomic systems is interacting with an electromagnetic field, as is the case in quantum-electronic devices.

Spectral line width. In the first part of this section the various relaxation processes for bringing a set of atomic systems to their thermal equilibrium energy distribution was discussed. We shall assume that as a result of these relaxation processes a two-level system will approach thermal equilibrium with a characteristic relaxation time γ. The amount of spontaneously emitted radiation above its equilibrium value will be proportional to the amount by which N_2 exceeds its equilibrium value. If $(E_2 - E_1) \gg kT$, then N_2 is virtually unpopulated in thermal equilibrium. Therefore, the intensity of the spontaneously emitted radiation will vary as follows:

$$I = I_0 e^{-\gamma t}. \quad (9.31)$$

If the electric field of the wave is proportional to the square root of the intensity and has a fundamental frequency $\omega_0 = E/\hbar$,

$$\mathcal{E} = \mathcal{E}_0 e^{-\gamma t/2} \cos \omega_0 t. \quad (9.32)$$

A Fourier analysis of the above wave yields a Lorentz shape for the frequency distribution of the intensity of the wave:

$$I(\omega) = \frac{\text{const}}{(\omega - \omega_0)^2 + (\gamma/2)^2}. \quad (9.33)$$

Therefore the line width bears a definite relationship to the lifetime of the state; the longer the lifetime the narrower the width, in accordance with the uncertainty principle.

9.3 Operation of devices based on stimulated emission

Within the last 10 years a number of devices have been developed which are based on the stimulated emission of radiation from atomic systems. The possibility of utilizing this principle was proposed independently by Townes, Weber, and Basov and Prokhoroff about 15 years ago. The first operating device was developed by a group under Townes and was given the name *maser* (Microwave Amplification through Stimulated Emission of Radiation). Later devices were developed which operated at other than microwave frequencies but were also based on the principle of stimulated emission. Those devices, involving amplification through stimulated emission at optical frequencies, have been called LASERS.

We have discussed so far the existence of various energy levels in atomic systems, and the fact that transitions between energy levels can involve interactions with electromagnetic radiation. In this and the following sections we shall consider the actual configuration of devices that are able to achieve amplification of electromagnetic radiation through the interaction of that radiation with excited electronic states through the process of stimulated emission. In principle, two requirements must be met to obtain this "maser action." First, one must have an ensemble of atomic systems between two of whose states a population inversion exists and can be maintained. This condition must be met if stimulated emission is to exceed resonance absorption, which must be the case if the particle system is to give up energy to the radiation field and hence amplify it. Second, one must obtain a sufficiently strong interaction between the field and the particles to allow stimulated emission from the particle systems to overcome any energy losses in the system and also produce amplification. We shall first consider techniques for causing population inversion and then consider the problem of coupling.

Population inversion: Two-level system. In a two-level system the condition of population inversion means that the higher-energy level E_2 is more densely populated than the lower-energy E_1 ($N_2 > N_1$). This is obviously not true for any state in thermal equilibrium, at least for positive absolute temperatures. Also it is impossible to achieve this condition simply by applying a strong electromagnetic field of the proper frequency. By considering Eq. (9.27) one can see that in equilibrium ($dN_1/dt = dN_2/dt = 0$) the largest we can make the ratio N_2/N_1 is 1, because $W_{12} = W_{21}$ and $w_{12} < w_{21}$. However, several techniques have been devices for obtaining the condition $N_2 > N_1$ for periods of time long enough to be useful.

In the ammonia maser a state of population inversion is attained by using an inhomogeneous electric field to physically separate the molecules that are in the higher-energy state from those in the lower-energy state. Once this separation is achieved, the beam of molecules in the higher-energy state is directed into a microwave cavity, where they can then interact with the field.

The process of adiabatic rapid passage has been used to obtain a population inversion in a two-level solid-state maser. This involves sweeping a paramagnetic sample through its magnetic resonance frequency under the proper conditions. The details of this process are too involved to be treated here.

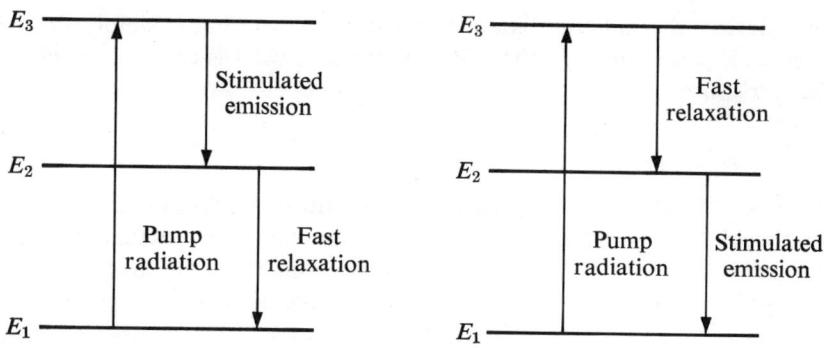

Fig. 9.1
Energy-level diagram of a three-level maser and the possible modes of operation

Population inversion: Three-level system. It was shown that for the two-level system population inversion could not be obtained by pumping with an external electromagnetic field of the frequency of the transition. However, if one goes to a multilevel system it is possible to obtain population inversion between two of the states of the system. For simplicity let us consider the three-level system of Fig. 9.1. If radiation of frequency $\nu_{31} = (E_3 - E_1)/h$ is used to pump the ensemble to saturation, then $N_3 = N_1$ and we must have either the condition that $N_2 > N_1$ or $N_3 > N_2$ for all but one special case when all three populations are equal. Therefore, a state of population inversion exists between E_1 and E_2 or E_2 and E_3. Many effects come into play in deciding which pair of levels has the inversion, but if $E_2 - E_1 \gg kT$, it is usually true that $N_3 > N_2$. Obviously, similar arguments can be used to show the possibility of obtaining population inversion between two states in a system containing four or more levels. Some of the schemes will be discussed in more detail later in connection with specific devices.

Coupling. To maintain efficient maser action one must not only have a fairly high degree of population inversion between two energy levels of an ensemble of atomic systems, but one must have good coupling between the particle system and the electromagnetic field that is to be amplified. To achieve this, one wants to make the transition probability per unit time as large as possible and to keep the field in contact with the active medium for as long a time as possible. These two goals are accomplished by placing the active material containing the atomic systems inside a region that can trap the electromagnetic radiation. Such an enclosure is termed a *resonant cavity* and will be discussed in more detail in connection with devices designed for specific frequencies of operation. Because the resonant cavity tends to restrict the frequency range over which a device can operate, in some devices the radiation is not trapped in a cavity but is simply slowed down as it passes through the active material in what is called a *traveling-wave structure*.

9.4 The solid-state maser

The two-level ammonia maser was described previously, and it was mentioned that population inversion was obtained by actual physical separation of the higher-energy molecules from those of lower energy. This is not possible in the solid-state maser, because in the usual case the active material consists of paramagnetic ions formed from transition-group elements, such as chromium or gadolinium, embedded in an inert lattice matrix. Ruby has proved to be the most useful material. It cosists of Cr^{3+} ions in an aluminum oxide lattice. The Cr^{3+} ions replace Al^{3+} ions in the lattice. The lowest states are a pair of Kramer's doublet states separated by an energy corresponding to a frequency of 11.46 gigacycles. The degeneracy of the doublets may be removed by application of an external magnetic field. A diagram of the four energy levels as a function of the magnetic field for the magnetic field parallel to the crystal axis is shown in Fig. 9.2. As the direction of the external magnetic field is changed, relative to the crystal axis, the diagram of the energy levels as a function of magnetic field changes. Also, the transition probabilities between the various levels depend on the direction and magnitude of the magnetic field. To obtain the optimum operation of the maser at the desired frequency one needs to choose the

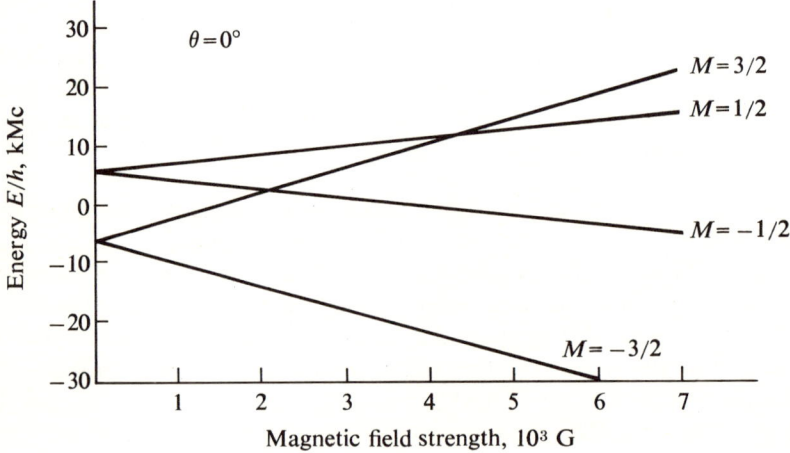

Fig. 9.2

Energy-level diagram of Cr^{3+} in ruby as a function of the external magnetic field for the field parallel to the crystalline C axis

proper field and direction of application. Because of the field dependence of the energy levels, a maser using these ions is tunable, an advantage over the ammonia maser. Only three of the four levels shown for Cr^{3+} are needed for maser operation, although a four-level push-pull operation has been successfully utilized in which the spacing between levels 1 and 3, and 2 and 4, is equal, and the system is pumped at the frequency corresponding to this energy difference. Maser action takes place between levels 2 and 3, which are being emptied and filled, respectively, by the pump radiation.

It is the magnetic field of the pump and signal radiation which interacts with the paramagnetic ions, so it is the magnetic dipole approximation which must be used to calculate the transition probabilities. Also, for this reason, it is desirable to place the active material in that region of the resonant cavity where the magnetic fields of the pump and signal are strongest. The cavity needs to be resonant at both the pump and signal frequency. A typical microwave cavity designed for this purpose is shown in Fig. 9.3. The use of a microwave cavity permits the build-up of large pump and signal fields at the resonant frequencies. This increases the transition probabilities, which are proportional to the square of the magnetic field in the magnetic dipole approximation. However, the cavity limits the bandwidth of the amplifier. As mentioned before, a traveling-wave structure allows increased bandwidth without reducing the field-ion interaction too much.

The microwave photon energy is quite small, about 6.6×10^{-24} J at

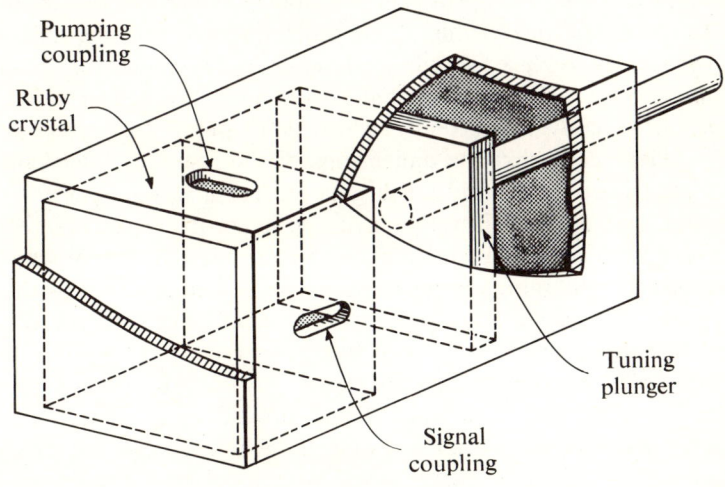

Fig. 9.3
Typical microwave maser cavity

10 gigacycles, so the amount of stored energy available for maser action in a sample of pink ruby if all the ions were excited would be

$$E = \tfrac{1}{2}N \times 6.6 \times 10^{-24} \text{ J}, \tag{9.34}$$

where N is the number of ions in the sample. In pink ruby the density of Cr^{3+} ions is approximately 2.3×10^{19} ions/cm³. Therefore, the energy density available for amplification is about 70 μJ/cm³. For this reason the maser is not useful for producing large amounts of microwave power. However, it can be used to amplify very weak microwave signals up to the microwatt level and introduce very little noise in the process.

The noise in the output of a maser generally comes from two sources: (1) the spontaneous emission of radiation by the active ions in the upper energy state, and (2) the thermal radiation from the walls of the cavity. At microwave frequencies the spontaneous emission is quite small, as was pointed out in Sec. 9.2. Because most solid-state masers are operated at liquid helium temperatures (4°K), to increase the spin-lattice relaxation time the thermal radiation in the cavity is quite low. For these reasons thermal noise can be reduced to the point where measured noise temperatures of less than 2°K may be achieved. A detailed analysis of noise and noise in maser amplifiers can be found in the book by Siegman.

9.5 The laser

The principle of amplification through stimulated emission has been extended to devices that operate in the optical range. However, in this region the problems and techniques are quite different from the microwave range. In the first place, spontaneous emission becomes the dominant relaxation process. However, excited states can be formed in solids, with lifetimes of several milliseconds, which are useful for laser action. Another difference is that the dimensions of the resonant cavity and the active material are necessarily much greater than a wavelength at optical frequencies, and consequently the cavity design problems are different. A further complication is that no sources of coherent monochromatic light are available for pumping the ions into an excited state. Even with these operation differences, the basic principles of operation of the laser, which is the general name given to devices of the maser type which operate at optical frequencies, are the same as the maser.

Ironically, ruby, the material that had proved most useful for maser operation, was also the material that yielded the first successful laser action, although entirely different energy levels were involved. The energy level diagram that is appropriate for laser action in ruby is shown in Fig. 9.4. The splitting of the ground state into two Kramer's doublets is not shown,

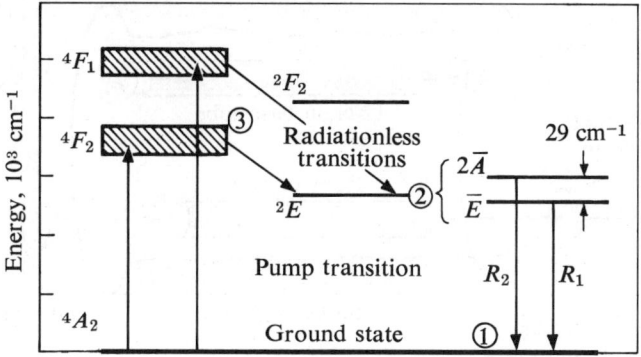

Fig. 9.4
Energy-level diagram of Ct^{3+} in ruby showing levels appropriate for laser action; the symbols designating the levels come from group theory

because the vertical scale now covers a much larger energy range than that of Fig. 9.2. The laser is pumped between the ground state and the broad $^4F_1$ and $^4F_2$ levels by a high-intensity flash tube. These levels are quite broad, and hence absorb the light from the flash tube, which has a broad spectral output, quite efficiently. Very quickly (10^{-8} sec) the ions make the transition from the absorption bands to the pair of states designated ^2E which have a lifetime of several milliseconds. The energy given up by this transition goes directly to the lattice and no electromagnetic radiation is emitted. The laser action takes place between the lower of the two ^2E states and the ruby ground state. This radiation has a wavelength of 6943 Å. Because of the large separation of the levels involved in the laser action, cooling of the crystal is not necessary, but a somewhat higher efficiency of operation is obtained with cooling to liquid-nitrogen temperatures.

A diagram of the usual configuration of a solid-state optically pumped laser is shown in Fig. 9.5. A cylindrical ruby rod is put at one focus of an elliptical reflector and a linear flash tube at the other. This configuration focuses the pumping light from the flash tube onto the ruby rod. The resonant cavity for the output light is formed by putting plane or spherical mirrors, which are reflective at the output wavelength, at the ends of the ruby rod. Although the cavity is many wavelengths long and has no sides, it serves as a sufficiently high Q cavity to allow build-up of laser action in the crystal. In its usual form the ruby laser is operated in a pulsed fashion with the light from the flash tube lasting only 1 or 2 msec but being of sufficient intensity to pump enough ions into the ^2E state so that a condition of population inversion exists between that state and the ground state. In this condition the laser could act as an amplifier for any external signal of the

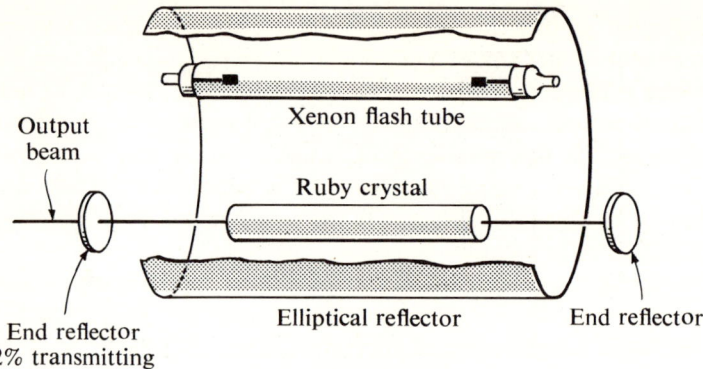

Fig. 9.5
Schematic diagram of one form of the ruby laser

proper frequency introduced into the cavity. However, the laser is generally used as an oscillator rather than an amplifier, and it is triggered into oscillation by its own spontaneous emission. The output light is extracted through one of the end reflectors, which can have a transmitivity of 2 or 3 per cent.

From the description so far given of the laser, we have a device into which we can dump a lot of electrical energy, which causes a flash tube to produce a lot of light, which is partially absorbed by a ruby, which emits some light. The whole process has an efficiency of less than 1 per cent. In terms of ordinary light sources this is quite inefficient. However, the light from the ruby has some very special properties, which make it worth accepting such poor efficiency. Because of the nature of the stimulated emission process the phase of the emitted light is the same as that of the light which stimulated the emission. For this reason there is a definite phase relationship among the radiations at all points, just as is the case for radiation from a single source or antenna. This property of light is termed *coherence*. The spectral width of the radiation is also much narrower than the natural line width because of the frequency-selective properties of the stimulated emission.

Let us now look at what the intensity of the light is. The density of stored energy available for stimulated emission is approximately 10^4 times what it is at microwave frequencies, so it is therefore about 0.7 J/cm^3. For ruby with a volume of 10 cm^3, this corresponds to a total energy of 7 J. If all this energy were dumped in 10^{-4} sec, the power output would be 70 kw.

To obtain even higher power outputs, a technique called *Q switching* has been developed to yield a single giant pulse from the laser. During the initial part of the flash from the flash tube one of the end reflectors is tilted a bit, so that the Q of the resonant cavity is quite low and laser action cannot occur; but a large population inversion is achieved. Suddenly the mirror is tilted back into position to restore the cavity to its normal Q. The laser then emits all its energy in one giant pulse, within a period of less than 10^{-8} sec. In this manner peak output power in the vicinity of 10^9 W have been achieved. Many variations of this technique have been developed but they all involve spoiling the Q of the cavity in one way or another, until a high degree of population inversion is attained.

From the above we see that the fundamental properties which distinguish light from a laser from light from other sources are as follows: (1) Laser light possesses a high degree of coherence, (2) it can be generated with high intensity, (3) it can be produced with a very narrow line width, and (4) it can be produced with a very narrow beam width. These four properties are responsible for the numerous advantages lasers have over other light sources. The fact that light is of a frequency of about 10^{15} Hz means that if only a modulation of frequency of 0.1 per cent were achieved, one would have an available bandwidth of 10^{12} Hz. Because the amount of information which may be transmitted is proportional to the bandwidth, the laser promises to be a means of transmitting and storing large amounts of information.

The coherency of the radiation as well as the mode patterns produced by the form of the cavity result in the output beam having very directional properties. If the beam were coherent and uniform over the cross section of the cylindrical ruby and the axial modes of oscillation produced plane waves, then the beam spread would be due only to diffraction effects. The beam width for a rod of d-cm diameter would be from diffraction theory proportional to λ/d, and would therefore be quite small in the visible range. Generally the measured beam spread is about 100 times this theoretical width, indicating that the laser rods tend to radiate from small fibers along the rod. Indeed such behavior has been directly observed.

The plane-wave character of the output, as well as its monochromaticity, make it easy to focus the laser output to a very small spot with very simple optics. It is possible to focus it to a diameter of less than 1μ. Enormous power densities, which can vaporize any substance, can be so obtained. Also extremely intense electric fields may be produced which can induce nonlinear effects in dielectric materials. This discussion merely mentions a few of the many possible applications of laser radiation.

9.6 Semiconductor lasers

A type of laser was developed in 1962 which achieved laser action from the recombination of electrons and holes in semiconductor diodes. This process is the inverse of the process described in Chapter 5, in which electromagnetic radiation is absorbed in a photovoltaic cell and produces hole-electron pairs.

The first condition to be met if we are to have laser action is that a condition of population inversion exists. This can be achieved in a forward-biased P-N junction, where both the P and N regions have been doped to degeneracy. By being "doped to degeneracy" is meant that the impurity concentration is so great that the Fermi level E_f lies in the conduction band in the N-type region and in the valence band in the P-type region. The energy-band diagram for such a junction is shown in Fig. 9.6. The rates at

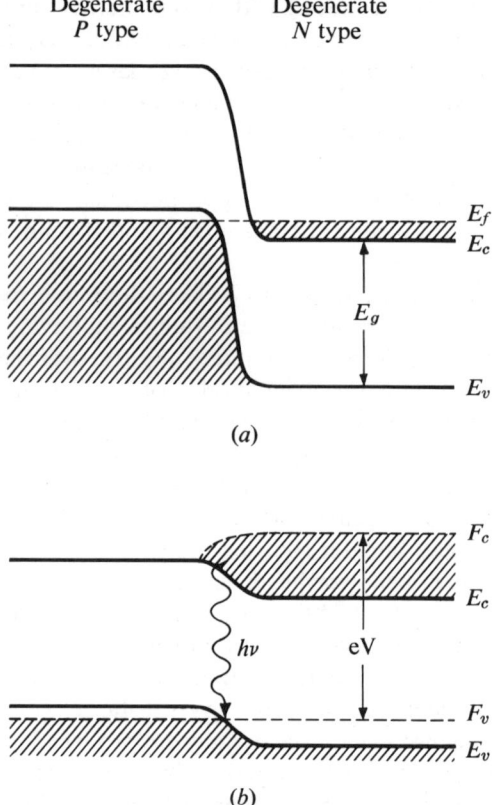

Fig. 9.6
Energy diagram of a P-N junction doped to degeneracy as used in a semiconductor laser. (*a*) No bias. (*b*) Applied forward voltage $V = F_c - F_v > h\nu$

which quanta are being absorbed and emitted are given by the following expressions:

$$\left(\frac{dn}{dt}\right)_{abs} = B_{vc} N_v N_c \rho(\nu_{vc}) p_v (1 - p_c), \quad (9.35)$$

$$\left(\frac{dn}{dt}\right)_{em} = B_{cv} N_v N_c \rho(\nu_{vc}) p_c (1 - p_v), \quad (9.36)$$

where $B_{vc} = B_{cv}$ = Einstein B coefficient used in Eq. (9.20),
N_v = density of states in the valence band,
N_c = density of states in the conduction band,
$\rho(\nu_{vc})$ = density of radiation of the frequency W_{vc} corresponding to a direct transition from the valence to the conduction band,
p_c = probability of occupation of a state in the conduction band,
p_v = probability of occupation of a state in the valence band.

Obviously, under conditions of no applied voltage, the system will be in thermodynamic equilibrium, and there will be no population inversion. Therefore, bias must be applied and the equilibrium distribution of occupation probabilities must be disturbed. We shall assume that the electrons in each of the two bands are in thermal equilibrium among themselves and that the population distribution in each band can be described by a so-called *quasi Fermi level*, which is approximately constant across the junction:

$$P_c = \frac{1}{1 + \exp[(E - F_c)/kT]}, \quad (9.37)$$

$$P_v = \frac{1}{1 + \exp[(E - F_v)/kT]}. \quad (9.38)$$

For emission to exceed absorption one must have the condition

$$\left(\frac{dn}{dt}\right)_{em} > \left(\frac{dn}{dt}\right)_{abs}. \quad (9.39)$$

This leads to the condition that

$$F_c - F_v > h\nu. \quad (9.40)$$

Because of the close relationship between $E_c - E_v$ and the applied voltage, one can say that this inequality implies that the forward applied voltage must be greater than the energy of the emitted radiation.

We have assumed in the preceding analysis that the emission process resulted from direct recombination of an electron from the conduction band with a hole in the valence band. However, either one or both of the terminal states in the radiative recombination process may be localized donor

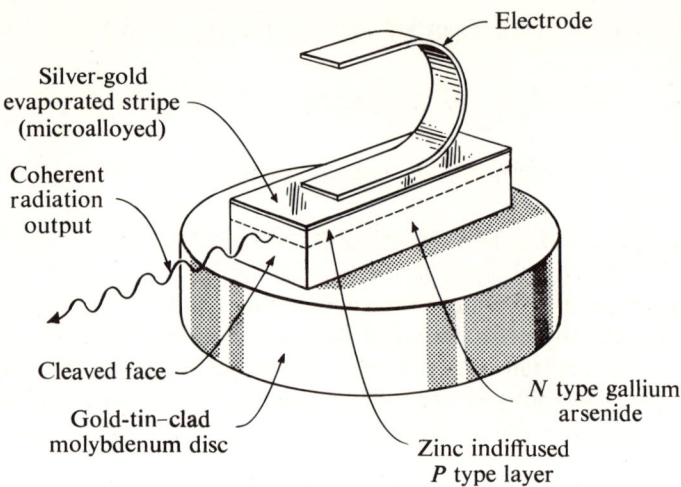

Fig. 9.7
GaAs laser in a typical mounting

or acceptor levels. So-called *"indirect" recombination processes* involving phonon interactions apparently are not involved in any of the transitions from which laser action may be obtained.

The resonant cavity essential for laser operation is formed by cleaving two parallel faces on the edges of the junction as shown in Fig. 9.7. These two faces, as a result of their reflectivity caused by the high index of refraction of the diode material, form a Fabry-Perot interferometer. The radiation is emitted from both of the parallel faces.

A gallium arsenide, gallium phosphide mixture $Ga(As_xP_{1-x})$, has proved to be an extremely efficient material for obtaining laser action. Wavelengths in the visible, 6300 Å, have been obtained with lasers made of this material. Most semiconductor lasers operate in the infrared, however. Other materials with which laser action has been achieved are straight GaAs, $(In_xGa_{1-x})As$, $In(As_xP_{1-x})$, InSb, PbTe, and PbSe.

The semiconductor laser possesses the advantages of small size, high efficiency, the property of converting electrical energy directly to coherent light, and ease of modulation of intensity through modulation of the forward current. Greatest efficiency is usually obtained by low-temperature operation. Diode lasers operating in the 4°K range have produced up to 6 W output power at close to 50 percent efficiency.

Problem 1

Calculate the probability per unit time that a transition will take place from the $n = 1$, $l = 0$ ground state of the hydrogen atom to the $n = 2$, $l = 1$ state in a broad-band electromagnetic field whose electric vector per unit frequency is 10^2 (V/m)/Hz.

Problem 2

What is the spontaneous emission transition probability from the $n = 2$, $l = 1$ state of the hydrogen atom?

Problem 3

Consider a three-energy-level system such as that shown in Fig. 9.1. Assume $A_{21} = A_{32} = 0$. The other parameters are $E_{12} = 0.0005$ eV and $E_{23} = 0.0003$ eV, $w_{21} = 2w_{32}$, $T = 4°$K. If the system is pumped at a frequency corresponding to E_3 until saturation is achieved ($N_1 \approx N_3$), which set of two levels would be suitable for laser operation?

Problem 4

Consider a laser cavity formed by two infinite parallel planes, one perfectly reflecting and the other with a reflectivity of 98 percent. The planes are separated by a distance d. What is the Q of the cavity for axial-plane-wave modes of oscillation of wavelength 6943°A?

Problem 5

By Fourier analysis determine the frequency spectrum of the following function of time:
$$E = E_0 e^{-\gamma t/2} \cos \omega_0 t.$$

Problem 6

Consider a system of active ions in a completely inverted state in a cavity of $Q = Q_0$ tuned to the transition frequency. Assume that no pumping is done after complete inversion of the ions is achieved. What will be the general behavior of the electric field as a function of time assuming Q_0 large enough to allow the build-up of a pulse?

Bibliography

Azaroff, L. V., and J. J. Brophy, *Electronic Processes in Materials* (McGraw-Hill, 1963).
Beam, W. R., *Electronics of Solids* (McGraw-Hill, 1965).
de Groot, S. R., *Thermodynamics of Irreversible Processes* (North-Holland, 1960).
Dekker, A. J., *Electrical Engineering Materials* (Prentice-Hall, 1959).
Dunlap, W. C., Jr., *An Introduction to Semiconductors* (Wiley, 1957).
Gibson, A. F., R. E. Burgess, and P. R. Aigrain, eds., *Progress in Semiconductors*, Vol. 3 (Wiley, 1958).
Jaeger, J. C., *An Introduction to Laplace Transformation* (Methuen, 1955).
Joffe, A. F., *Semiconductor Thermoelements and Thermoelectric Cooling* (Infosearch Limited, 1957).
Katz, H. W., *Solid State Magnetic and Dielectric Devices* (Wiley, 1959).
Kittel, C., *Introduction to Solid State Physics* (Wiley, 1956).
Landau, L. D., and E. M. Lifschitz, *Electrodynamics of Continuous Media* (Addison-Wesley, 1960).
Lengyel, B. A., *Lasers* (Wiley, 1962).
Levine, S. N., *Quantum Physics of Electronics* (Macmillan, 1965).
Megaw, H. D., *Ferroelectricity in Crystals* (Methuen, 1957).
Nussbaum, A., *Semiconductor Device Physics* (Prentice-Hall, 1962).
Putley, E. H., *Hall Effect and Related Phenomena* (Butterworth, 1960).
Reitz, J. R., and F. J. Milford, *Foundations of Electromagnetic Theory* (Addison-Wesley, 1960).
Ramey, R. L., *Physical Electronics* (Wadsworth, 1961).
Rogers, W. E., *Introduction to Electric Fields* (McGraw-Hill, 1954).
Rojansky, V., *Introductory Quantum Mechanics* (Prentice-Hall, 1938).
Seitz, F., *The Modern Theory of Solids* (McGraw-Hill, 1940).
Seitz, F., and D. Turnbull, eds., *Solid State Physics, Vol. 1* (Academic, 1955).
Seitz, F., and D. Turnbull, eds., *Solid State Physics, Vol. 5* (Academic, 1957).
Siegman, A. E., *Microwave Solid-State Masers* (McGraw-Hill, 1964).
Smith, R. A., *Semiconductors* (Cambridge, 1964).
Sokolnikoff, I. S., and E. S. Sokolnikoff, *Higher Mathematics for Engineers and Physicists*, 2nd ed. (McGraw-Hill, 1941).

Spenke, E., *Electronic Semiconductors* (McGraw-Hill, 1958).
Starkey, B. J., *Laplace Transforms for Electrical Engineers* (Philco Optical Laboratories, 1956).
Tralli, N., *Classical Electromagnetic Theory* (McGraw-Hill, 1963).
Van der Ziel, A., *Solid State Physical Electronics* (Prentice-Hall, 1957).
Vuylsteke, A. A., *Elements of Maser Theory* (Van Nostrand, 1960).
Walsh, J. B., *Electromagnetic Theory and Engineering Applications* (Ronald, 1960).

Index

Adiabatic demagnetization, 299
Ampère's law, 260
Antiferromagnetism, 294

Band, conduction, 115
  valence, 115
Biot-Savart law, 259
Bloch, F., 104
Bohr-Sommerfeld atom, 1
Boltzmann constant, 41
Boltzmann factor, 48
Boltzmann transport equation, 345
Brillouin theory, 277
Brillouin zone, 109

Classical theory, of paramagnetism, 280
  of specific heats, 67
Clausius-Mosotti relationship, 230
Conduction electrons, 57, 88
Conductivity, electrical, 306, 332
  photo-, 153
  thermal, 307, 337, 343
Contact potential, 188
Covalent bonding, 139
Curie constant, 283
Curie point, 232
Curie temperature, 279, 296
Curie-Weiss law, 274, 285

de Broglie's hypothesis, 9
Debye's approximation, 77
Degeneracy, threefold, 16
Delta function, 107
Density of states, 16, 52
Depletion region, 180
Dielectric constant, 230
Dielectric losses, 247
Diffusion current, 166
Diffusion length, 176
Dipole moment, electric, 207, 244
  magnetic, 255
  permanent magnetic, 265

Domains, ferroelectric, 288
  ferromagnetic, 279
Donors, 140
Drift current, 166
Drift velocity, 321
Dulong-Petit value, 69

Effective mass, 111
Eigenvalues, 12
Einstein coefficient, 360
Einstein relation, 167
Einstein theory of specific heats, 69
Electrochemical potential, 163
Electron, gas, 39, 92, 117, 119
  spin, 36
  volt, 3
Emission, spontaneous, 361
  stimulated, 363
Energy levels, 356
Ettinghausen effect, 348
Exchange interactions, 287
Excitons, 159
Exclusion principle, 36, 49, 89

Fermi-Dirac statistics, 49, 89, 331
Fermi level, 51, 131, 146, 331
Fermi temperature, 95
Ferrites, 298
Ferroelectric domains, 288
Ferroelectric energy conversion, 249
Ferroelectricity, 231
Figure of merit, thermoelectric, 318
  thermomagnetic, 352
Fluorescence, 161

Galvanomagnetic effects, 320
Gamma functions, 46
Gas kinetics, 39
Gauss theorem, 207
Group velocity, 7

INDEX

Hall effect, 322
Hall mobility, 324
Harmonic force, 4
Harmonic oscillator, 67
Holes, 128
Hooke's constant, 4
Hydrogen atom, 24, 31
Hund's rule, 268

Insulators, 115, 127
Irreversible processes, 303

Kelvin's thermodynamic relationships, 311
Kronig-Penny model, 103

Langevin-Debye equation, 223
Langevin functions, 222, 236, 273
Larmor frequency, 263
Lasers, 364, 368
   semiconductor, 371
Lattice vibration, 67, 96
Legendre equation, 27
Legendre polynomials, 28
Lifetime of charge carriers, 171
Lorentz force, 213, 324
Lorentz method, 227
Luminescence, 153

Magnetic flux, 255
Magnetic number, 26
Magnetic susceptibility, 261
Magnetism, dia-, 269
   ferro-, 279
   para-, 272
Magnetization, 258
Magnetoresistance, 329
Maser, 363
   solid-state, 366
Matthiessen rule, 127
Mean free path, 336
Metals, 114
   free-electron model, 116

Neel temperature, 297
Nernst effect, 347
Nernst-Ettinghausen generator, 349
Nuclear magnetic moments, 269

Onsager's theorem, 303

Peltier effect, 303, 310
Periodic table, 36
Permeability, 259
Phase velocity, 7
Phonons, 88
Phosphorescence, 161
Photocells, 194
Photosensitivity, 156
Photovoltaic cells, 194
Planck's constant, 4, 70
Planck's law, 361
P-N junction, capacitance of, 185
   optical effects in, 177
   theory of, 177
Polarization, electronic, 212, 241
   ionic, 243
   orientational, 217, 245
   residual, 232
   spontaneous, 232
Population-inversion, 364
Potential box, 12
Potential well, 17
Probability density, 24
Probability function, 34

Quantum electronics, 355
Quantum numbers, 34
   magnetic, 26, 36
   orbital angular momentum, 27, 34, 265
   principal, 31, 34
   spin, 267
Quantum oscillator, 21, 71

Recombination, 171
Recursion relation, 23
Richardson-Dushman equation, 61
Righi-Leduc effect, 349
Rutherford's atom, 1

Schrödinger equation, 9
   time-dependent, 11
   time-independent, 12
Seebeck effect, 308
Semiconductors, 114
   extrinsic, 138
   intrinsic, 127
   N-type, 139
   P-type, 142
Spectral line width, 363
Spherical coordinates, 25
Spin-lattice relaxation, 361

Statistics, classical, 42
   Fermi-Dirac, 49
   Maxwell-Boltzmann, 125
Surface work function, 57

Thermionic emission, 57
Thompson effect, 310
Transition probabilities, 357

Traps, 157

Velocity, group, 7
   phase, 7
Velocity distribution function, 42

Wave function, normalization of, 12
Wiedemann-Franz ratio, 116, 122